REVIT MEP 2025 FOR NOVICES

(Learn by Doing)

CADSoft Technologies

Excellence in Technology

CADSoft Technologies

Revit MEP 2025 for Novices: Learn By Doing

CADSoft Technologies

ISBN: 978-1-64057-152-5

NOTICE TO THE READER

Publisher does not warrant or guarantee any of the products described in the text or perform any independent analysis in connection with any of the product information contained in the text. Publisher does not assume, and expressly disclaims, any obligation to obtain and include information other than that provided to it by the manufacturer.

The reader is expressly warned to consider and adopt all safety precautions that might be indicated by the activities herein and to avoid all potential hazards. By following the instructions contained herein, the reader willingly assumes all risks in connection with such instructions.

The publisher makes no representation or warranties of any kind, including but not limited to, the warranties of fitness for particular purpose or merchantability, nor are any such representations implied with respect to the material set forth herein, and the publisher takes no responsibility with respect to such material. The publisher shall not be liable for any special, consequential, or exemplary damages resulting, in whole or part, from the reader's use of, or reliance upon this material.

www.cadsofttech.com

DEDICATION

To students, who are dedicated to learning new technologies

THANKS

To employees of CADSoft Technologies
for their cooperation

Online Training Program Offered by CADSoft Technologies

CADSoft Technologies provides effective and affordable virtual online training on various software packages including Computer Aided Design, Manufacturing and Engineering (CAD/CAM/CAE), computer programming languages, animation, architecture, and GIS. The training is delivered 'live' via Internet at any time, any place, and at any pace to individuals as well as the students of colleges, universities, and CAD/CAM/CAE training centers. The main features of this program are:

Training for Students and Companies in a Classroom Setting

Highly experienced instructors and qualified Engineers at CADSoft Technologies conduct the classes under the guidance of Prof. Sham Tickoo of Purdue University Northwest, USA. This team has authored several textbooks that are rated "one of the best" in their categories and are used in various colleges, universities, and training centers in North America, Europe, and in other parts of the world.

Training for Individuals

CADSoft Technologies with its cost effective and time saving initiative strives to deliver the training in the comfort of your home or work place, thereby relieving you from the hassles of traveling to training centers.

Training Offered on Software Packages

CADSoft Technologies provides basic and advanced training on the following software packages:

CAD/CAM/CAE: *CATIA, Creo Parameteric, SOLIDWORKS, Autodesk Inventor, Solid Edge, NX, AutoCAD, AutoCAD LT, Customizing AutoCAD, AutoCAD Electrical, and ANSYS*

Architecture and GIS: *Autodesk Revit (Architecture/Structure/MEP), AutoCAD Civil 3D, AutoCAD Map 3D, Autodesk Navisworks, Bentley Staad.Pro, Oracle Primavera P6, MS Project, ArcGIS, and Raster Design.*

Animation and Styling: *Autodesk 3ds Max, 3ds Max Design, Autodesk Maya, Autodesk Alias, Pixologic ZBrush, and CINEMA 4D*

Computer Programming: *C++, VB.NET, Oracle, AJAX, and Java*

For more information, please visit the following link:

Table of Contents

Chapter 2: Getting Started with an MEP Project

Chapter 3: Creating Building Envelopes

Chapter 4: Creating Spaces and Zones and Performing System Analysis

Chapter 5: Creating an HVAC System

Chapter 6: Creating an Electrical System

Chapter 7: Creating Plumbing Systems

Chapter 8: Creating Fire Protection System

Chapter 9: Creating Construction Documents

Chapter 10: Creating Families and Worksharing

This page is intentionally left blank

Preface

Autodesk Revit 2025

Autodesk Revit is a Building Information Modeling software designed for Architects, Structural Engineers, MEP Engineers, Designers, and Contractors. The software has the capability to design the 3D model of a building with its various components, annotate the model with 2D drafting elements, and access building information from the building model's database.

Autodesk Revit also has the capability of executing 4D BIM with tools to plan and track various stages in the building's lifecycle, from concept to construction and later demolition.

Revit was developed by Charles River Software, in Newton, Massachusetts on October 31, 1997. Leonid Raiz and Irwin Jungreis were the key developers of Revit, which was designed specifically with Architecture in mind. The Charles River Software company was later renamed to Revit Technology Corporation. In 2002, Autodesk purchased Revit from Revit Technology Corporation for their building solutions and infrastructure group.

The MEP discipline of Revit was introduced to the users in 2006 as Revit MEP. This software is primarily used in designing the Mechanical, Electrical, and Plumbing and Piping systems, which are the three disciplines of building services. In Revit, you can create the plans, elevations, sections, schedules, and 3D models of a building project that can be easily accessed and shared between different users.

The **Revit MEP 2025 for Novices** explains the concepts and principles of MEP in Revit through practical examples, tutorials, and exercises. This enables the users to harness the power of BIM with Autodesk Revit for their specific use. In this textbook, the author explains in detail the procedure of evaluating HVAC cooling and heating loads and the usage of tools required for designing HVAC, electrical, and plumbing design. In addition, in this textbook, you will learn tools and concepts for creating families and process to document the final drawings.

In this textbook, special emphasis has been laid on the concepts of space modeling and tools to create systems for all disciplines (MEP). Each concept in this textbook is explained using the detailed description and relevant graphical examples and illustrations. The accompanying tutorials and exercises, which relate to the real world projects, help you understand the usage and abilities of the tools available in Autodesk Revit. Along with the main text, the chapters have been punctuated with tips and notes to make the concepts clear, thereby enabling you to create your own innovative projects.

The main features of this textbook are as follows:

- **Project-based Approach**
  The author has adopted the project-based approach and the learn-by-doing theme throughout the textbook. This approach guides the users through the process of creating the designs given in the tutorials.

- **Real-World Designs as Projects**
  The author has used real-world building designs and architectural examples as projects in this textbook so that the users can correlate them to the real-time designs.

- **Tips and Notes**
  Additional information related to various topics is provided to the users in the form of tips and notes.

- **Learning Objectives**
  The first page of every chapter summarizes the topics that are covered in that chapter.

Symbols Used in the Textbook

Note
The author has provided additional information to the users about the topic being discussed in the form of Notes.

Tip
Special information on various techniques is provided in the form of Tips that will increase the efficiency of the users.

Formatting Conventions Used in the Textbook

Please refer to the following list for the formatting conventions used in this textbook.

- Names of tools, buttons, options, browser, palette, panels, and tabs are written in boldface.
  Example: The **Duct** tool, the **Modify** button, the **HVAC** panel, the **Systems** tab, **Properties** palette, **Project Browser**, and so on.

- Names of dialog boxes, drop-downs, drop-down lists, list boxes, areas, edit boxes, check boxes, and radio buttons are written in boldface.
  Example: The **Options** dialog box, the **Wire** drop-down in the **Electrical** panel of the **Systems** tab, the **Name** edit box in the **Name** dialog box, the **Chain** check box in the **Options Bar**, and so on.

- Values entered in edit boxes are written in boldface.
  Example: Enter **4”** (**100mm**) in the **Offset** edit box.

Naming Conventions Used in the Textbook

Tool

If you click on an item in a panel of the ribbon and a command is invoked to create/edit an object or perform some action, then that item is termed as **tool**.

For example:
Duct tool, **Air Terminal** tool, **Isolated** tool
Filled Region tool, **Trim/Extend to Corner** tool, **Rotate** tool

If you click on an item in a panel of the ribbon and a dialog box is invoked wherein you can set the properties to create/edit an object, then that item is also termed as **tool**, refer to Figure 1.

For example:
Load Family tool, **Duct** tool, **Wall** tool
Plumbing Fixture tool, **Visibility/Graphics** tool

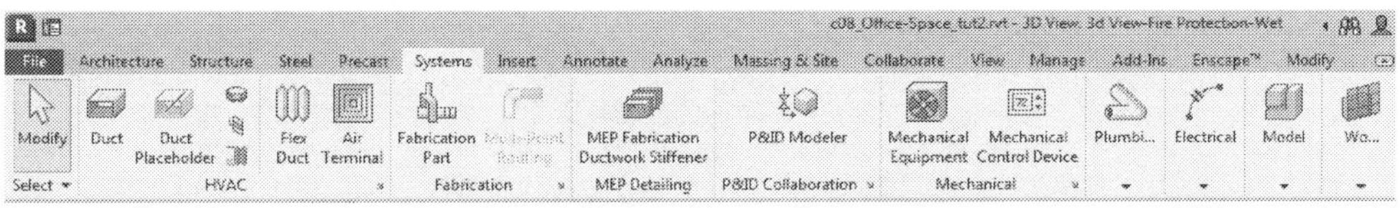

Figure 1 *Tools in the ribbon*

Button

The item in a dialog box that has a 3d shape like a button is termed as **button**. For example, **OK** button, **Cancel** button, **Apply** button, and so on. If the item in a ribbon is used to exit a tool or a mode, it is also termed as button. For example, **Modify** button, **Finish Editing System** button, **Cancel Editing System** button, and so on; refer to Figure 2.

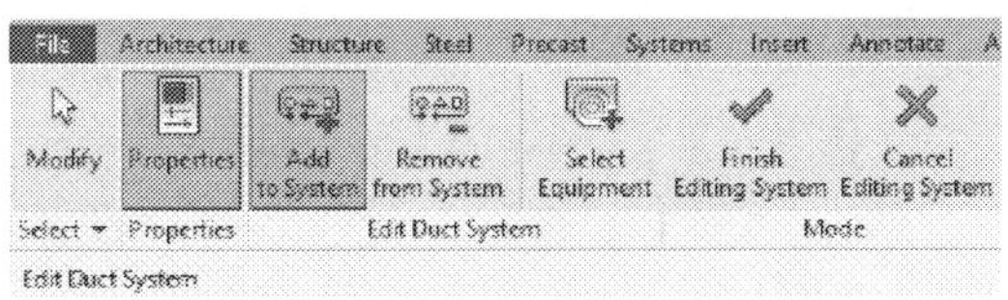

Figure 2 *Choosing the* ***Finish Editing System*** *button*

Dialog Box

In this textbook, different terms are used for referring to the components of a dialog box. Refer to Figure 3 for the terminology used.

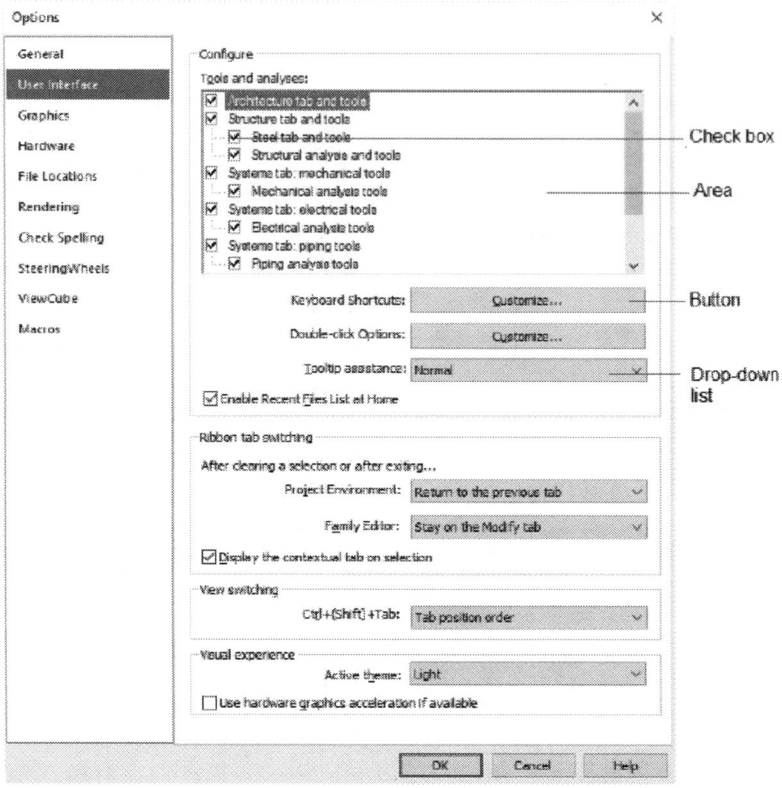

Figure 3 *Components of a dialog box*

Drop-down

A drop-down is the one in which a set of common tools are grouped together for creating an object. You can identify a drop-down with a down arrow on it. These drop-downs are given a name based on the tools grouped in them. For example, **Wall** drop-down, **Component** drop-down, **Region** drop-down, and so on; refer to Figure 4.

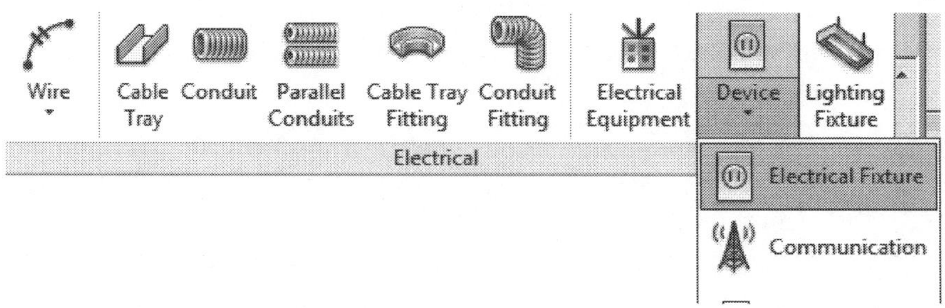

Figure 4 *Choosing a tool from the drop-down*

Drop-down List

A drop-down list is the one in which a set of options are grouped together. You can set various parameters using these options. You can identify a drop-down list with a down arrow on it. For example, **Type Selector** drop-down list, **Units** drop-down list, and so on; refer to Figure 5.

Options

Options are the items that are available in shortcut menus, drop-down lists, dialog boxes, and so on. For example, choose the **Zoom In Region** option from the shortcut menu displayed on right-clicking in the drawing area; refer to Figure 6.

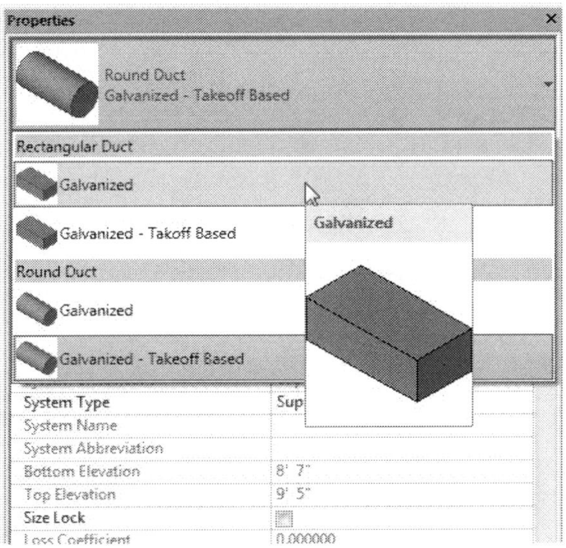

*Figure 5 Selecting an option from the **Type Selector** drop-down list*

Figure 6 Choosing an option from the shortcut menu

Some New and Enhanced Features in Revit 2025

In Revit 2025, some of the enhanced and new features are discussed next:

Measure in 3D

In Revit 2025, you can measure elements while working in a 3D view.

View on Sheet Indicator

The project browser has an option to indicate which views have been placed on sheets. If a view is placed on a sheet then a blue colored square icon appears next to the view name in the **Project Browser**.

Swap Views on Sheets

Views that are placed on sheets can now be swapped for other views. TTo swap a view, select it in the sheet and then select the sheet to swap the view with, from the ribbon.

Filter by Sheet for Schedules

The filter by sheet will look at the objects that are visible on the sheet that the schedule is placed on. As a schedule can be placed on multiple sheets, the filter will adapt to each sheet and list only the objects visible in the views placed on each sheet.

Cut CAD Models

In Revit 2025, there is an option to have imported or linked CAD models cut with the view settings. When a model is linked, a check box in its instance properties can be used to enable cutting in views.

Performance Improvement

In Revit 2025, there have also been improvements in performance for working with large models.

Tag Groups and Linked Models

In Revit 2025, groups and linked models have standard parameters applied and they have the ability to add custom parameters. These parameters can also be tagged and included in schedules.

Duplicate Material Assets

In Revit 2025, you can now duplicate materials and assets simultaneously.

Fill Pattern in 3D Families

In Revit 2025, while working in a 3D family, you can create fill patterns on selected views for added detailing value.

Pick Working Plane

In Revit 2025, you can select your working planes using the **Pick Plane** tool.

Chapter 1

Introduction to Autodesk Revit 2025 for MEP

Learning Objectives

After completing this chapter, you will be able to:

- *Understand the basic concepts and principles of Revit for MEP*
- *Understand various terms used in Revit for MEP*
- *Describe the parametric behavior of Revit*
- *Start the Revit 2025 program*
- *Understand the interface of Revit 2025*
- *Access the Revit 2025 Help*

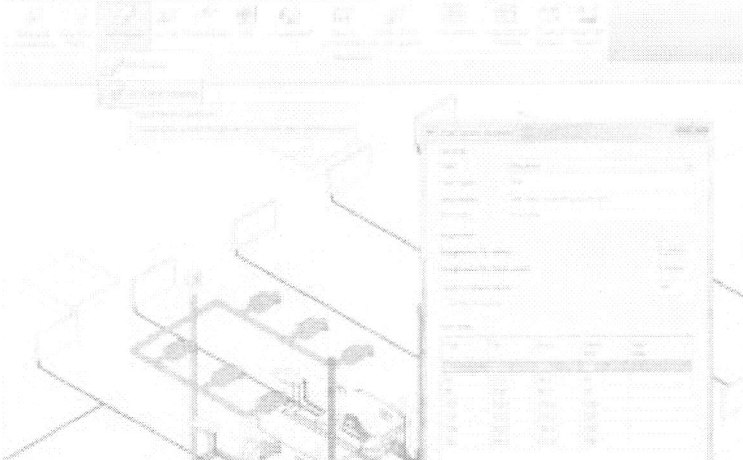

INTRODUCTION TO Autodesk Revit FOR MEP

Autodesk Revit is a Building Information Modeling software developed for professionals in the AEC (Architecture, Engineering, and Construction) industry. Revit is used by Architects, Structural Engineers, MEP Engineers, Designers and Contractors for a building project.

The MEP functionality in Revit was introduced in 2006 as a separate software, Revit MEP. It was specifically built for MEP engineers and designers. Since then, it has become very popular in the Building Information Modeling (BIM) workflow. This software provides engineers and designers with tools for the analysis, modeling, and design of various building elements and systems for MEP (Mechanical, Electrical, and Plumbing) services. Since 2017 release of Autodesk Revit, Revit MEP has been discontinued as an individual software and the MEP functionality of this software is now available in the Revit software.

Revit is a BIM software that helps users to coordinate the documentation of MEP designs with other engineering disciplines. Its integrated parametric modeling technology is used to create the information model of a project and to collect and coordinate information across all its representations. In Autodesk Revit, drawing sheets, 2D views, 3D views, sectional view, callout details, and schedules directly represent the same building information model (BIM) as the real one does. Autodesk Revit (for MEP) is developed with an approach to bring the Mechanical, Electrical, and Plumbing engineers together under the BIM framework and make the building services system efficient and interoperable with other systems. In Revit, a designer can not only work with various pre-designed elements of different MEP disciplines but can also model customized elements and add parameters to them. This helps in modeling complex designs with various permutations. Different disciplines of Revit (for MEP) are briefly described next.

Mechanical Discipline

In the Mechanical discipline, you can develop an HVAC (Heating, Ventilation, and Air Conditioning) system, keeping in view the energy requirements of that building. The study of the energy requirements of the building is very essential for developing an efficient and cost effective design. In mechanical discipline, you can design the whole ducting network with the ventilation layout plan. You can also route the piping or ducting networks manually or generate routing solutions by using various tools in this software. In this discipline, you can also develop a Fire Suppression System.

Electrical Discipline

While working with the electrical discipline, you can design an electrical system. In this system, you can add various lighting fixtures, switches, alarms, communication devices, and more as per the requirement of the project. You can also add panels and prepare panel schedules and perform the load analysis. Further, you can connect the devices and fixtures through logical circuits.

Plumbing Discipline

In this discipline, you can design a plumbing system for a project. In the plumbing system, you can add plumbing fittings, accessories, and fixtures as per the requirement of a project. In addition, you can also design fire fighting system for a building and add fire safety components to the system.

Autodesk Revit AS A BUILDING INFORMATION MODELER (BIM)

The history of computer aided design and documentation dates back to the early 1980s when architects and engineers began using this technology for documenting their projects. Realizing its advantages, information sharing capabilities were developed, especially to share data with other consultants. This led to the development of object-based CAD systems in the early 1990s. Before the development of these systems, objects such as HVAC components, pipes, plumbing fixtures, electrical fixture, and more were stored as a non-graphical data with the assigned graphics. These systems arranged the information logically but were unable to optimize its usage in a building project. Realizing the advantages of the solid modeling tools, the mechanical and manufacturing industry professionals began using the information modeling CAD technology. This technology enabled them to extract data based on the relationship between model elements.

The Building Information Modeling (BIM) provided an alternative approach to building design, construction, and management. This approach, however, required a suitable technology to implement and reap its benefits. In such a situation, the use of parametric technology with the Building Information Modeling approach was envisaged as an ideal combination. In 1997, a group of mechanical CAD technologists began working on a new software dedicated to the building industry. They developed a software that was suitable for creating MEP projects. This led to the development of Autodesk Revit.

Autodesk Revit is a design and documentation platform in which a digital MEP model is created using the parametric elements such as HVAC system, mechanical equipment, plumbing network, fire fighting, and so on. All MEP elements have inherent characteristics, and therefore, they can be tracked, managed, and maintained by using computer.

BASIC CONCEPTS AND PRINCIPLES

Autodesk Revit enables you to envisage and develop an MEP model with actual 3D parametric elements. It provides a new approach to MEP design and implementation process. It replicates the way MEP engineers conceive the structure of an MEP system. For example, the 2D CAD platforms mostly use lines to represent all elements, as shown in Figure 1-1. However, in Autodesk Revit, you can create the MEP model of a building project using 3D elements, such as HVAC components, pipes, plumbing fixtures, electrical fixtures, as shown in Figure 1-2.

Using these 3D elements, you can visualize the MEP project with respect to its scale, volume, and proportions. This enables you to study design alternatives and develop superior quality design solutions. Autodesk Revit automates routine drafting and coordination tasks and helps in reducing errors in documentation. This, in turn, saves time, improves the speed of documentation, and lowers the cost for the users.

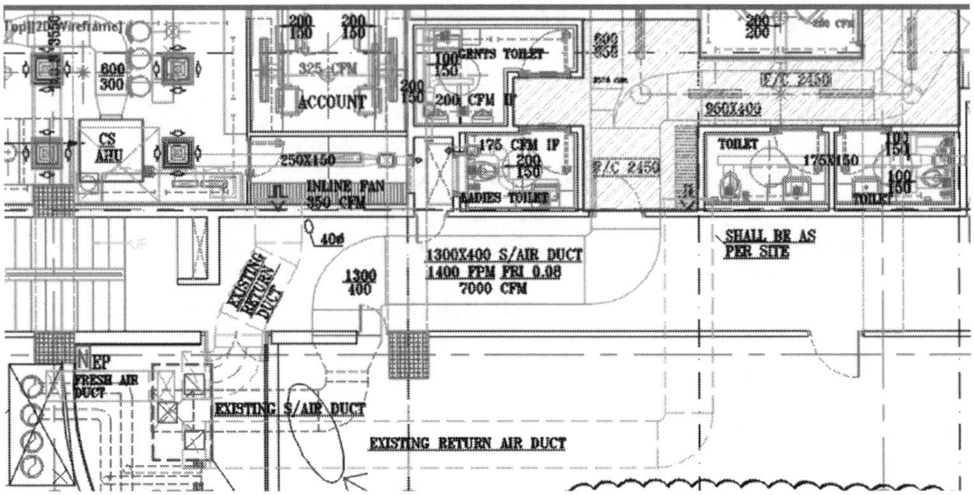

Figure 1-1 *CAD project created using 2D lines*

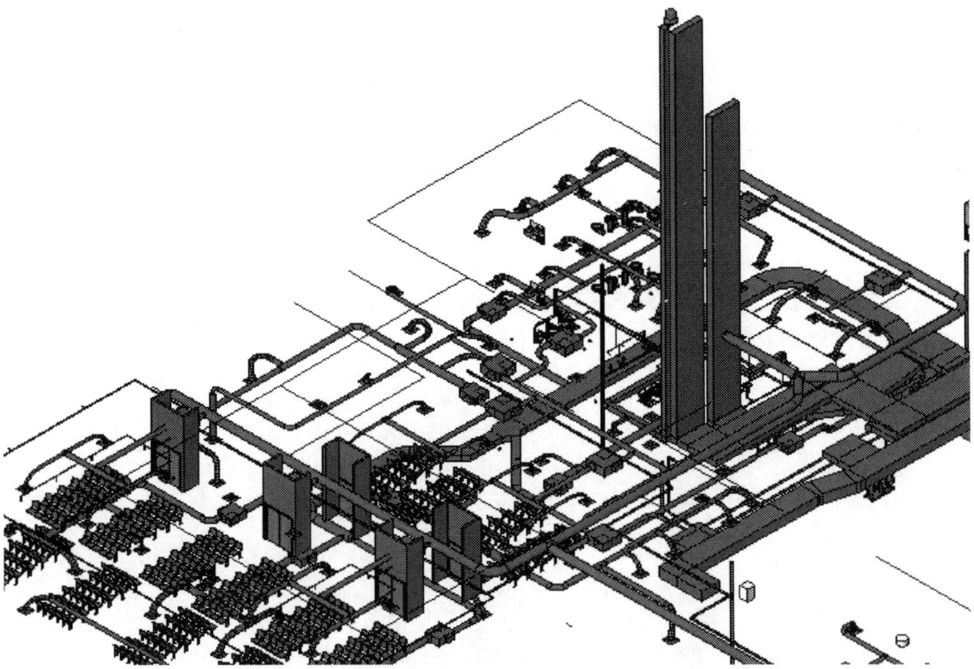

Figure 1-2 *An MEP project created using parametric elements*

Understanding the Parametric Building Modeling Technology

A project in Autodesk Revit is created using the inbuilt parametric building elements. The term 'parametric' refers to parameters that define relationship between various building elements. Some of these relationships are defined by Autodesk Revit itself and others by the users. For example, the relationship between air terminals and ceilings are defined by MEP and the relationship between connectors and ducts are defined by the users.

In an MEP project, each element has inbuilt bidirectional associativity with many other elements. These elements together form an integrated building information model. This model contains all data needed for the design and development of the project. You can then use this data to create project presentation views such as ceiling plans, sections, elevations, and so on for documentation. As you modify the model while working in certain views, Autodesk Revit's parametric change engine automatically updates other views. This capability is, therefore, the underlying concept in Autodesk Revit.

Autodesk Revit's parametric change engine enables you to modify design elements at any stage of the project development. As changes in the model are reflected immediately and automatically in the project, the time and effort required in coordinating the changes in other views is saved. This feature provides immense flexibility in the design and development process along with an error-free documentation.

Autodesk Revit also provides a variety of in-built parametric element libraries that can be selected and used to create a building model. It also provides you with the flexibility to modify the properties of these elements or to create your own parametric elements, based on the project requirement.

Terms Used in Autodesk Revit for MEP

Before working with Autodesk Revit, it is important to understand the basic terms used for creating a building model. Various terms in Autodesk Revit such as project, level, category, family, type, and instance are described next.

Autodesk Revit Project

A project in Autodesk Revit is similar to an actual project. In an actual project, the entire documentation such as drawings, 3D views, specifications, schedules, cost estimates, and so on are inherently linked and read together. Similarly, in Autodesk Revit, a project not only includes the digital 3D MEP model but also its parametrically associated documentation. Thus, all the components such as the building model and its standard views, MEP drawings, and schedules together form a complete project. A project file contains all the project information such as building and MEP elements used in a project, drawing sheets, schedules, cost estimates, 3D views, renderings, and so on. A project file also stores various settings such as environment, lighting, and so on. As the entire data is stored in the same file, so it becomes easier for Autodesk Revit to coordinate the database.

Levels in a Building Model

In Autodesk Revit, a building model is divided into different levels. These levels may be understood as infinite horizontal planes that act as hosts for different elements such as roof, floor, ceiling, and so on. Each element that you create belongs to a particular level.

Subdivisions of Elements into Categories and Subcategories

Apart from MEP elements, an Autodesk Revit project also contains other associated elements such as annotations, imported files, links, and so on. These elements have been divided into the following categories:

Model Category	:	Consists of various MEP elements such as HVAC elements, ducts, air terminals, diffusers, pipes, plumbing fixtures, electrical conduits, and others used in creating systems
Annotation Category	:	Consists of annotations such as dimensions, text notes, tags, symbols, and so on
Datum Category	:	Consists of datums such as levels, grids, reference planes, and so on
View Category	:	Consists of interactive project views such as the architectural, mechanical, and plumbing floor plans, elevations, sections, 3D views, and renderings

In addition to these four categories, other categories such as **Imported**, **Workset**, **Filter**, and **Revit Categories** can also exist if the project has imported files, enabled worksets, or linked Autodesk Revit projects, respectively.

Families in Autodesk Revit

Another powerful concept in Autodesk Revit is family. A family is described as a set of elements of the same category that are grouped together based on certain common parameters or characteristics. Elements of the same family may have different properties, but they all have common characteristics. For example, **Rectangular Diffuser - Round Connection** is an air diffuser family and it contains different sizes of air diffusers. Family files have the *.rfa* extension. You can load additional MEP component families from the libraries provided in Autodesk Revit package.

Families are further divided into certain types. A type or family type, as it is called, is a specific size or style of a family. For example, **Rectangular Diffuser - Round Connection 24x24 - 10 Neck** in Metric (**M_Rectangular Diffuser - Round Connection 600x600- 230 Neck)** is an air diffuser type. Family and family types can also be used to create new families using the **Family Editor**.

Instances are the actual usage of model elements in an MEP model or annotations in a drawing sheet. A family type, created at a new location, is identified as an instance of the family type. All the instances of the same family type have the same properties. Therefore, when you modify the properties of a family type, the properties of all its instances also get modified. The family categorization of Revit elements is given below:

Model Category	: Air diffuser
Family	: **Rectangular Diffuser - Round Connection**
Family type	: **Rectangular Diffuser - Round Connection 24x24 - 10 Neck**
Instance	: Particular usage of a family type

The hierarchy of service elements in Autodesk Revit plays an important role in providing flexibility and ease in managing a change in a building model. Figure 1-3 shows the hierarchy of categories and families in a typical Autodesk Revit project.

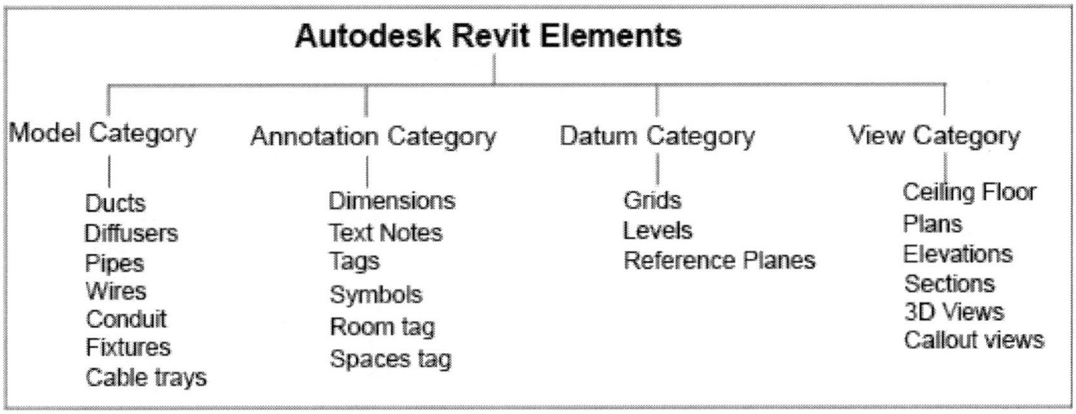

Figure 1-3 *Hierarchy of Autodesk Revit MEP categories and families*

Creating an MEP Model Using Parametric Elements

Another classification of categories of elements followed in Autodesk Revit is based on their usage. Autodesk Revit uses five classes of elements: Host, component, annotation, view, and datum. Hosts are the element categories that form the basic system of an MEP model and include model elements such as ducts, pipes, cables, and more. Components are the elements that are added to host elements or act as stand-alone elements such as air terminals, diffusers, and conduits. Annotations are the 2D, view-specific elements such as dimensions, tags, text notes, and so on that add content to the project documentation. Views represent various orientations of a building model such as plans, elevations, sections, 3D views, and so on. Datum refers to the reference elements that assist you in creating a building model, which include grids, levels, reference planes, and so on.

There is no specific methodology available for creating a services model in Autodesk Revit. It provides you with the flexibility of generating the MEP model based on the project requirement, design complexity, and other factors. However, the following steps describe a general procedure that may be followed for creating an MEP model using the built-in parametric elements provided in Autodesk Revit.

In Revit, you can start designing a project for individual discipline (Mechanical, Electrical, or Plumbing) by selecting the specific template. For example, to design a project for a mechanical discipline, you can select the *Mechanical-Default.rte* (*Mechanical-Default_Metric.rte*) template file. Alternatively, you can start a project to work in all the disciplines by selecting the *Systems-Default.rte* (*System-Default_Metric.rte*) template file.

Once you have started a project, you need to copy the levels of the architectural model to the current project or create additional levels as per the requirement. Next, you can start with any of the disciplines by activating the specific view from the **Project Browser**. For example, to start with mechanical discipline, activate the desired mechanical plan view under **Views** (**Discipline**) **> Mechanical** node from the **Project Browser**.

In Revit, there are specific workflows for each discipline as per the requirement of the project. The workflow for disciplines generally includes analysis, design, and documentation. For mechanical discipline, you need to analyze spaces to design an appropriate HVAC system. Then, based on the analysis, you need to place air terminals, equipment, and design ducts for the system. Figure 1-4 shows an example of a mechanical system.

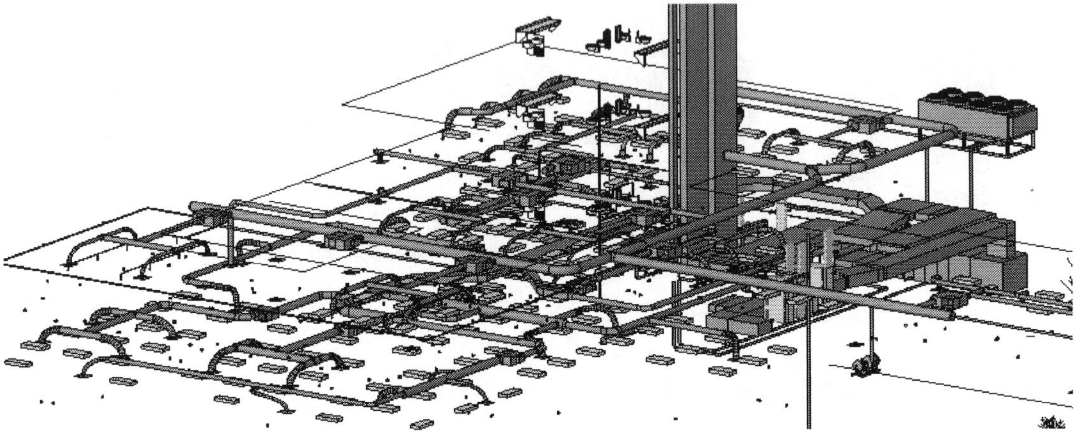

Figure 1-4 *Mechanical system with its elements*

Visibility/Graphics Overrides, Scale, and Detail Level

Autodesk Revit enables you to control the display and graphic representation of a single element or the element category of various elements in the project views. This is done by using the visibility and graphics overrides tools. You can select a model category and modify its linetype and detail level. This can also be done for various annotation category elements and imported files. These settings can be done for each project view based on its desired representation. You can also hide an element or an element category in a view using the **Hide in view** and **Isolate** tools. You can override the graphic representation of an element or an element category in any view using the **Visibility/Graphics** tool.

The scale is another important concept in an Autodesk Revit project. You can set the scale for each project view by selecting it from the available list of standard scales such as **1/16"=1'0"**, **1/4"=1'0"**, **1"=1'0"**, **1/2"=1'0"** for Imperial system or **1: 50**, **1: 100**, **1: 200**, **1: 500** for Metric system. As you set a scale, Autodesk Revit automatically sets the detail level that is appropriate for it. There are three detail levels provided in an Autodesk Revit project: **Coarse**, **Medium**, and **Fine**. You can also set the detail level manually for each project view. Each detail level has an associated linetype and the detail lines associated with it. The details of annotations, such as dimensions, tags, and so on, are also defined by the selected scale.

Extracting Project Information

A single integrated building information is used to create and represent a building project. You can extract project information from a building model and create area schemes, schedule, and cost estimates, and then add them to the project presentation.

Autodesk Revit also enables you to export the extracted database to the industry standard Open Database Connectivity (ODBC) compliant relational database tables. The use of the building information model to extract database information eliminates the error-prone method of measuring building spaces individually.

Creating an MEP Drawing Set

After creating the building model, you can easily arrange the project views by plotting them on the drawing sheets. The drawing sheets can also be organized in a project file based on the established CAD standards followed by the firm. In this manner, the project documentation can easily be transformed from the conceptual design stage to the design development stage and finally to the construction document stage. The project view on a drawing sheet is only a graphical representation of the building information model. Therefore, any modification in it is immediately made in all associated project views, keeping the drawing set always updated.

Creating an Unusual Building Geometry

Autodesk Revit also helps you conceptualize a building project in terms of its volume, shape, and proportions before working with actual building elements. This is done by using the **Massing** tool, which enables you to create quick 3D models of buildings and conduct volumetric and proportion study on overall masses. It also enables you to visualize and create an unusual building geometry. The same massing model can then be converted into a building model with individual parametric building elements. It provides continuity to the generation of building model right from sketch design to its development. You can also create various custom MEP elements as per the project requirement and then load them to the project.

Flexibility of Creating Special Elements

Autodesk Revit provides a large number of in-built family types of various model elements and annotations. Each parametric element has the associated properties that can be modified based on the project requirement.

Autodesk Revit also enables you to create the elements that are designed specifically for a particular location. The in-built family editor enables you to create new elements using family templates. This provides you with the flexibility of using in-built elements for creating your own elements. For example, using the furniture template, you can create a reception desk that is suitable for a particular location in the design.

Creating Services Layouts

Autodesk Revit provides you with an extensive in-built library of MEP elements that can be used to add elements such as ducts, air terminals, diffusers, conduits, and so on to a project. This helps MEP consultants to include these service elements in the basic architectural building model and check for inconsistency, if any.

Working on Large Projects

In Autodesk Revit, you can work on large projects by linking different building projects together. For a large project that comprises of a number of buildings, you can create individual buildings as separate projects and then link all of them into a single base file. The database recognizes the linked projects and includes them in the project representation of the base file.

For example, while working on a large educational institution campus, you can create separate project files for academic building, administration area, gymnasium, cafeteria, computer center, and so on, and then link them into the base site plan file. In this manner, large projects can be subdivided and worked upon simultaneously.

Working in Large Teams and Coordinating with Consultants

In Autodesk Revit, worksets enable the division of the MEP model into small editable sets of disciplines such as Mechanical, Electrical, and Plumbing. The worksets can be assigned to different teams working on the same project and then their work can easily be coordinated by sharing the files in the central file location. The effort required to coordinate, collaborate, and communicate the changes between various worksets is taken care of by the computer. Various consultants working on a project can be assigned a workset with a set of editable elements. They can then incorporate their services and modify the associated elements.

For example, a high rise commercial building project can be divided into different worksets with independent teams working on different disciplines such as Mechanical, Electrical, Plumbing, Architecture, Structure, and so on. The structural consultants can be assigned to the exterior skin and the core workset, in which they can incorporate structural elements. Similarly, the rest of the teams can work independently on different worksets.

STARTING Autodesk Revit 2025

You can start Autodesk Revit by double-clicking on the Revit 2025 icon on the desktop. Alternatively, choose **Autodesk > Revit 2025** from the **Start** menu (for Windows 10); the user interface screen is displayed, as shown in Figure 1-5.

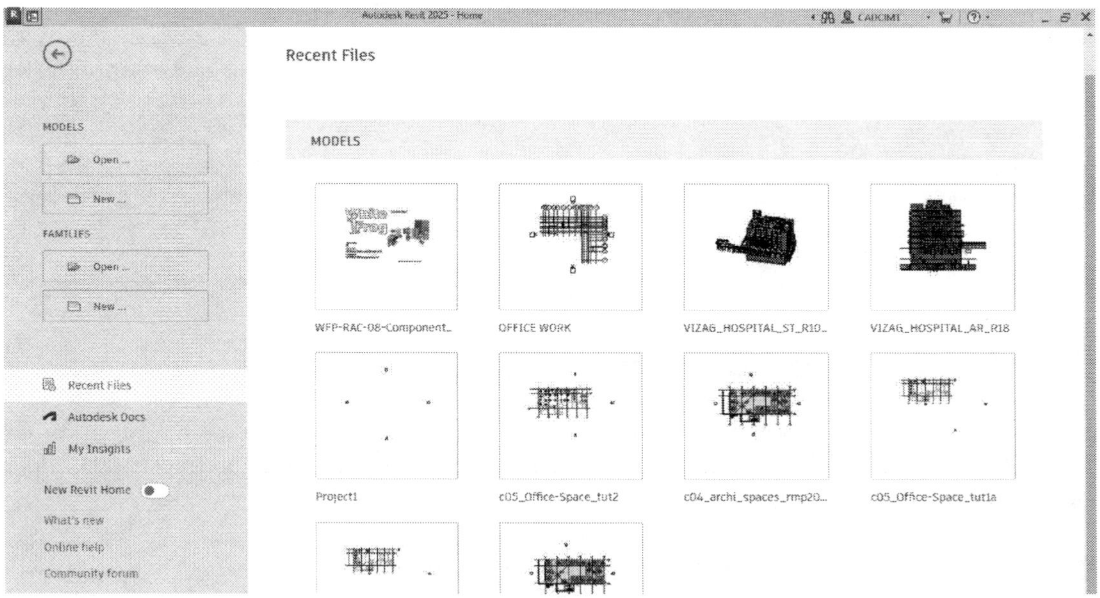

***Figure 1-5** The interface of Autodesk Revit 2025*

Note
The path for starting Autodesk Revit depends on the operating system being used.

The screen interface has two sections: **MODELS** and **FAMILIES**. The options in the **MODELS** section are used to open a new or an existing project. The options in the **FAMILIES** section are used to open a new or an existing family.

In the interface, choose the **Autodesk Docs** option to browse the projects saved in the **Autodesk Construction Cloud** platform. Also, in the interface, you can choose the **Recent Files** option to toggle between the projects saved in the **Autodesk Docs** and Revit platform.

In the interface, you can choose the **What's new** option to get information regarding the new features introduced in Revit 2025. Also, you can choose the **My Insights** option where a range of data-driven insights about their Revit usage are shown through interactive cards. These cards come equipped with icons that provide positive or negative feedback, aiming to enhance the software's ability to tailor the displayed insights based on user preferences.

In the **MODELS** section, choose the **Open** option; the **Open** dialog box will be displayed. Browse to the desired location in the dialog box and select the file. Now, choose the **Open** button to open the file. The **New Revit Home** slider allows you to seamlessly switch to the redesigned home screen in Revit 2025.

To open a new project file, choose the **New** option from the **MODELS** section. Alternatively, choose **New > Project** from the **File** menu; the **New Project** dialog box will be displayed. In this dialog box, make sure that the **Project** radio button is selected, and then choose the **OK** button; a new project file will open and the interface screen is activated.

USER INTERFACE

In Autodesk Revit, the user interface consists of the Ribbon, Drawing area, Properties palette, Status Bar, and the View Control Bar, as shown in Figure 1-6. In Autodesk Revit, all the tools are grouped in several panels in the ribbon.

The ribbon, which contains task-based tabs and panels, streamlines the structural workflow and optimizes the project delivery time. In Autodesk Revit, when you select an element in the drawing area, the ribbon displays a contextual tab that comprises of tools corresponding to the selected element. The interface of Autodesk Revit is similar to the interfaces of many other Microsoft Windows-based programs. The main components in the Revit interface are discussed next.

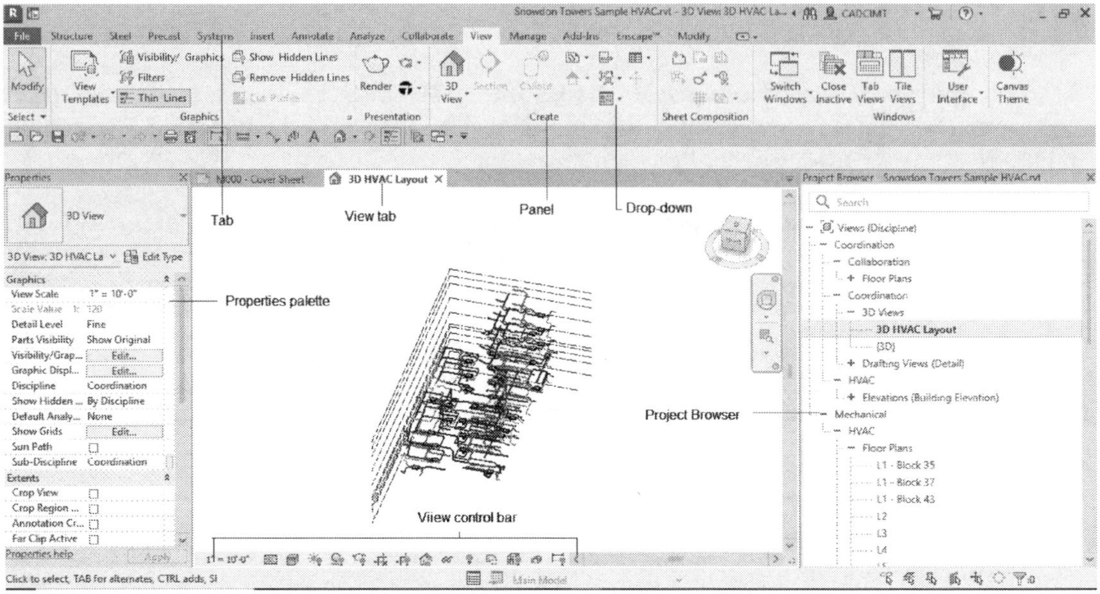

Figure 1-6 The Autodesk Revit 2025 user interface screen

Title Bar

The Title Bar, docked on the top portion of the user interface, displays the program's logo, name of the current project, and the view opened in the viewing area. **Project 1- Floor Plan: Level 1** is the default project name and view displayed.

Ribbon

The ribbon, as shown in Figure 1-7, is an interface that is used to invoke tools. When you open a file, the ribbon is displayed at the top in the screen. It comprises of task-based tabs and panels, refer to Figure 1-7, which provide all the tools necessary for creating a project. The tabs and panels in the ribbon can be customized according to the need of the user. This can be done by moving the panels and changing the view states of the ribbon (the method of changing the ribbon view state is discussed later in this chapter). The ribbon has three types of buttons: general, drop-down, and split. These buttons can be used from the panels.

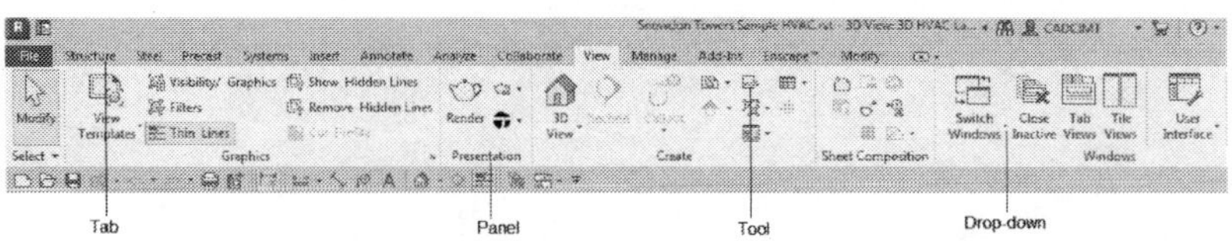

Figure 1-7 Different components of a ribbon

In the ribbon, you can move a panel and place it anywhere on the screen. To do so, press and hold the left mouse button on the panel label in the ribbon, and then drag the panel to a desired place on the screen. After using the tools of the moved panel, place the panel back to the ribbon. To do so, place the cursor on the moved panel and choose the **Return Panels to Ribbon** button from the upper right corner of this panel, as shown in Figure 1-8; the panel will return to the ribbon.

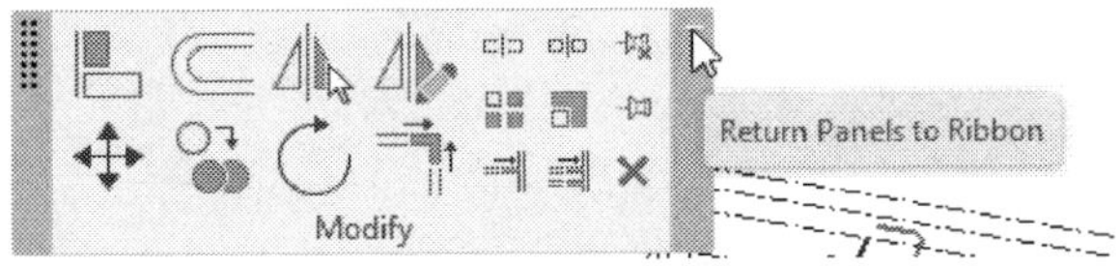

Figure 1-8 *Choosing the **Return Panels to Ribbon** button*

> **Tip**
> *Tooltips appear when you place the cursor over any tool icon in the ribbon. The name of the tool appears in the box, assisting you in identifying each tool icon.*

Changing the View States of the Ribbon

The ribbon can be displayed in three view states by selecting any of the following four options: **Minimize to Tabs**, **Minimize to Panel Titles**, **Minimize to Panel Buttons**, and **Cycle through All**. To use these options, click on the down arrow located on the right of the **Modify** panel; the arrow will be highlighted. Next, click on the down arrow; a flyout will be displayed, as shown in Figure 1-9.

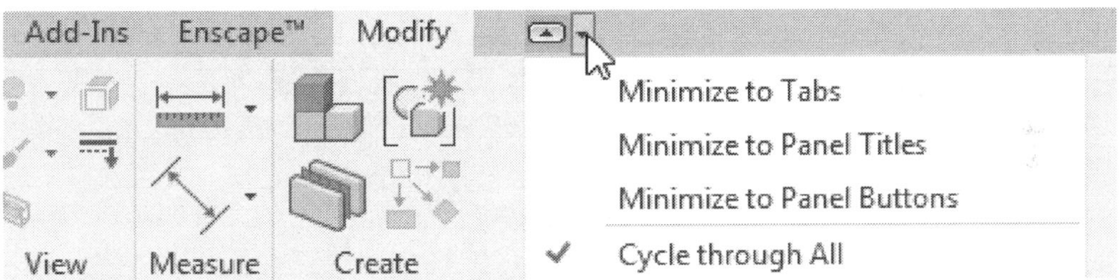

Figure 1-9 *Various options in the flyout for changing the view state of the ribbon*

From this flyout, you can choose the **Minimize to Tabs** option to display only the tabs in the ribbon. If you choose the **Minimize to Panel Titles** option, the ribbon will display the titles of the panels along with the tabs. You can choose the **Minimize to Panel Buttons** option to display panels as buttons along with tabs in the ribbon.

> **Note**
> *If the ribbon is changed to a different view state, then on placing the cursor over the first arrow on the right of the **Modify** tab, the **Show Full Ribbon** tooltip will be displayed. Click on the arrow; the full ribbon will be displayed.*

The following table gives description of various tabs in ribbon.

Tab	Description
Architecture	Contains tools for creating an architectural model
Structure	Contains tools for creating structural model
Systems	Contains tools for creating an MEP model
Precast	Contains tools related to precast structures
Steel	Contains tools for the modification of the steel structure
Insert	Contains tools for inserting or managing secondary files such as raster image files and CAD files
Annotate	Contains tools for documenting a building model such as adding texts and dimensions
Analyze	Contains tools for analyzing the structural model
Massing & Site	Contains tools for creating massing and site elements
Collaborate	Contains tools for collaborating the project with other team members (internal and external)
View	Contains tools for managing and modifying the current views, switching views, and so on.
Manage	Contains tools for specifying the project and system parameters and settings
Add - Ins	Contains add in links for interoperability of BIM software
Modify	Contains tools for editing elements in the model

Contextual Tabs in the Ribbon

These tabs are displayed when you choose certain tools or select certain elements. They contain a set of tools or buttons that relate only to a particular tool or element.

For example, when you invoke the **Duct** tool, the **Modify | Place Duct** contextual tab is displayed. This tab has the following panels: **Select**, **Properties**, **View**, **Measure**, **Geometry**, **Clipboard**, **Create**, **Modify**, **Tag**, and **Placement Tools**. The **Select** panel contains the **Modify** tool. The **Properties** panel contains the **Properties** button and the **Type Properties** tool. The **Mode** panel has tools that are used to load model families or to create the model of a window in a drawing.

The other panels, apart from those discussed above, contain the tools that are contextual and are used to edit elements when placed in a drawing or selected from a drawing for modification.

Application Frame

The application frame helps you manage projects in Autodesk Revit. It consists of **File** menu, **Quick Access Toolbar**, **InfoCenter**, and **Status Bar**. These are discussed next.

File Menu

The **File** menu contains the tools that provide access to tools such as **Open**, **Close**, and **Save**, as shown in Figure 1-10. Alternatively, press ALT+F to display tools in the **File** menu.

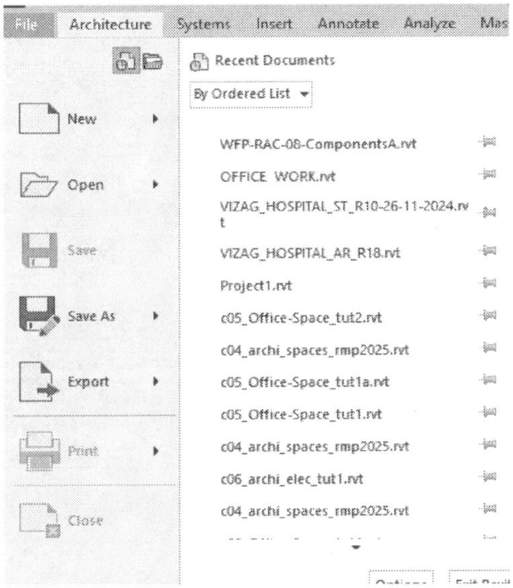

Figure 1-10 The ***File*** *menu*

Quick Access Toolbar

The **Quick Access Toolbar**, shown in Figure 1-11, contains the options to undo and redo changes, open and save a file, create a new file, and so on.

By default, the **Quick Access Toolbar** contains the options such as **Open**, **Save**, **Redo**, **Undo**, and so on. You can customize the display of the **Quick Access Toolbar** by adding more tools and removing the unwanted tools. To add a tool or a button from the panel of the ribbon to the **Quick Access Toolbar**, place the cursor over the button; the button will be highlighted. Next, right-click; a flyout will be displayed. Choose **Add to Quick Access Toolbar** from the flyout displayed; the highlighted button will be added to the **Quick Access Toolbar**. The **Quick Access Toolbar** can be customized to reorder the tools displayed in it. To do so, choose the down arrow next to the **Switch Windows** drop-down in the **Quick Access Toolbar**, refer to Figure 1-11; a flyout will be displayed. Choose the **Customize Quick Access Toolbar** option located at the bottom of the flyout; the **Customize Quick Access Toolbar** dialog box will be displayed. Use various options in this dialog box to customize the display of toolbar and choose the **OK** button; the **Customize Quick Access Toolbar** dialog box will close and the tools in the **Quick Access Toolbar** will be reordered.

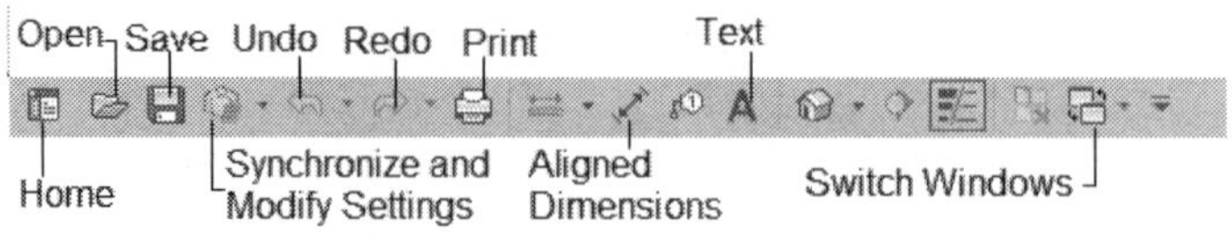

Figure 1-11 The ***Quick Access Toolbar***

InfoCenter

You can use **InfoCenter** to search for information related to Revit (Help) to display the **Subscription Center** panel for subscription services and product updates, and to display the **Favorites** panel to access saved topics. Figure 1-12 displays various tools in **InfoCenter**.

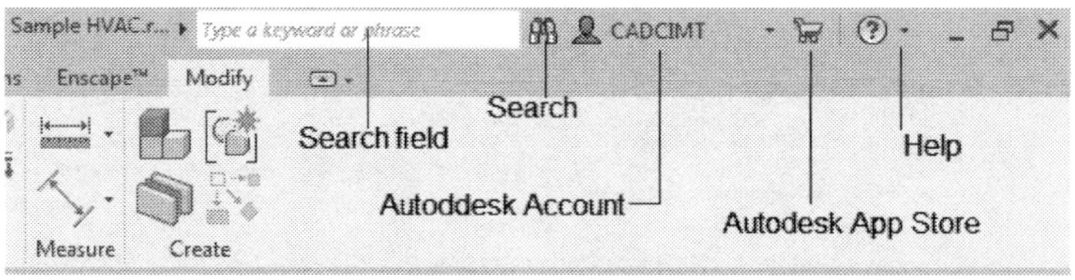

Figure 1-12 The ***InfoCenter***

Status Bar

The Status Bar is located at the bottom of the interface screen. When the cursor is placed over an element or a component, the Status Bar displays the name of the family and the type of the corresponding element or components. It also displays prompts and messages to help you use the selected tools.

View Control Bar

The **View Control Bar** is located at the lower left corner of the drawing window, as shown in Figure 1-13. It can be used to access various view-related tools. The **Scale** button shows the scale of the current view. When you can choose this button, a flyout containing standard drawing scales is displayed. From this flyout, you can then select the scale for the current view. The **Detail Level** button is used to set the detail level of a view. You can select the required detail level as **Coarse**, **Medium**, and **Fine**. Similarly, the **Visual Style** button enables you to set the display style. The options for setting the display style are: **Wireframe**, **Hidden Line**, **Shaded**, **Consistent Colors**, **Shaded**, and **Raytrace**.

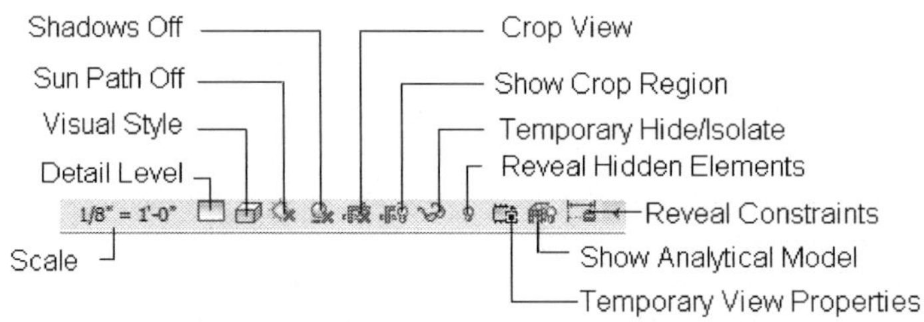

Figure 1-13 The ***View Control Bar***

Options Bar

The **Options Bar** provides information about the common parameters of a component type. It also displays options for creating or editing them. The options displayed in the **Options Bar** change according to the type of component being created and selected for editing. Figure 1-14 displays the options in the **Options Bar** to create a structural column.

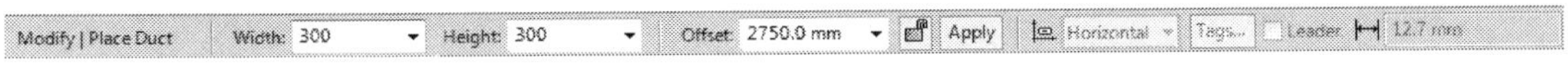

***Figure 1-14** The **Options Bar** with different options to create a structural column*

Type Selector

The **Type Selector** drop-down list is located in the **Properties** palette for the currently invoked tool. On invoking the **Duct** tool, the properties of the duct will be displayed in the **Properties** palette. In this palette, you can use the **Type Selector** drop-down list to select the required type of the beam. The options in the **Type Selector** drop-down list keep changing, depending upon the current function of the tool or the elements selected. The **Type Selector** drop-down list can also be used to specify the type of an element or component while placing that element or the component in a drawing by using the **Place a Component** tool. You can also use this drop-down list to change the type of a selected element.

Drawing Area

The Drawing Area is the actual modeling area where you can create and view the building model. It covers the major portion of the interface screen. You can draw building components in this area. The position of the pointing device is represented by the cursor. The Drawing Area also has the standard Microsoft Windows functions and buttons such as close, minimize, maximize, scroll bar, and so on. These buttons have the same function as that of the other Microsoft Windows-based programs.

PROJECT BROWSER

The **Project Browser** is located below the ribbon. It displays project views, schedules, sheets, families, and groups in a logical, tree-like structure, as shown in Figure 1-15 and helps you to open and manage them. To open a view, double-click on the name of the view, or drag and drop the view in the Drawing Area. You can close the **Project Browser** or dock it anywhere in the Drawing Area.

Note

*If the **Project Browser** is not displayed on the screen, choose the **View** tab from the ribbon and then select the **Project Browser** check box from **View > Windows > User Interface** drop-down.*

The **Project Browser** can be organized to group the views and sheets based on the project requirement. For example, while working on a large project with a number of sheets, you can organize the **Project Browser** to view and access specific sheets.

Note

*In the **Project Browser**, you can expand or contract the view listing by selecting the '+' or '-' sign, respectively. The current view in the drawing window is highlighted in bold letters. The default project file has a set of preloaded views.*

In the **Project Browser**, you will find a Search bar at the top. This feature enables you to search seamlessly within the views, schedules, sheets, families, groups, Revit Links, and any other elements loaded into the project. The search functionality is highly responsive, as it initiates the search as you begin typing, ensuring a swift and smooth experience.

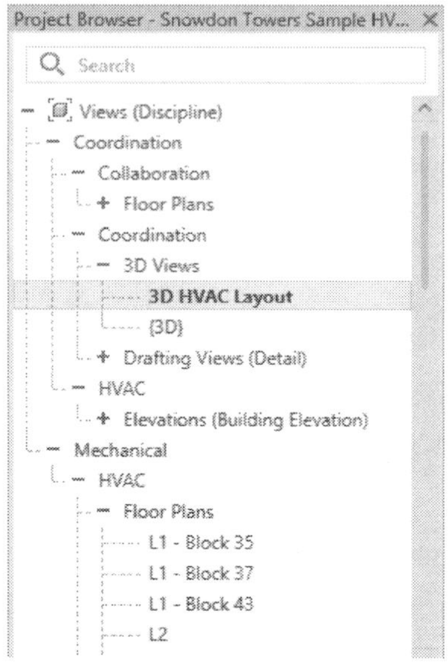

Figure 1-15 The ***Project Browser***

Keyboard Accelerators

In Autodesk Revit, accelerator keys have been assigned to some of the frequently used tools. These keys are shortcuts that you can type from the keyboard to invoke the corresponding tool. The accelerator key corresponding to a tool appears as a tooltip when you move the cursor over the tool.

Properties Palette

The **Properties** palette, as shown in Figure 1-16, is an interface without model, which displays the type and element properties of various elements and views in a drawing.

The **Properties** palette is dockable and resizable, and it supports multiple monitor configurations. The **Properties** palette is displayed in the Revit interface by default and it shows the instance properties of the active view.

Tip

As you become accustomed to use Autodesk Revit, you will find these Keyboard Accelerators quite useful because they save the effort of browsing through the menus.

When you select an element from a drawing, the **Properties** palette displays its instance properties. You can also access the type properties of the selected element from the **Properties** palette. To do so, choose the **Edit Type** button from the palette; the **Type Properties** dialog box will be displayed. In this dialog box, you can change the type properties of the selected element.

In the **Properties** palette, you can assign a type to the selected element in a drawing using the options from the **Type Selector** drop-down list. In Revit, you can toggle the display of the **Properties** palette in its interface. Choose the **Properties** button in the **Properties** panel of the **Modify** tab to hide it. Similarly, you can choose the **Properties** button to display the palette if it is not visible in the interface.

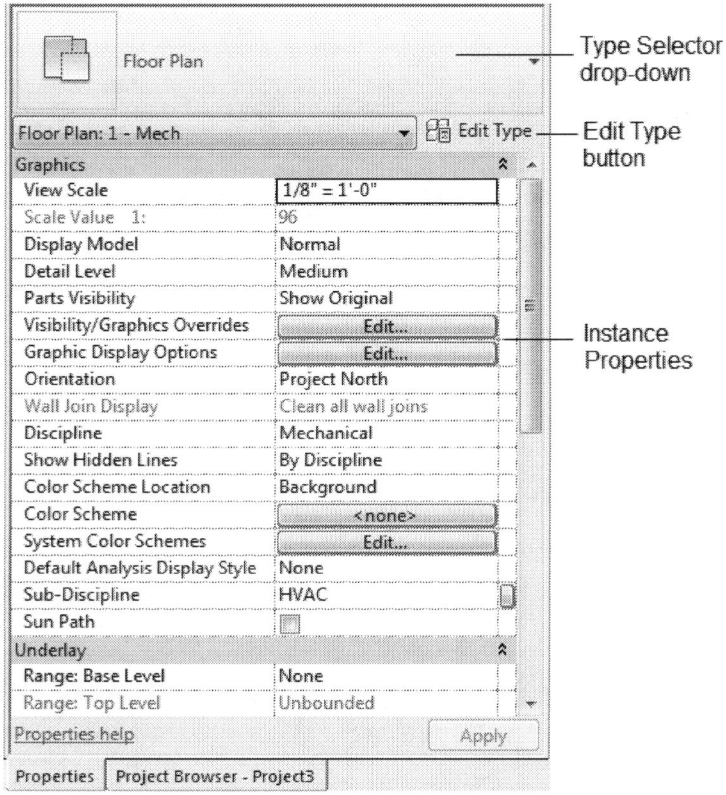

*Figure 1-16 The **Properties** palette*

DIALOG BOXES

Some Autodesk Revit tools, when invoked, display a dialog box. A dialog box is an interface for accessing, specifying, and modifying the parameters related to that tool. For example, when you choose **Save As > Project** from the **File** menu, the **Save As** dialog box is displayed, as shown in Figure 1-17.

A dialog box consists of various parts such as dialog label, radio buttons, text or edit boxes, check boxes, slider bars, image box, buttons, and tools, which are similar to other windows-based programs. Some dialog boxes contain the **Browse** button, which displays another related dialog box. There are certain buttons such as **OK**, **Cancel**, and **Help**, which appear at the bottom of most of the dialog boxes. The names of the buttons imply their respective functions.

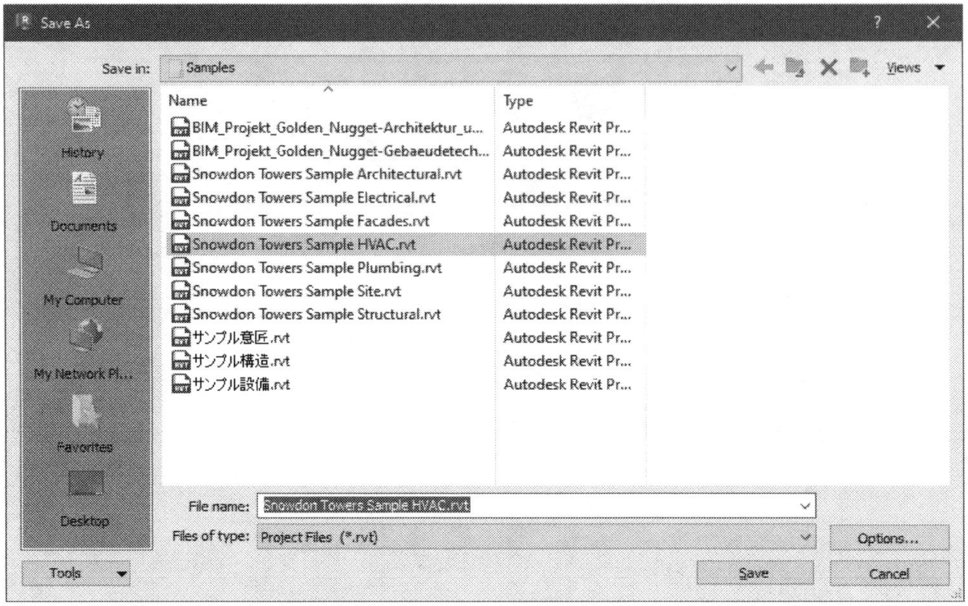

*Figure 1-17 The **Save As** dialog box*

MULTIPLE DOCUMENT ENVIRONMENT

The multiple document environment feature allows you to open more than one project at a time in a single Autodesk Revit session. This is very useful when you want to work on different projects simultaneously and make changes with reference to each other.

Sometimes, you may need to incorporate certain features from one project into the other. With the help of multiple document environment feature, you can open multiple projects and then use the **Cut**, **Copy**, and **Paste** tools from the **Clipboard** panel of the **Modify (type of element)** tab to transfer the required components from one project to another. These editing tools can also be invoked by using the CTRL+C and CTRL+V keyboard shortcuts.

To access the opened projects, click on the **Switch Windows** drop-down in the **Windows** panel of the **View** tab; the options for the names of different opened project files will be displayed, as shown in Figure 1-18.

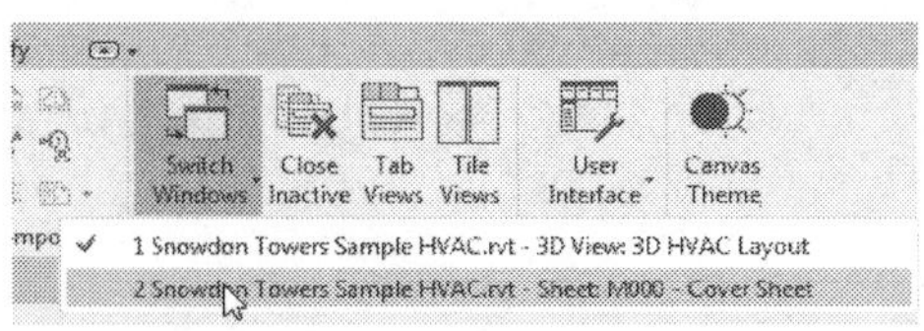

***Figure 1-18** Selecting an option from the **Switch Windows** drop-down*

Like other Microsoft Windows-based programs, you can select and view the opened projects using the **Tile Views** option from the **Windows** panel of the **View** tab. The tile view of projects is shown in Figure 1-19.

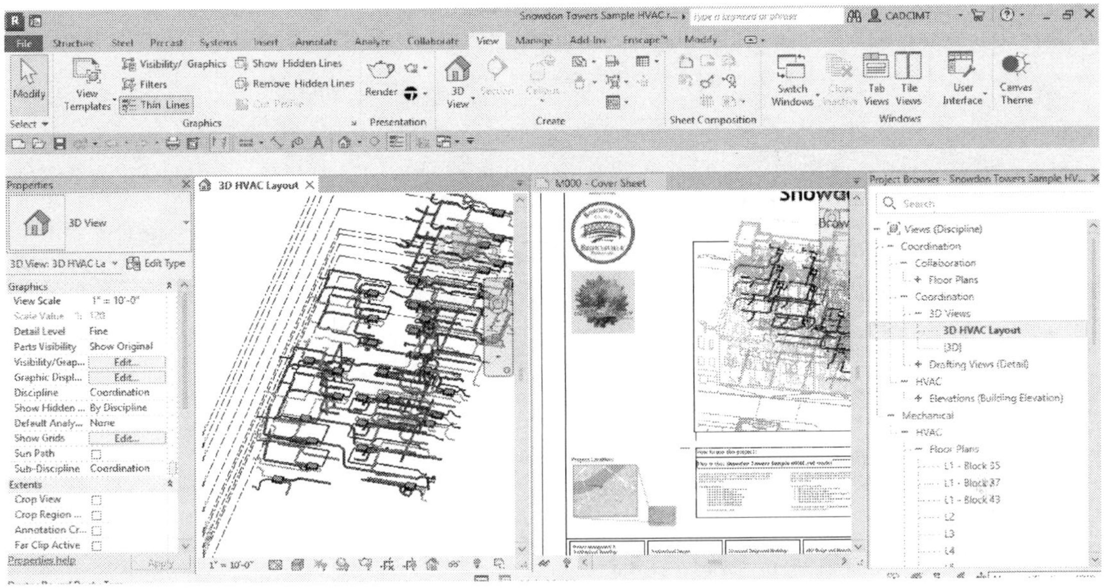

Figure 1-19 The tiled view of projects

INTEROPERABILITY OF Autodesk Revit

The models or geometries created in Autodesk Revit can be easily exported to AutoCAD and AutoCAD Architecture in the DWG file format. This enables structural engineers to collaborate with Architects. One of the important aspects of the job of a structural engineer is to collaborate and share information with the rest of the design team including the architect. To facilitate this requirement, Revit 2025 follows a wide range of industry standards and supports various CAD file formats such as *DWF*, *DGN*, *DWG*, *DGN*, *IFC*, *SKP*, and *SAT*. For image files, it supports *JPG*, *TIFF*, *BMP*, *PNG*, *AVI*, *PAN*, *IVR*, and *TGA* file formats. Besides these, the formats that are supported by Revit include *ODBC*, *HTML*, *TXT*, *XML*, *XLS*, and *MDB*. Autodesk Revit is compatible with any CAD system that supports the DWG, DXF, or DGN file format. Revit can import the models and geometries as ACIS solids. This enables engineers to import models from AutoCAD Architecture and AutoCAD MEP (Mechanical, Electrical, and Plumbing) software and to link the 3D information to Revit. This feature makes Autodesk Revit 2025 an efficient, user-friendly, and compatible software.

BUILDING INFORMATION MODELING AND Autodesk Revit 2025

Building Information Modeling (BIM) is defined as a design technology that involves the creation and use of coordinated, internally consistent, and computable information about a building project in design and construction. Using BIM, you can demonstrate the entire life cycle of a building project starting from the process of construction, facility operation, and information about quantities and shared properties of elements. BIM enables the circulation of virtual information model from the design team to contractors and then to the owner, thereby updating them about the changes made in the model at each stage. The ability to keep information up-to-date and make it available in an integrated digital environment enables the architects, owners, builders, and engineers to have a clear vision of the project before the commencement of actual construction. It also enables them to make better and faster decisions to improve the quality and profitability of projects. Autodesk Revit 2025 is a specially designed platform based on BIM.

In Revit, the analytical and physical representations of an MEP model are created simultaneously. These representations are different views of a computable building model that contains necessary information for a third-party analysis application which is done with the help of a common modeling interface. You can use Revit API to move data directly from the Revit building information model to the analysis software. Further, you can bring back the analysis reports while keeping the analysis, design, and documentation synchronized. Revit's parametric model represents a building as an integrated database of coordinated information. In Revit, a change anywhere is a change everywhere. A change made in your project at any stage is reflected in the entire project, and also due to the parametric behavior of elements, the project is updated automatically. Also, the integration of Revit with the available in-built commercial tools such as solar studies, material takeoffs, and so on greatly simplifies the project design and reduces the time consumed for analysis, thereby enabling faster decision making.

AUTODESK CONSTRUCTION CLOUD

Autodesk Construction Cloud™ is used as a cloud-based construction management and collaboration solution developed by Autodesk. Autodesk Construction Cloud solution connects building data, project work flows, and project teams throughout the entire building life cycle, from design to operations. Autodesk Construction Cloud is built on an integrated platform and common data environment that will empower general contractors, specialty trades, designers and owners to drive better business outcomes. Using the Autodesk Construction Cloud solution, project teams will now have a comprehensive construction management platform in which all data will be located centrally. As such, it will simplify collaboration, predict project changes, and will provide guidance that will be driven by data for improvement of the organization. The Autodesk Construction Cloud offers three main products namely Autodesk Build, Autodesk Quantity, and Autodesk BIM Collaborate. As part of the unified Autodesk Construction Cloud, every product is reinforced by Autodesk Docs, Insights, and Administration.

Autodesk Docs cloud platform supports the common data environment for every product of Autodesk Construction Cloud. It provides the users with uninterrupted navigation of data across the project teams and throughout the project workflow. Autodesk Docs is a single, cloud-based platform for streamlining document management by storing and managing all project documents and data from design through construction in a single platform.

Insights platform is used to deliver analytical report from the data collected and analyzed from the project. It is also used to export that data to the user with the application of artificial intelligence to identify and mitigate risk.

Administration platform is used for administrating the data in the project for better practices.

Autodesk Revit HELP

In Autodesk Revit, you can access various help topics online by using the **Autodesk Revit 2025** help page. You can access this page by choosing the **Help** tool from the **InfoCenter**, refer to Figure 1-20. In this page, different areas such as **Learn about Revit**, **Resources**, and others are displayed. You can click on the required link from these areas to get the related information. In the **Learn about Revit** area, various help options related to Autodesk Revit are available. You can click on the required option to display the help page corresponding to the option. The **Resources** area contains various learning resources. You can click on the desired option from this area to get the information related to it.

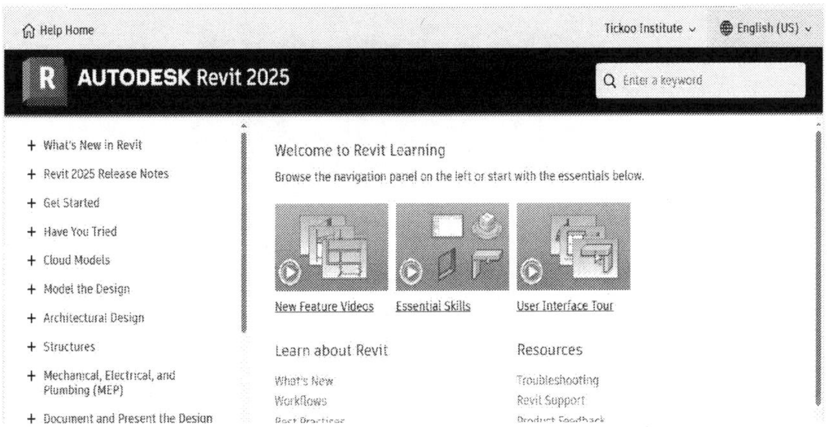

Figure 1-20 The AUTODESK Revit 2025 page

EXPORT IFC ELEMENT MAPPING

In Revit 2025, the mapping of IFC for different components within the model can be accomplished by utilizing the options available under the **IFC Parameters** section in the **Properties** palette of the selected component.

To map a component, go to the **Properties** palette of the chosen component and locate the **Export to IFC As** parameter. Click on the field corresponding to it and choose the browse button; the **Select Export As IFC Entity** dialog box will open. In this dialog box, you have the option to select the specified IFC schema from the **IFC Schema Version** drop-down list, where the default selection is the **IFC4** schema. You can use the **Search** text box to find the desired component type and associate it with the corresponding IFC category. Within the left box, situated below the **Search** text box of the **Select Export As IFC Entity** dialog box, select the **IFC** type, and in the right box, specify the related IFC category. For instance, as demonstrated in Figure 1-21, input **pipe** in the **Search** text box, select the **ifcpipesegment** radio button in the left box, and in the right box corresponding to the category, select the **RIGIDSEGMENT** radio button. Next, choose the **OK** button to close the **Select Export As IFC Entity** dialog box. The parameters will be displayed in the instance properties of the selected pipe.

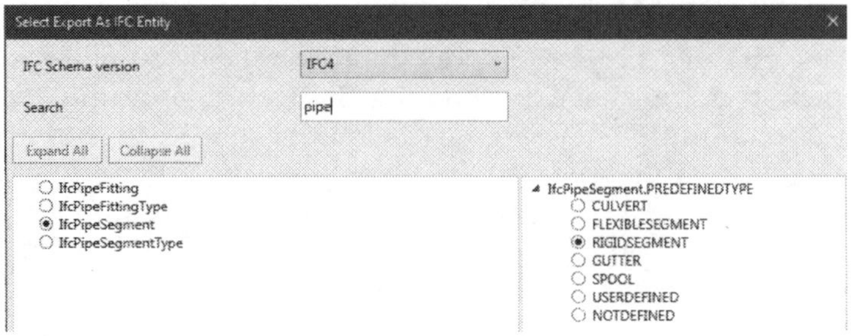

***Figure 1-21** The partial view of the **Select Export As IFC Entity** dialog box*

Chapter 2

Getting Started with an MEP Project

Learning Objectives

After completing this chapter, you will be able to:

- *Understand the essentials of an MEP project*
- *Start a new project*
- *Set the units of various measurement parameters in a project*
- *Use project templates*
- *Understand the concept of snaps, dimensions, and object snaps*
- *Save a project*
- *Use the Options dialog box*
- *Close and exit an MEP project*
- *Open an existing project*

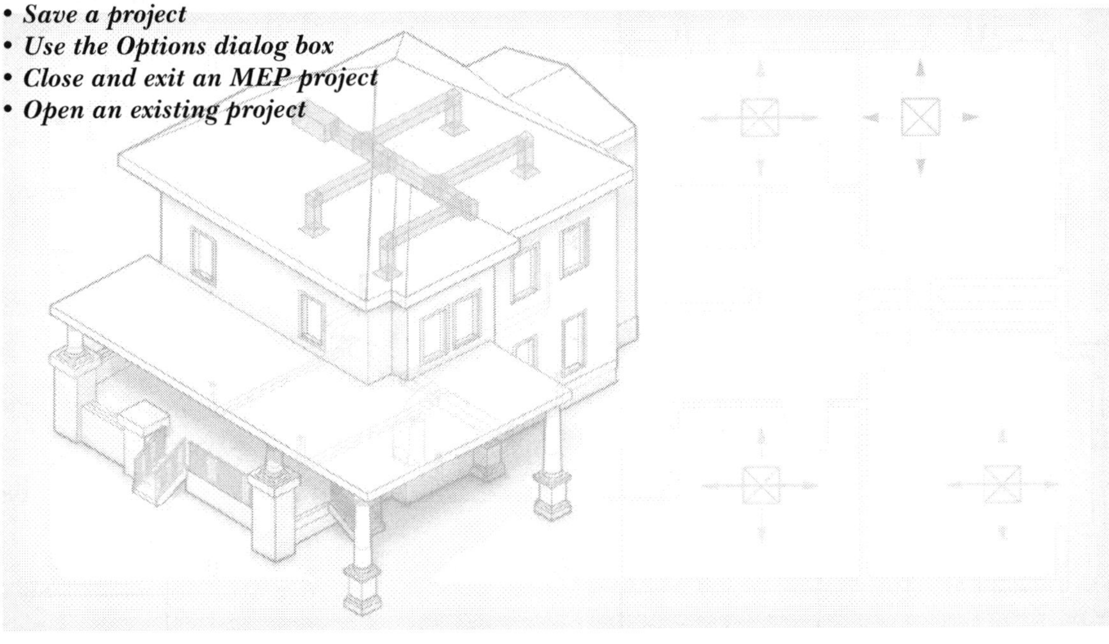

OVERVIEW OF A BIM-MEP PROJECT

In BIM environment, a project is delivered as an integrated model comprising elements of various disciplines such as Architecture, Mechanical, Electrical, Plumbing, Structure, and Coordination. This means that the project file that you create will contain all the information related to building design from geometry to various construction documentations such as schedules and legends. Generally, this information includes the details of the building elements like walls, doors, windows, beams, columns, ducts, pipes, equipment, fixtures, and others. The information in the project also includes different views of the project, working drawings created from the building elements, and the documentation related to design of the model.

In Revit, the project file in which you will create the MEP project is based on a Revit Template File (*.rte*). This template provides initial settings for the project such as its units, material used, and display settings. You can customize the default settings of a project as required. The basic template file has predefined information and settings for a project.

Generally, each organization has its own standard of working in a building project. Based on the standard, a user can customize the template and then save it for further use.

Tip

It is recommended to follow a slower approach to set up a project and give more time to create the standard template for practice and to organize the structure of the required components in a project. This helps in carrying out the project smoothly and efficiently.

ESSENTIALS FOR AN MEP PROJECT

Before starting up an MEP Project, there are some essential tips that are recommended to be followed for smooth functioning of a project. The essential tips are as follows:

Arrange a BIM Project Kick-off Meeting. It is recommended to arrange a BIM project kick-off meeting for the people involved in all disciplines and the BIM modeler. This meeting is essential as it brings all the people involved in the project in a single platform and provides everyone an opportunity to share the information about the expectation of the clients and the firms from the building model. Also, the information is shared on the expected Level of Development (AIA Document E202) of the building model.

Establish a common Project Settings and Project Goal. It is required to establish project settings and goals before starting up a new project in Revit. Following are the project settings and goals that you need to establish for the MEP project: File Structure, Shared Coordinates, and project milestones.

Communicate with the Architects. Since the architects have been using Revit for a long time, it is required to communicate with those who are involved in the project regarding the design.

STARTING A NEW MEP PROJECT

File Menu: New > Project
Shortcut Key: CTRL+N

To start a new MEP project in Revit 2025, choose **Start > Autodesk > Revit 2025 > Revit 2025** (for Windows 10) or double-click on the Revit 2025 icon available on desktop; the Autodesk Revit interface will be displayed. Next, choose **New > Project** from the **File** menu, as shown in Figure 2-1; the **New Project** dialog box will be displayed, as shown in Figure 2-2.

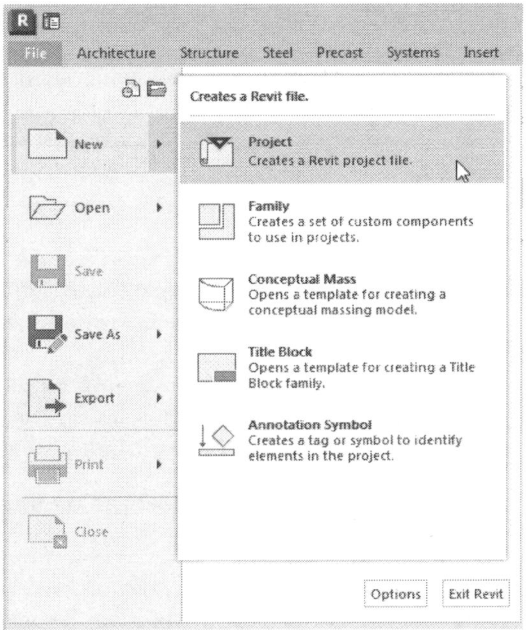

Figure 2-1 *Choosing the* ***Project*** *option from the* ***File*** *menu*

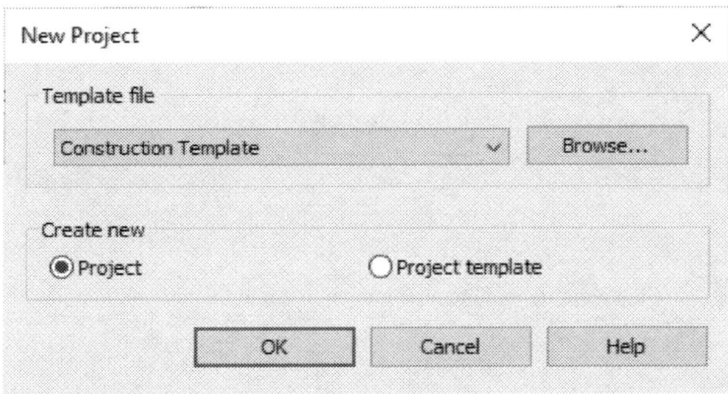

Figure 2-2 *The* ***New Project*** *dialog box*

In this dialog box, the **Construction Template** option is selected by default in the drop-down list in the **Template file** area. As a result, the new project will adopt the settings of the *Construction-Default* template file. Alternatively, you can select any of the following options from the drop-down list in the **Template file** area: **Architectural Template**, **Structural Template**, **Mechanical Template**, and others.

Note

The selection of the option from the drop-down list depends on the MEP discipline or discipline that you are going to work within the project.

Tip

For an MEP project, it is recommended to start with the ***Systems.rte*** *template file. To use this file, you need to select the* ***Systems Template*** *option from the drop-down list in the* ***Template file*** *area of the* ***New Project*** *dialog box. The* ***Systems.rte*** *template file provides a useful set up for all the disciplines such as Mechanical, Electrical, and Plumbing.*

A template file has various project parameters saved in it such as units, views, and so on. When you apply the template file to a new project, it will adopt the same parameters as that of the template file. The difference between a template file format and a project file format is that the former has a *.rte* extension, whereas the latter has a *.rvt* extension. You can either select any of the template files provided in Revit or create your own template file. You can also save any project file as a template file.

You can select a file as template by choosing the **Browse** button in the **Template file** area of the **New Project** dialog box. On doing so, the **Choose Template** dialog box will be displayed. In this dialog box, you can browse to the specified file location and select the desired template file. After selecting the template file, choose the **Open** button; the **Choose Template** dialog box will be closed and the selected file will be displayed as an option in the drop-down list that is displayed in the **Template file** area of the **New Project** dialog box.

In the **New Project** dialog box, you can select the **Project** radio button to create a new project. Alternatively, you can select the **Project template** radio button to start with a new project template.

After specifying the various options in the **New Project** dialog box, choose the **OK** button; the **New Project** dialog box will be closed and the Revit project interface will be displayed with the applied settings.

Tip

Revit 2025 introduces a new multi-discipline template, expanding from the earlier options of discipline-specific and systems (for example MEP) templates. In earlier releases, some tools in Revit had issues functioning correctly based on the chosen template, due to varying settings and required loaded families for specific commands and workflows. The new all-discipline template addresses this problem, offering a solution to avoid such inconsistencies.

PROJECT UNITS

Ribbon: Manage > Settings > Project Units
Shortcut Key: UN

Units are important parameters of a project as they provide a standard of measurement for different entities. While installing Revit, you are prompted to set the Imperial (feet and inches) or Metric (meter) unit as the default unit system. Setting a default unit system helps you start your project with a specific type of unit. To set units, choose the **Project Units** tool from the **Settings** panel; the **Project Units** dialog box will be displayed, as shown in Figure 2-3. Under the **Units** column in this dialog box, you can specify various units that are relevant to the building project.

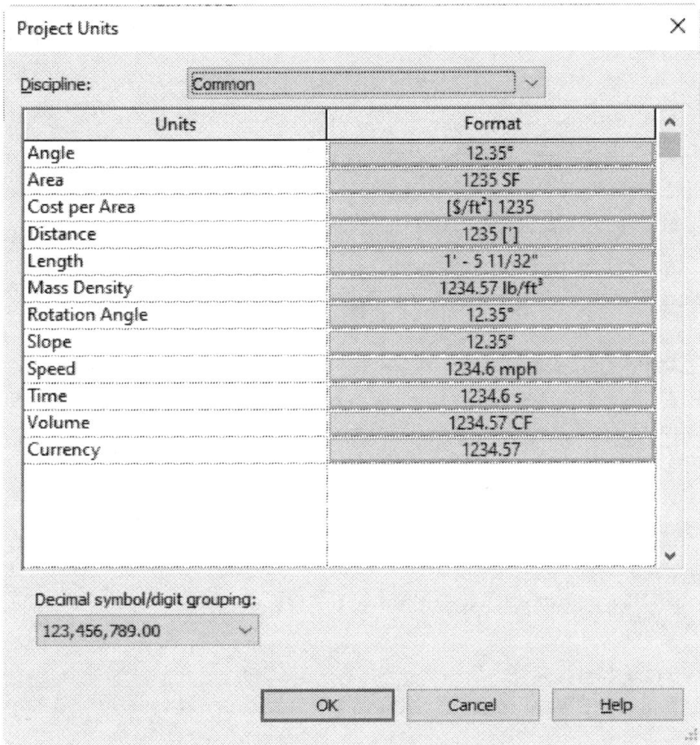

*Figure 2-3 The **Project Units** dialog box*

In the **Project Units** dialog box, units are grouped into seven disciplines: **Common**, **Structural**, **HVAC**, **Electrical**, **Piping**, **Infrastructure**, and **Energy**. Each discipline has a set of measurement parameters. You can select any of these disciplines from the **Discipline** drop-down list in the **Project Units** dialog box. The **Format** column in this dialog box displays the current unit format for the corresponding parameter in the **Units** column. You can preview and select the possible digit grouping and decimal separators from the **Decimal symbol/digit grouping** drop-down list, which is at the lower left corner of the dialog box. Some of the disciplines that are used in MEP projects are discussed next.

Note

*The values for different parameters displayed in the **Format** column of the **Project Units** dialog box may differ depending upon the type of unit system, Imperial or Metric, selected for the project. In this textbook, the Imperial unit system has been used in the tutorials and illustrations.*

Common Unit Type

The **Common** unit type used in an MEP project includes the parameters such as length, volume, angle, slope, and so on. In the **Project Units** dialog box, the **Common** option is selected by default in the **Discipline** drop-down list, refer to Figure 2-3. The **Common** unit type used in Revit is similar to that used in other Revit platforms. Moreover, the settings of the parameters of common units are similar to those used in other CAD programs. The methods of setting various parameters under the **Common** unit types are :-

Setting Length Units

In an MEP project, you can set the length unit by clicking the Length field in the **Project Units** dialog box and choosing from options like meters, millimeters, or feet. You can also define rounding values and select a unit symbol, with certain options like custom rounding becoming active based on your selections.

Setting Area Units

To set area units in an **MEP** project, click the **Area** field in the **Project Units** dialog box and choose a unit like square feet, square meters, or acres from the **Units** drop-down list. You can also adjust rounding, rounding increment, and unit symbol as needed.

Setting the Volume Units

To set volume units, click the **Volume** field in the Format column of the **Project Units** dialog box. In the **Format** dialog box, choose a unit like cubic feet, cubic meters, or liters from the **Units** drop-down list, then click **OK** to apply it.

Setting the Angle Units

You can specify the unit for the angle measurement by selecting the required option from the **Units** drop-down list in the **Format** dialog box for the **Angle** parameter.

Setting the Slope Units

To set slope units, click the **Slope** field in the **Project Units** dialog box to open the **Format** dialog box. From the **Units** drop-down list, select options like Ratio:12, Rise/12”, **Decimal degrees**, or **Percentage**. By default, Rise/12” is selected if **Imperial** units were chosen during installation.

Setting the Currency Units

In Revit, you can set the currency unit by clicking the **Currency** field in the **Project Units** dialog box. In the **Format** dialog box, choose the desired currency symbol and rounding value from the respective drop-down lists.

Setting the Mass Density Units

In Revit, you can set the unit for mass density by clicking the **Mass Density** field in the **Project Units** dialog box. Choose a unit like kilogram per cubic meter or pound per cubic foot from the **Units** drop-down list, and adjust rounding and symbol settings as needed.

Setting the Rotation Angle Units

To set the rotation angle unit in **Revit**, click the **Rotation Angle** field in the Project Units dialog box. In the Format dialog box, choose a unit like **Degrees, Radians,** or **Gradians**, and adjust rounding and symbol settings as needed.

Setting the Cost per Area Units

In Revit, you can set the unit for cost per area by clicking the **Cost** per Area field in the **Project Units** dialog box. Choose from options like cost per square foot or square meter, and adjust rounding, increment, and unit symbol settings as needed.

Setting the Speed Units

In Revit, you can set the speed unit by clicking the **Speed** field in the **Project Units** dialog box and selecting a unit like miles per hour, feet per second, or meters per second from the Units drop-down list. You can also adjust rounding, rounding increment, and unit symbol settings as needed.

Setting the Time Units

In Revit, you can set the time unit by clicking the **Time** field in the **Project Units** dialog box and choosing from options like seconds, minutes, or hours in the **Format** dialog box. You can also adjust rounding, rounding increment, and unit symbol settings from their respective drop-down lists. By default, **Seconds** is selected if the **Imperial** unit system was chosen during installation.

HVAC Unit Type

HVAC units are commonly used while working in the **Mechanical** discipline of an **MEP Project**. Some of the frequently used **HVAC** units are **Density**, **Power**, **Pressure**, **Velocity**, **Air Flow**, and more. In Revit, you can set the **HVAC** units in the **Project Units** dialog box. To do so, select the **HVAC** option from the **Discipline** drop-down list in the **Project Units** dialog box, as shown in Figure 2-4. Some of the important **HVAC** units are discussed next.

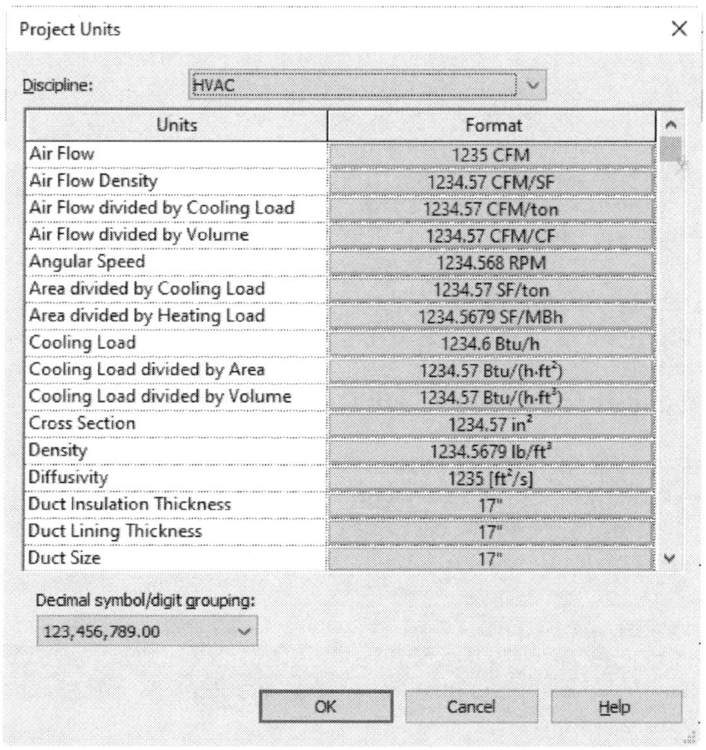

Figure 2-4 *The **Project Units** dialog box with the **HVAC** option selected in the **Discipline** drop-down list*

Setting the Unit for Power
To set the power unit for HVAC in Revit, click the Power field in the Project Units dialog box and choose from options like **Watts, Kilowatts**, **BTU (British Thermal Units) per second**, **BTU (British Thermal Units) per hour**,. The default is BTU/hr for Imperial and Watts for Metric.

Setting the Unit for Pressure
To set the pressure unit in an **HVAC** system in Revit, click the Pressure field in the Project Units dialog box to open the Format dialog box. Then, select the desired unit from the Units drop-down list. This defines how static pressure is measured in the system.

Setting the Unit for Air Flow
In Revit, you can set the Air Flow unit by clicking the **Air Flow** field in the **Project Units** dialog box and selecting from options like **CFM**, liters per second, or cubic meters per second. The default is Cubic feet per minute for **Imperial** units and Liters per second for Metric units.

Setting the Unit for Heating Load
To set the **Heating Load** unit in **Revit**, click the corresponding field in the **Project Units** dialog box to open the **Format** dialog box. Choose from units like **Watts**, **Kilowatts**, or **BTU/hr**. By default, **BTU/hr** is selected for **Imperial** units, while **Watts** is default for **Metric** units.

Electrical Unit Type
Electrical units are commonly used while working in the **Electrical** discipline of an **MEP Project**. Some of the frequently used **Electrical** units are **Current**, **Electrical Potential**, **Frequency**, **Illuminance**, and more. To specify these units, invoke the **Project Units** dialog box and then select the **Electrical** option from the **Discipline** drop-down list. Some of the frequently used **Electrical** units are discussed next.

Setting the Unit for Illuminance
To set the unit for **Illuminance** in Revit, click the **Illuminance** field in the **Project Units** dialog box and choose either **Footcandles** or Lux from the Units drop-down list. **Footcandles** is the default for **Imperial** units, while **Lux** is default for **Metric**.

Setting the Unit for Electrical Potential
In **Revit**, you can set the unit for **Electrical Potential** by clicking its field in the **Project Units** dialog box, which opens the **Format** dialog box. From there, choose units like **Volts**, **Kilovolts**, or **Millivolts**, with **Volts** being the default for both **Imperial** and **Metric** systems.

Setting the Unit for Demand Factor
In **Revit**, you can set the unit for the **Demand Factor** by clicking its field in the **Project Units** dialog box. In the **Format** dialog box, choose between **Percentage** and **Fixed**, with **Percentage** being the default for both **Imperial** and **Metric** systems.

Piping Unit Type
Piping units include units for Density, Flow rate, Pressure, Velocity, and so on. In Revit, you can set the piping units in the **Project Units** dialog box. To do so, select the **Piping** option from the **Discipline** drop-down list in the **Project Units** dialog box. Some of the piping units are discussed next.

Setting the Unit for Flow

To set the flow unit in Revit, click the Flow field in the **Project Units** dialog box and choose a unit like **US** gallons per minute or cubic meters per hour from the **Units** drop-down list. By default, **US** gallons per minute is selected for **Imperial** units, and **Liters** per second for Metric.

> **Tip**
> *In Revit, you can also select the piping flow unit as **Liters per minute**.*

Setting the Unit for Velocity

To set the unit for velocity in Revit, click the **Velocity** field in the **Project Units** dialog box and choose a unit like **Feet** per second or **Meters** per second from the **Units** drop-down list. The default is Feet per second for **Imperial** and **Meters** per second for **Metric** units.

Setting the Unit for Pipe Size

To set the unit for pipe size in Revit, click the **Pipe Size** field in the **Project Units** dialog box and choose from options like fractional inches, decimal inches, or millimeters. The default unit is **Fractional** inches for **Imperial** and **Millimeters** for **Metric** settings.

PROJECT TEMPLATES

Project templates are commonly known as template files. These files contain predefined settings for projects for the display of annotations, graphics, and so on. In a project, these files contain predefined settings for units, mechanical components, electrical components, and piping components. Similarly, for the display of graphics, the template files contain predefined settings for materials, line styles, line weights, line patterns, and various symbols relevant to the **MEP** project.

When you install the **Revit** software, you will find in-built templates that are saved with *.rte* as file extension. You can also create your own template based on the project requirement. In Revit, any new template-based project inherits all families, settings (such as units, fill patterns, line styles, line weights, and view scales), and geometry from the template.

The use of template file is quite extensive and it helps in reducing the cycle time of a project. In the following sections, you will learn to create custom templates and then to use them in a structural project.

Creating a Custom Project Template

In Revit, there are various methods to create a custom project template. The common method is to open an existing template file and modify its settings based on the project requirement and then save it as a different template file. You can also create a custom project template by starting a blank project file, defining all settings such as naming the viewports, creating levels, adding grids, and others, and then saving it as a template (.rte) file. For certain projects, you can create a template file which includes geometry that can be used repeatedly as a base for the new projects. For example, if you have defined geometry for a hospital project and want to include this geometry whenever you start a new project, you can save the file that includes this geometry as a template. Each time you open a project with this template, the geometry will be included.

Creating a New Template from a Blank Project File

In Revit, you can create a new template from a blank project by selecting **New** > **Project** and choosing None in the **Template** file area. After selecting the project type, you'll be prompted to choose **Imperial** or **Metric** units. Once the blank project opens, customize the settings as needed, then save it as a template using **Save As** > **Template** from the File menu.

Creating a New Project Template from an Existing Project Template

To create a new project using an existing template in **Revit**, go to **File** > **New** > **Project**, then select the **Project** template option. Click **Browse** to locate and select the desired template file, then click **Open**. Finally, choose **OK** to start a new project with all settings inherited from the selected template.

Settings for the Project Template

While creating a project template in **Revit**, you can predefine settings like project name, number, and client details, which are useful for publishing or plotting. You can also set units, snaps, line styles, fill patterns, and family settings. Additionally, you can customize system and loaded families, set up project views, visibility/graphics, and print settings.

Setting the Project Information

When creating a project template in **Revit**, you can set project information using the **Project Information** tool from the **Manage** tab. In the **Project Information** dialog box, you can enter details such as organization name, building name, author, issue date, project status, address, project name, and number. To update the address, use the browse button to open the **Edit Text** dialog box and input the location. Once all details are entered, click **OK** to save them. Additionally, you can add custom parameters—either local or shared—using the **Project Parameters** tool. In the **Project Parameters** dialog box, click **New Parameter** to define attributes such as name, discipline, type, and group. For instance, to add an **MEP Consultant** parameter, select the **Common** discipline and assign it to the Text type under the "**Other**" group. After saving, the new parameter will appear in the **Project Information** dialog box.

Setting the Project Location

In an MEP project, you must define the geographical location, including latitude, longitude, and coordinate system. To do this, use the **Location** tool from the **Project Location** panel in the **Manage** tab, which opens the **Location** and **Site** dialog box with **Location** and **Site** tabs. The **Location** tab allows you to define the location using an **Internet Mapping Service** or a **Default City List**, entering a city name or coordinates manually. The Site tab lets you name and manage multiple location settings; you can duplicate, delete, or make a location current. The active location will be marked with "(current)" next to its name.

Transferring Project Standards

In Revit, you can transfer project standards from another project to your current template file using the **Transfer Project Standards** tool. These standards include system family types, line weights, styles, patterns, materials, and view templates. To do this, open the source project, go to the **Manage** tab, and select **Transfer Project Standards**. In the Select Items To **Copy** dialog box, choose the source project from the drop-down list, then select or deselect the checkboxes for the standards you want to copy. You can use the **Check All** or **Check None** buttons for quick selection. After choosing the desired standards, click **OK**. If there are duplicates, you'll be prompted to either overwrite or only transfer new standards. Once confirmed, the selected standards will be applied to your template.

Note

*If you invoke the **Transfer Project Standards** tool without any opened files in the current session, the **Revit** message box informing that there are no open projects to transfer project from will be displayed. Choose the **Close** button in this message box and then open a project in the current session from which you want to transfer the project standard to the current project.*

Setting the Browser Organization

While creating a project template in Revit, you can organize the Project Browser using the **Browser Organization** tool from the Windows panel of the **View tab**. In the **Browser Organization** dialog box, which includes **Views**, **Sheets**, and **Schedules** tabs, you can edit or create browser organizations by selecting or deselecting checkboxes, editing existing ones, or creating new ones using the **New** button. The Browser Organization Properties dialog box lets you apply filters and define grouping and sorting for views, sheets, and schedules. You can specify view properties, filter values, and sorting options to customize how project elements are displayed in the browser.

LINKING REVIT MODELS AND SHARING COORDINATES

In a Design Build project, you need to link architectural and structural models to the **MEP** model for coordination and interference checks. Revit allows you to link models similar to **AutoCAD Xrefs** and manage them through the **Project Browser**. To maintain alignment, use the **Acquire Coordinates** tool to share coordinates between the host and linked models. You can link files using the **Link Revit** tool in the Insert tab and choose positioning options to place the models accurately.

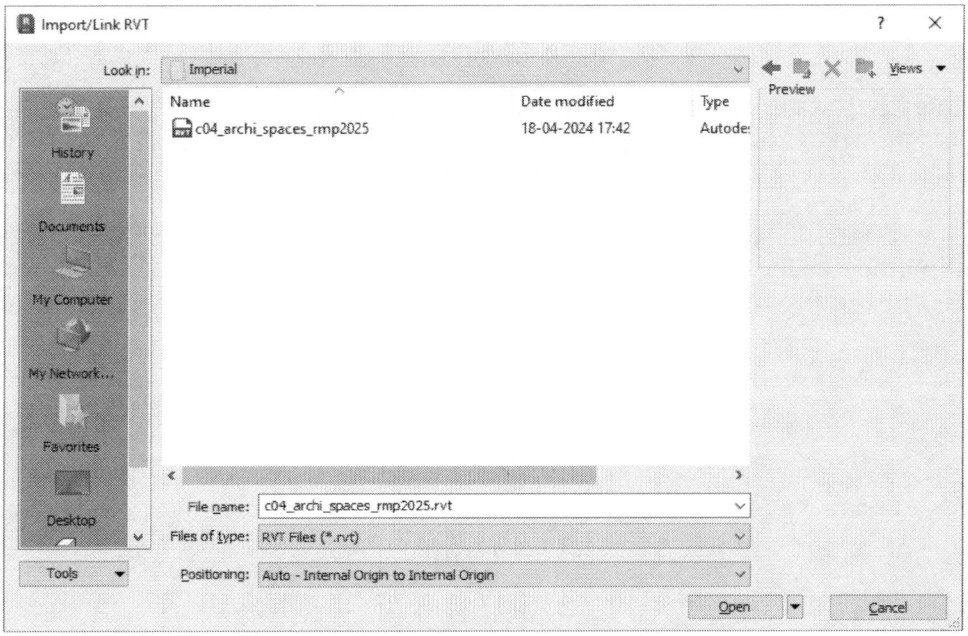

***Figure 2-5** The **Import/Link RVT** dialog box*

Managing the Linked Revit Models in the Project Browser

The linked Revit models are listed in the **Project Browser** under the **Revit Links** head. You can access the linked model files and also link a new file from the **Project Browser**. To do so, select the **Revit Links** sub-node in the **Project Browser** and right-click; a shortcut menu will be displayed. Choose **New Link** from the shortcut menu; the **Import/Link RVT** dialog box will be displayed. Select the file to be linked from this dialog box. Similarly, you can access the link manager by choosing **Manage Links** from the shortcut menu and manage the links. You can also open the linked model in the project by dragging it from the **Project Browser** and dropping it in the project view.

Converting Linked Models to Groups - Binding Links

You can convert a linked Revit model into a group in the host project by using the **Bind Link** tool. This process, called binding, makes the linked model part of the host project, simplifying edits. After selecting the linked model, use the **Bind Link** tool, choose the elements to include (like levels and grids), and confirm the action. The model will then appear as a group under the Groups section in the Project Browser.

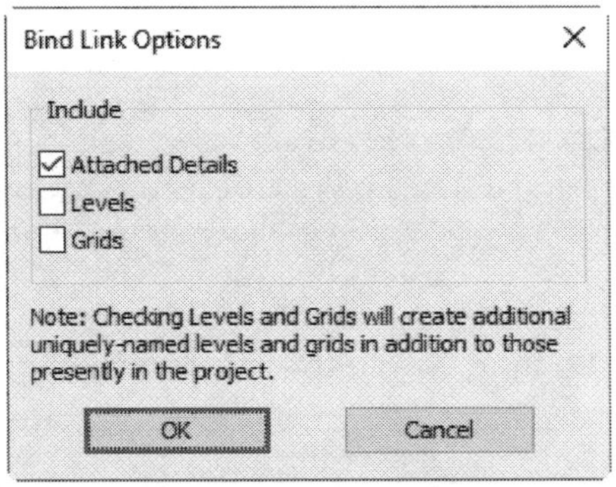

*Figure 2-6 The **Bind Link Options** dialog box*

Controlling the Visibility of Linked Models

You can control the visibility, detail level, and display settings of linked and nested Revit models within a host project using the **Visibility/Graphics** tool in the View tab. In the **Visibility/Graphic Overrides** dialog box, the **Revit Links** tab lets you manage the visibility of linked files and their components. You can adjust settings like halftone, display, and filters, and use the **RVT Link Display Settings** dialog box to choose between host view, linked view, or custom settings. These options help ensure consistent display and coordination across linked and nested models.

Managing Links

In an MEP project, you can manage linked models using the **Manage Links** tool from the **Insert** tab. The **Manage Links** dialog box displays details like the status of linked files, reference type, saved path, and position information. You can choose to show or hide nested links, check if coordinates are saved, and set the path type—keeping it relative is recommended for easier file tracking if project folders are moved.

Including Elements of Linked Models in Schedules

In Revit, you can include elements from linked models—like walls, doors, and windows—into project schedules. To do this, open the existing schedule from the **Project Browser**, go to its **Properties Palette**, and select Edit under the Fields parameter. In the **Schedule Properties** dialog box, check the "Include elements in linked files" option, apply changes, and click **OK**.

Copying Linked Model Elements

You can copy the elements of the linked model to the host model. To do so, move the cursor over the linked model and press **TAB** to highlight the required elements, in case the file has a nested link. Click on the element when it is highlighted and choose the **Copy to Clipboard** tool from the **Clipboard** panel of the **Modify | RVT Links** tab. Next, open the host file in which you want to paste the element and choose the **Paste from Clipboard** tool from the **Clipboard** panel. You can place the element by clicking at the required location in the drawing area.

Copying and Monitoring Linked Model Elements

After linking a Revit Architecture project, it's important to monitor changes to avoid conflicts between architectural and MEP elements. Using the **Copy/Monitor** tool from the **Collaborate** tab, you can copy elements from the linked model and establish relationships between corresponding elements to track changes. Monitored elements display an eye symbol, and once done, you can finish the process by clicking the **Finish** button.

SNAPS TOOL

Ribbon:	Manage > Settings > Snaps

The **Snaps** tool is one of the important tools used to snap elements in an **MEP** model. This tool is used to make the cursor snap or jump depending on the preset increments or on the specific object properties of elements such as endpoint and midpoint of elements. When you invoke the **Snaps** tool from the **Settings** panel of the **Manage** tab, the **Snaps** dialog box is displayed, as shown in Figure 2-7.

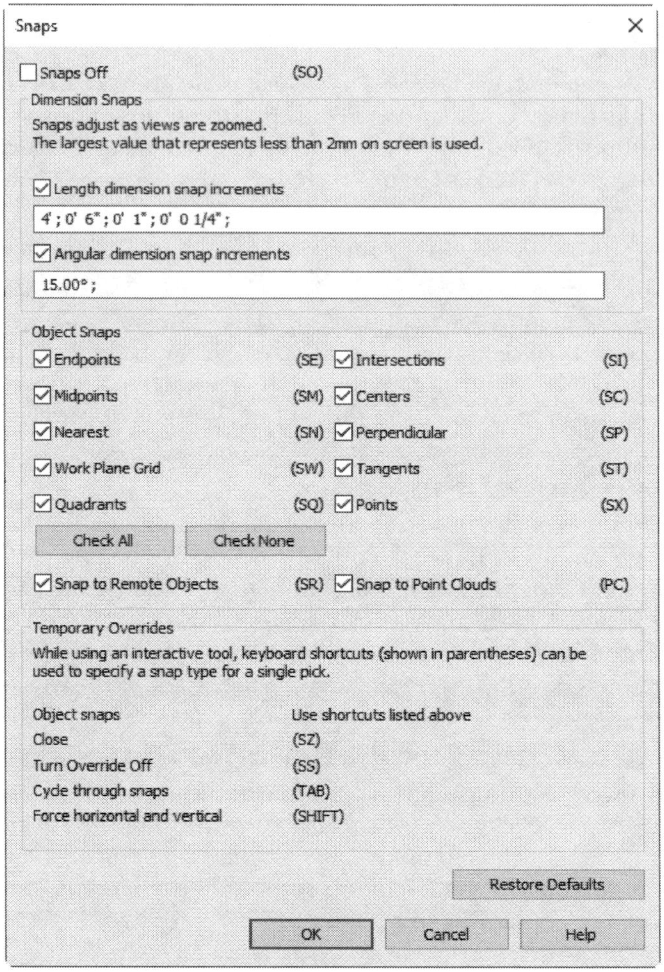

Figure 2-7 *The* ***Snaps*** *dialog box*

This dialog box has three areas, **Dimension Snaps**, **Object Snaps**, and **Temporary Overrides**. These areas are discussed next.

Note

The settings specified in this dialog box will be applied to all the projects opened in the session and will not be saved

.

Dimension Snaps Area

In the Dimension Snaps area, you can set length and angle snap increments to control how the cursor moves when placing elements in a project. By default, the **Length** snap option is enabled, allowing you to define custom values like 5', 2', or 4" to suit your layout needs. Similarly, you can enable and set angular snap increments, which is especially helpful for projects with radial geometry.

Note

The reason for specifying multiple length and angle snap increment values in the edit boxes is that the priority of increments may change on changing the zoom level of the drawing. Therefore, when you zoom in the drawing, Revit will use smaller increment values whereas on zooming out, the larger increment values will be used.

Object Snaps Area

The Object Snaps area allows the cursor to snap to key geometric points like endpoints, midpoints, or perpendiculars, making it easier to place elements accurately. When enabled, snaps activate automatically as the cursor nears an element, improving precision without manually selecting exact points. This enhances both accuracy and efficiency in creating and dimensioning drawings.

Note

The object snapping works only with the objects that are visible on the screen. A tooltip with a name as that of the object snap is also displayed when you bring the cursor close to a snap point.

Temporary Overrides Area

The Temporary Overrides options let you override snap settings for a single use without changing the main settings. For example, if the **Endpoint** snap isn't enabled, you can temporarily activate it by typing the shortcut "**SE**" instead of opening the **Snaps** dialog box. The snap turns off automatically after use. You can toggle between snap options at the same spot using the **TAB** key. **Holding** the **SHIFT** key lets you draw vertically or horizontally by restricting cursor movement to orthogonal directions, and releasing it returns the cursor to normal. To disable snapping completely, check the **Snaps Off** box.

SAVING AN MEP PROJECT

Before you close or exit a Revit session, it is recommended to save the project file. You can save a project file in a permanent storage device, such as a hard disk or a removable storage device like **CD** or **USB**. Also, you must save your work at regular intervals to avoid data loss due to any error in the computer's hardware or software.

Using the Save As Tool

In Revit, the Save As tool allows you to save your project file to a desired location by selecting **File > Save As > Project**, which opens the **Save As** dialog box where you can choose the location, name the file, and access frequently used folders. By clicking the **Options** button, you can open the **File Save Options** dialog box to customize file-saving preferences such as the number of backup files (three for non-workshared and twenty for workshared projects), and set a thumbnail preview image, typically the active view or sheet, to be displayed when opening the file. You can also choose to update the preview with the latest changes by selecting the **Regenerate if view/sheet is not up-to-date** checkbox.

Note

Revit updates the preview image continuously. Therefore, selecting the ***Regenerate if view/sheet is not up-to-date*** *check box can consume considerable resources.*

Using the Save Tool

Once a project has been saved using the **Save As** tool, you do not need to re-enter file parameters to save it again. To save a project to the hard disk, choose the **Save** option from the **File** menu. If you are saving the project for the first time, the **Save As** dialog box will be displayed, even if you invoke the **Save** tool. Alternatively, you can save your project by choosing the **Save** button from the **Quick Access Toolbar**. As you save your project file, Revit updates it automatically without prompting you to re-enter the file name and path.

THE OPTIONS DIALOG BOX

In Autodesk Revit 2025, you can configure global settings by using the **Options** dialog box. This dialog box can be invoked by choosing the **Options** button from the **File** menu. The **Options** dialog box, as shown in Figure 2-8, contains twelve tabs: **General**, **Colors, Graphics**, **File Locations**, **Hardware**, **Rendering**, **Check Spelling**, **SteeringWheels**, **ViewCube**, **User Interface**, **Cloud Model**, and **Macros**. These tabs are discussed next.

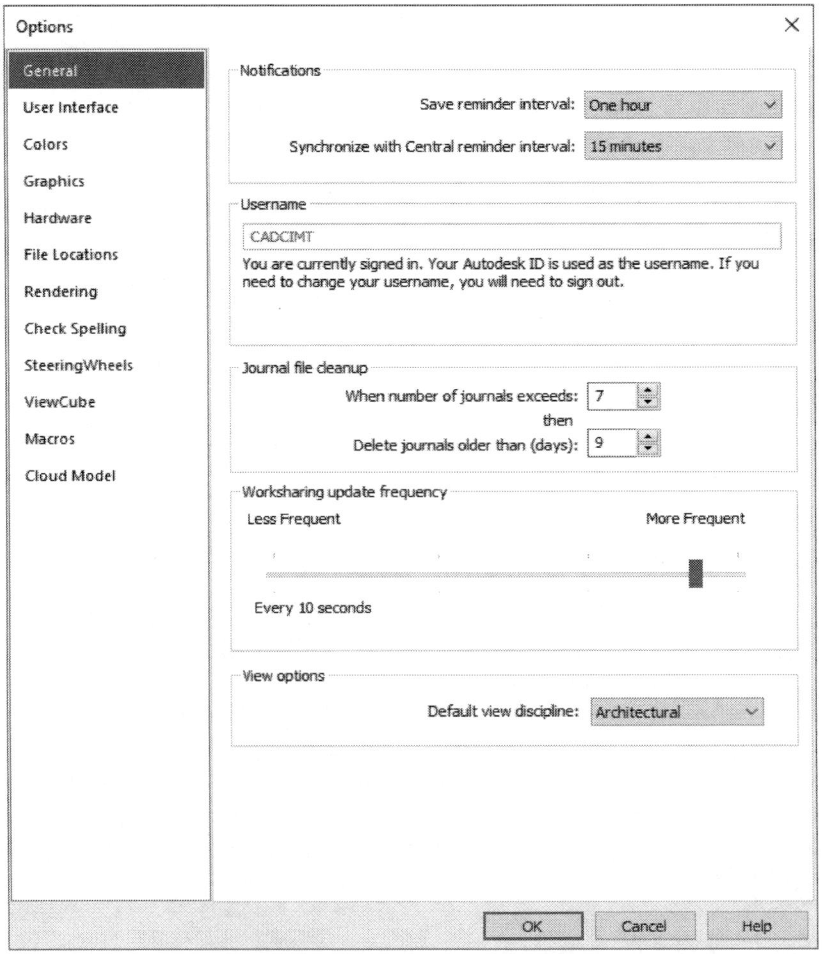

***Figure 2-8** The **Options** dialog box*

General Tab

The **General** tab is chosen by default and contains five areas: **Notifications**, **Username**, **Journal File cleanup**, **Worksharing update frequency**, and **View Options**. These areas are discussed next.

Notifications Area

Revit allows you to set reminders to save your work at regular intervals through the **Save reminder interval** drop-down in the Notifications area, with 30 minutes selected by default. You can also disable reminders by choosing **No reminders**. Similarly, the **Synchronize with Central reminder interval** drop-down lets you set how often you're reminded to sync your local file with the central file.

Username Area

In Revit, you can create a unique identification for a particular session. To do so, enter a name for the session in the edit box displayed in this area. The name entered in this edit box will be used for granting permissions for editing in a multiuser Revit environment. Note that if you sign in on **Autodesk**, the edit box will display the Autodesk ID with which you will sign in.

Journal File Cleanup Area

Journal files in Revit are text files that log every step during a session and help identify or recover from technical issues. These files are automatically saved each time Revit is closed and are stored by default in the **Journals** folder under the Revit installation path. Over time, the number of journal files can grow, so Revit provides a cleanup option in the **General** tab of the Options dialog box, where you can set limits based on file count and age to manage storage—for example, keeping only the most recent 15 files created within the last 30 days.

Worksharing Update Frequency Area

In this area, you can set the update frequency that indicates the time interval for updating the project in a worksharing environment. To specify the limits for worksharing, you can set the slider between the **Less Frequent** and **More Frequent** limits.

View Options Area

In the **View Options** area, you can specify the default view discipline to be used in the Revit project. To do so, select an option from the **Default view discipline** drop-down list in this area. In this drop-down list, the **Coordination** option is selected by default.

Graphics Tab

The **Graphics** tab in Revit allows you to configure your computer's display settings to enhance performance. By default, options like **Simplify display during view navigation** and **Allow navigation during redraw***are enabled to improve model navigation speed. The **Allow control for each view in Graphic Display Options** is also selected, enabling hardware acceleration, which helps render larger models faster and improves the experience when switching or navigating between 2D and 3D views using tools like the **ViewCube**, **Navigation** Bar, mouse wheel, or keyboard shortcuts.

Feature	Visual Styles			
	Hidden Line	**Shaded**	**Consistent Colors**	**Realistic**
Edges	-----------	Barred	Barred	Barred
Fill Patterns	Barred	Barred	Barred	Barred
Shadows	Barred	Barred	Barred	Barred
Structural Hidden Lines	Barred	Barred	Barred	Barred
Mechanical Hidden Lines	Barred	Barred	Barred	Barred

Colors Tab

The Colors tab in Revit allows you to customize the interface by setting the **UI** theme (Light, Dark, or System default), adjusting the canvas color scheme, and choosing specific colors for background, selection, pre-selection, and alert highlights. You can toggle these settings using the buttons in the **Canvas Colors** area and set transparency with the default **Semi-transparent** option to view elements behind selections. Additionally, in the Temporary Dimension Text Appearance section, you can adjust text size and choose between opaque or transparent backgrounds for temporary dimension text.

Hardware Tab

The Hardware tab in **Revit** allows you to configure your computer's display settings for better performance. By default, the **Use hardware acceleration (Direct 3D®)** checkbox is selected, enabling hardware accelerators that help display large models more quickly and improve the speed of switching between **2D** and **3D** views. This results in smoother navigation and overall display performance. However, if you encounter graphics issues or have an incompatible graphics card, you can disable this option.

File Locations Tab

The File Locations tab in Revit allows you to set or modify paths for frequently accessed files and directories, such as project files, templates, family templates, and point clouds. While default paths are set during installation, you can change them anytime using the corresponding **Browse** buttons. You can also set a customized template file location and define a default path for saving or opening projects through the **Default path for user files** field. Additionally, in the **Systems Analysis Workflows** area, you can add a system analysis workflow file (.osw) to run system analyses.

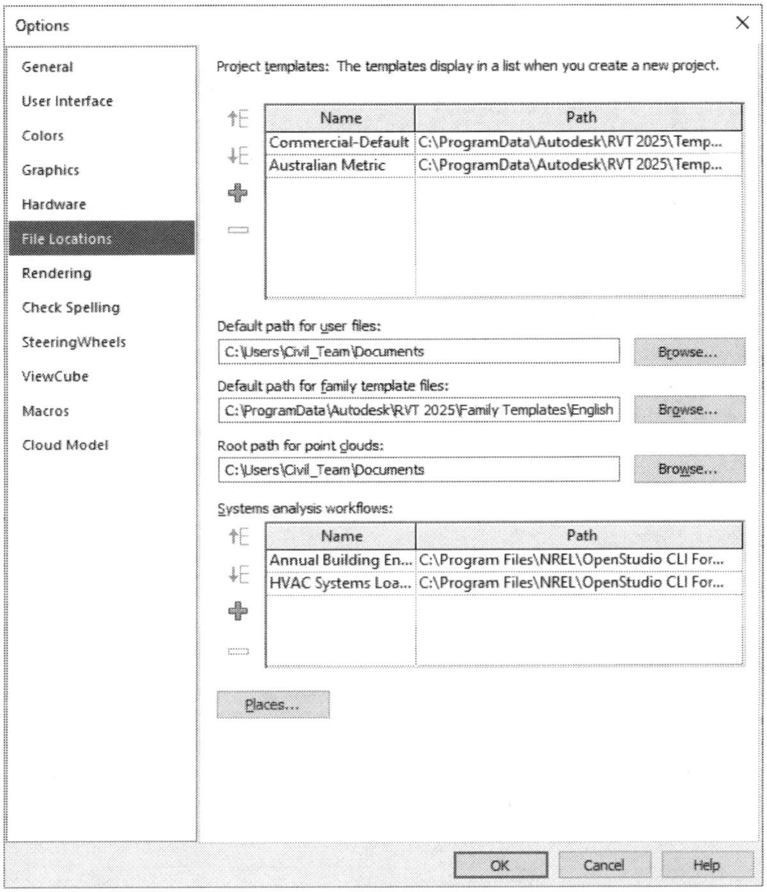

*Figure 2-9 The **Options** dialog box displaying the options in the **File Locations** tab*

Rendering Tab

Autodesk Revit uses the mental ray rendering engine, which relies on the Render Appearance Library to store material render appearances, default RPC content, and other rendering-related data. This read-only library is installed by default at **C:\Program Files (x86)\Common Files\Autodesk Shared\Materials\2025\assetlibrary_base.fbm**. In addition to this default location, users can add custom paths for external image files such as textures, bump maps, or custom colors not included with the software. To add these, click the **Add Value** button in the **Additional Render Appearance Paths** area, then either enter the path manually or use the **Browse** button to select the folder from the Browse for Folder dialog box.

Check Spelling Tab

Revit includes a spell check feature to identify and correct spelling errors in text. You can access its settings through the **Check Spelling** tab in the Options dialog box, where you can choose the main dictionary and enable additional dictionaries like the personal and building industry dictionaries, which include common industry terms not found in the default dictionary. You can also edit these dictionaries by adding or removing words. To run a spell check in your drawing, use the **Check Spelling** tool from the **Text panel** on the **Annotate** tab or press **F7**, then correct errors in the Spelling dialog box by selecting the right word and clicking **Change**.

SteeringWheels Tab

The **SteeringWheels** tab in the **Options** dialog box has options to control the text visibility, appearance of model, and behaviour of the operational tools of different types of SteeringWheels. This tab has seven different areas, as shown in Figure 2-10. These areas are discussed next.

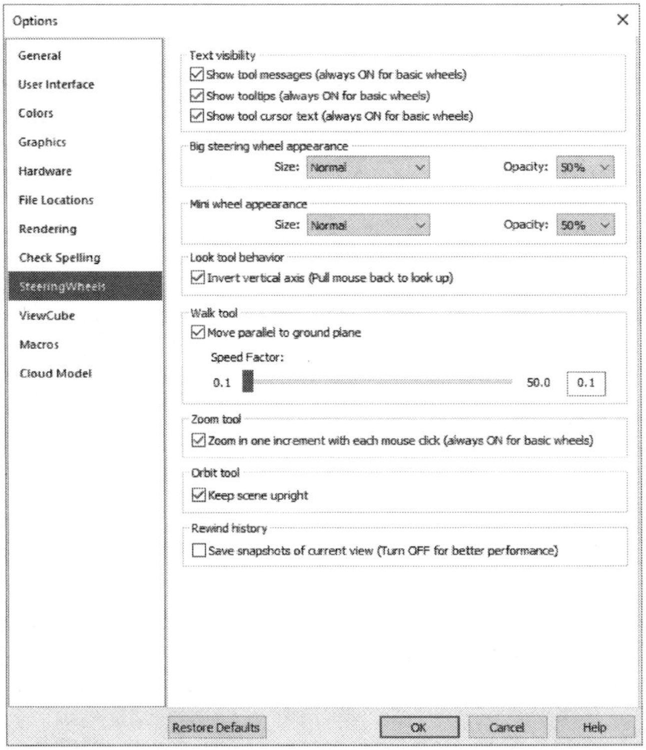

***Figure 2-10** The **Options** dialog box displaying the options in the **SteeringWheels** tab*

Text Visibility Area

You can control the display of tool messages, tooltips, and tool cursor of **SteeringWheels** by using the options in the **Text Visibility** area of the **SteeringWheels** tab. You need to select the **Show tool messages** check box to enable the visibility of tool messages in **SteeringWheels**. To display tooltips along with **SteeringWheels**, select the **Show tooltips** check box in this area. Similarly, to control the display of the cursor text when a tool is active, select the **Show tool cursor text** check box.

Big Steering Wheel Appearance and Mini Wheel Appearance Areas

The options in these areas are used to set the size and transparency of the **SteeringWheels**. To set the size of **SteeringWheels**, select the required option from the **Size** drop-down list in the corresponding areas and set its size to small, normal, or large. Similarly, you can set the transparency of **SteeringWheels** by selecting the required option from the **Opacity** drop-down list.

Look Tool Behavior and Walk Tool Areas

In the **Look Tool Behavior** area of the **SteeringWheels** tab, selecting the **Invert Vertical Axis** check box enables the view to move in the same direction as does the cursor.

The **Speed Factor** slider in the **Walk Tool** area is used to change the walk speed while using the **Walk** tool of the SteeringWheels. You can also select the **Move parallel to ground plane** check box in the **Walk Tool** area to constrain the angular movement of the walk to ground plane.

Zoom Tool and Orbit Tool Areas

Select the **Zoom in one increment with each mouse click** check box in the **Zoom Tool** area to enable the zooming operation with a single click.

In the **Orbit Tool** area, select the **Keep scene upright** check box to maintain perpendicularity between the sides of the model and the ground plane while using the **Orbit** tool.

ViewCube Tab

The **ViewCube** tab in the **Options** dialog box is used to edit the settings of the ViewCube. It has four different areas: **ViewCube Appearance**, **When Dragging the ViewCube**, **When Clicking on the ViewCube**, and **Compass**, as shown in Figure 2-11. These areas are discussed next.

ViewCube Appearance Area

This area is used to control the appearance and display of the ViewCube. This area controls the ViewCube's appearance and display. By default, the **Show the ViewCube** option is enabled; disabling it hides the ViewCube and deactivates related settings. You can adjust its on-screen position, size, and transparency using the available drop-down lists.

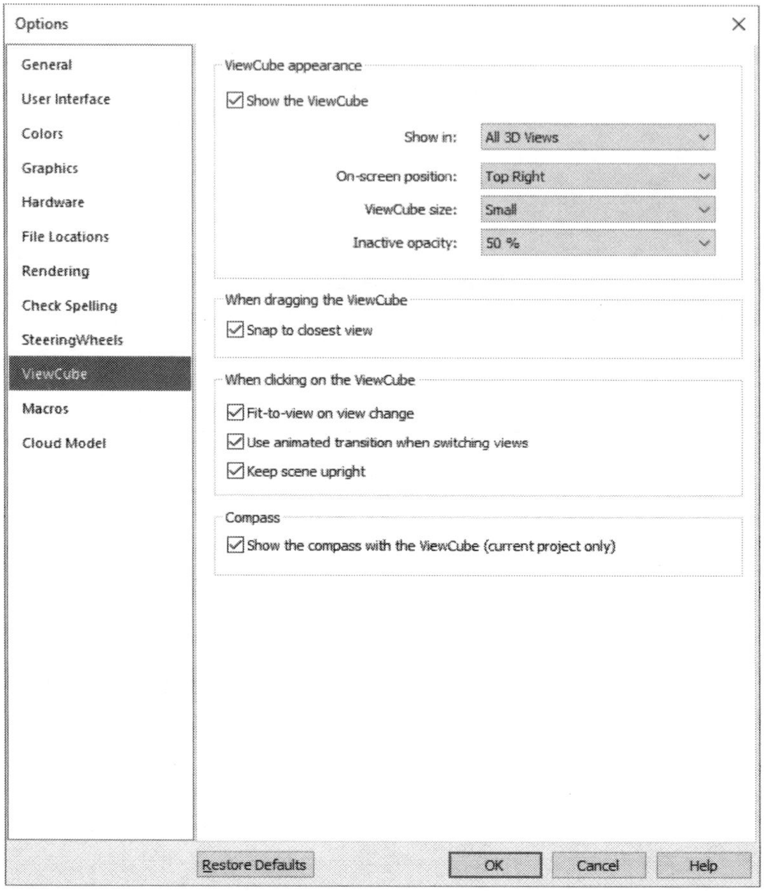

Figure 2-11 *The* ***Options*** *dialog box displaying the options in the* ***ViewCube*** *tab*

When Dragging the ViewCube Area

Select the **Snap to closest view** check box in this area to enable the snap to select the closest view in the ViewCube.

When Clicking on the ViewCube Area

The options in the **When Clicking on the ViewCube** area let you customize view transitions. Enabling **Fit-to-view on view change** fits the model to the screen when the view is changed. By default, **Use animated transition when switching views** is selected for smooth transitions, but you can disable it if preferred. The **Keep scene upright** option keeps the view aligned with the ground plane; clearing it allows full 360-degree rotation, which can be helpful when editing families.

Compass Area

In this area, the **Show the compass with the ViewCube** check box is selected by default. As a result, the compass along with the ViewCube is visible in the drawing. In the **ViewCube** tab, you can choose the **Restore Defaults** button to restore the default settings that were changed in its different areas.

User Interface Tab

The **User Interface** tab contains tools and options that will help to configure the user interface of the Revit software. In this tab, you can access various tools and options from the **Configure** and **Tab Switching Behavior** areas. In the **Configure** area, you can customize the display of the tabs and tools available in the user interface of Revit. This can be achieved by using various options displayed in the **Tools and analyses** list box. For example, you can clear the **Architecture tab and tools** check box to hide the **Architecture** tab in the Revit interface. In the **Configure** area you can specify the options for the display of Revit user interface. You can do so by selecting the **Dark** or **Light** option from the **Active theme** drop-down list in this area. Also, you can choose the **Customize** button corresponding to the **Keyboard Shortcuts** parameter to customize the use of shortcut keys in a project. In the **Configure** area, you can select an option from the **Tooltip assistance** drop-down list to set the extent of the tip that will displayed with the cursor when it is close to a tool. The options in this drop-down list are **None**, **Minimal**, **Normal**, and **High**. By default, the **Normal** option is selected in this drop-down list. Note that the tooltip will appear more frequently in your drawing if you select **High** in the **Tooltip assistance** drop-down list. In the **Configure** area, the **Enable Recent Files page at startup** check box is selected by default. As a result, the recent files will be displayed on starting the Autodesk Revit software. You can clear this check box if you do not want to display the recent files at the startup.

In the **Ribbon tab switching** area, you can specify the tab to be displayed once you clear a selection or exit a tool. In this area, the **Project environment** drop-down list contains two options: **Stay on the Modify tab** and **Return to the previous tab**. Select the **Stay on the Modify tab** option to display the options in the **Modify** tab after exiting a tool or clearing a selection. Alternatively, you can select the **Return to the previous tab** option to display the last used tab after exiting a tool or clearing a selection. In the **Tab Switching Behavior** area, the **Display the contextual tab on selection** check box is selected by default. As a result, the contextual tab is displayed once you select a tool from the Autodesk Revit interface. In the **View switching** area you can select an option from the **Ctrl+(Shift)+Tab** drop-down list to specify the function of the keys. By default, the **Tab position order** option is selected. You can select the **History order** option from the drop-down list to specify the functions of the keys to order the histories of the project.

Cloud Model Tab

The options in the **Cloud Modet** tab allow you to specify the cache folder's location for the cloud model. While working within the cloud model environment, it becomes feasible to edit the designated cache folder for the cloud model. This adjustment entails changing the cache folder associated with the cloud model on the local disk.

CLOSING AN MEP PROJECT

To close a project, choose the **Close** option from the **File** menu displayed. If you have already saved the latest changes, the project file will be closed. Otherwise, Revit will prompt you to save the changes using the **Save File** dialog box. You can save the changes by choosing the **Yes** button or discard them by choosing the **No** button. You can also choose the **Cancel** button to return to the interface and continue working on the project file. You can also use the **Close** button (X) in the drawing window to close the project.

EXITING AN MEP PROJECT

To exit a Revit session, choose the **Exit Revit** button from the **File** menu. Even if the project is open, you can choose the **Exit Revit** button to close the file and exit Revit. If the project has not been saved once, the **Save File** dialog box will be displayed on choosing the **Exit Revit** button. In this dialog box, if you choose the **No** button, all unsaved changes will be lost. You can also use the Close button (**X**) in the main Revit window (in the title bar) to end the Revit session.

OPENING AN EXISTING MEP PROJECT

In Autodesk Revit, there are several options available to open an existing project. These options are discussed next.

Opening an Existing Project Using the Open Tool

To open an existing project file, you can select **Open > Project** from the File menu, use the **Open** button on the Quick Access Toolbar, or press **CTRL+O**. This opens the **Open dialog box**, where you can browse to the desired folder using the **Look in** drop-down list. The **Preview area** shows a preview of the selected file to help identify it, even if you don't remember its name. You can change how files are displayed using the **Views menu**, and choose **Thumbnails** to preview file contents directly. On the left side of the dialog box, the **Places list** offers quick access to important folders, which you can customize. To add or remove folders, go to **File > Options > File Locations > Places**. In the **Places dialog box**, you can create new libraries using the **Add Value** button, assign a name, and choose a folder path. You can also reorder or delete libraries. After saving your changes, the new folders will appear in the Places list for easier access. Once you select a file, its name appears in the **File Name** box, and a preview is shown, confirming your selection before opening.

Note

*If you try opening an already opened file that has been modified in the Revit session, a message box will appear prompting you to close the file first and then reopen it. In case you open a file that has been created using an older version of Revit, the **Program Upgrade** message box will be displayed. This message box informs that the file is being upgraded to the latest file format and that this is a onetime process. Once the file is opened, it gets upgraded to Autodesk Revit 2025 version.*

Using the Windows Explorer to Open an Existing Project

Apart from using the **Open** tool from the Revit interface to open a file, you can also open files directly from the Windows Explorer by using the methods discussed next.

A file can be opened by double-clicking on its icon in the Windows Explorer. It opens the project file in the latest Revit session. If Revit is not running, double-click on the file icon to start Revit and then open the file.

Another method of opening a project file is by dragging the project file icon from the Windows Explorer and dropping it in the drawing window of the Revit interface. You can also select, drag, and drop more than one file in the drawing window. In this case, Revit prompts you to open the files in separate windows. Choose the **OK** button to open all files in the same Revit session.

TUTORIAL

Tutorial 1 Office Space

In this tutorial, you will create a project setup for the *Office-Space* project using the following parameters and project specifications: **(Expected time: 45 min)**

1. Template file:
   For Imperial **English-Imperial > Systems-Default**
   For Metric **English > Systems-Default_Metric**
2. Project Units: Refer to Figure 2-29
3. File name to be assigned:
   For Imperial *c02_Office-Space_tut1.rvt*
   For Metric *M_c02_Office-Space_tut1.rvt*

The following steps are required to complete this tutorial:

a. Start the Revit 2025 session.
b. Use **Systems-Default** (Imperial) or **Systems-Default_Metric** (Metric) as the template file for the project, refer to Figure 2-27.
c. Specify the project units, refer to Figure 2-29.
d. Specify the project information, refer to Figure 2-30.
e. Create the project parameter, refer to Figure 2-31.
f. Add the project parameter to project information.
g. Set the project location.
h. Set the browser organization.
i. Save the project as *c02_Office-Space_tut1.rvt* by using the **Save As** tool.
j. Close the project by using the **Close** tool.

Starting Autodesk Revit 2025

1. Choose **Start > All Programs > Autodesk > Revit 2025** from the taskbar (for Windows 10) or double-click on the Revit 2025 icon on the desktop; the Revit interface is displayed.

Opening a New Project

1. Choose **New > Project** from the **File** menu; the **New Project** dialog box is displayed, as shown in Figure 2-12.

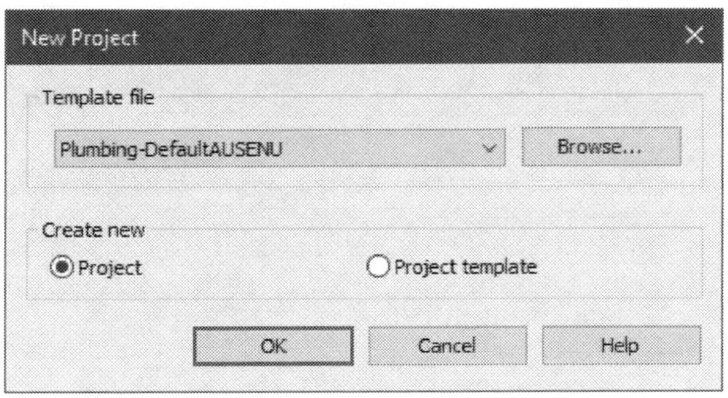

*Figure 2-12 The **New Project** dialog box*

Selecting the Template File

Before you start a Revit project, it is necessary to select a desired template file.

1. In the **New Project** dialog box, choose the **Browse** button from the **Template file** area; the **Choose Template** dialog box is displayed with a list of template files in the **English-Imperial** folder. Note that for Metric system, the dialog box will display a list of template files in the **English** folder.

2. In the **Choose Template** dialog box, select the **Systems-Default** template file from the list, as shown in Figure 2-13. For Metric system, select the **Systems-Default_Metric** template file from the list. Choose the **Open** button the **Choose Template** dialog box is closed.

3. Next, in the **New Project** dialog box, ensure that the **Systems-Default.rte** (for Metric system **Systems-Default_Metric.rte)** option is selected in the drop-down list located in the **Template file** area. Also, ensure that the **Project** radio button is selected in the **Create new** area. Now, choose the **OK** button; the **New Project** dialog box is closed and the *Systems-Default.rte* template file is loaded in the current file. For Metric system the *Systems-Default_Metric.rte* is loaded in the current file. Notice that the **Project Browser** now shows different levels and views that have already been created in the selected template file.

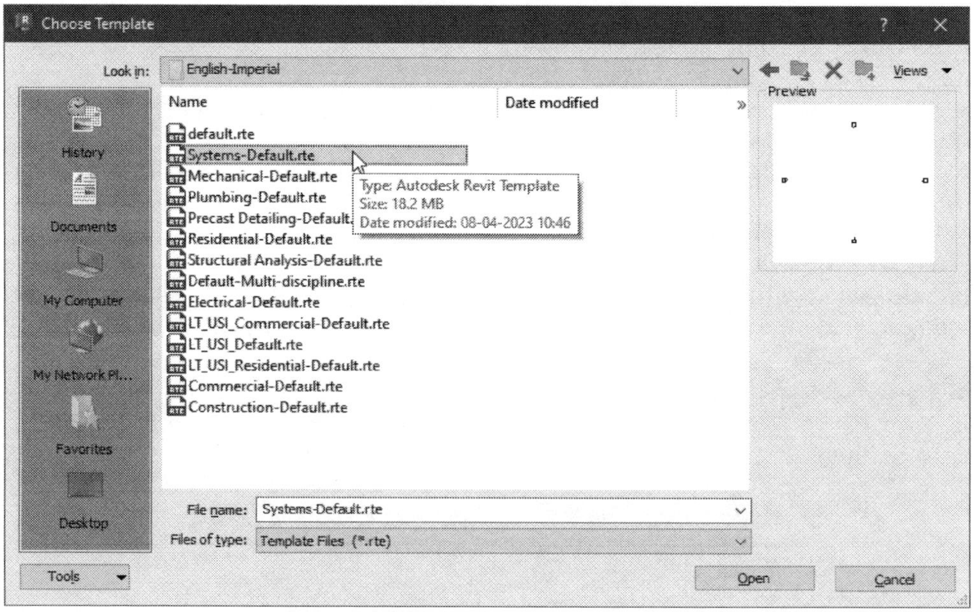

Figure 2-13 Selecting the ***Systems-Default.rte*** *file from the* ***Choose Template*** *dialog box*

Setting MEP Units

In this section, you will set the project units for the MEP project.

1. Choose the **Project Units** tool from the **Settings** panel of the **Manage** tab; the **Project Units** dialog box is displayed. Alternatively, you can type UN to invoke the **Project Units** dialog box.

2. In the **Project Units** dialog box, select the **HVAC** option from the **Discipline** drop-down list; various units required in the HVAC workflow are displayed in a table located below the **Discipline** drop-down list, refer to Figure 2-14.

3. Next, choose the button in the **Format** column corresponding to the **Density** parameter; the **Format** dialog box is displayed.

4. In the **Format** dialog box, select the **Pounds per cubic inch** option (for Metric system, select the **kilograms per cubic meter** option) from the **Units** drop-down list and then choose the **OK** button; the **Format** dialog box is closed.

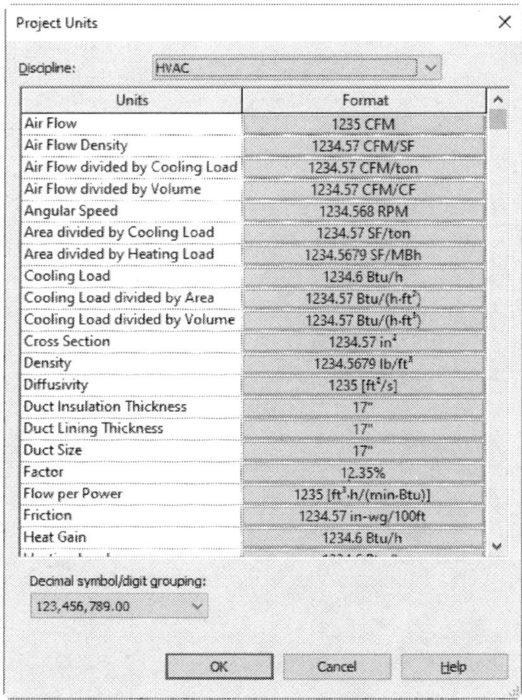

*Figure 2-14 The **Project Units** dialog box*

5. In the **Project Units** dialog box, repeat steps 2 to 4 and assign different units to the **Electrical**, **Piping**, and **Energy** disciplines. Refer to the table given next for the formats and the discipline of different units needed to be specified.

Discipline	Unit	Format (Imperial)	Format (Metric)
HVAC	Pressure	Pounds per square inch	Pascals
Electrical	Illuminance	Footcandles	Lux
Electrical	Apparent Power	BTU (British Thermal Units) per seconds	Watts
Piping	Density	Pounds per cubic inch	kilograms per cubic meter
Energy	Coefficient of Heat Transfer	BTU (British Thermal Units) per hour square foot degree Farenheit	Watts per square meter kelvin

6. In the **Project Units** dialog box, after specifying the various formats for different units, choose the **OK** button; the **Project Units** dialog box is closed.

Setting the Project Information

In this section, you will add the project information to the *Office-Space* project file.

1. To set the project information, choose the **Project Information** tool from the **Settings** panel of the **Manage** tab; the **Project Information** dialog box is displayed.

2. In the **Project Information** dialog box, specify the project parameters as given in Figure 2-15.

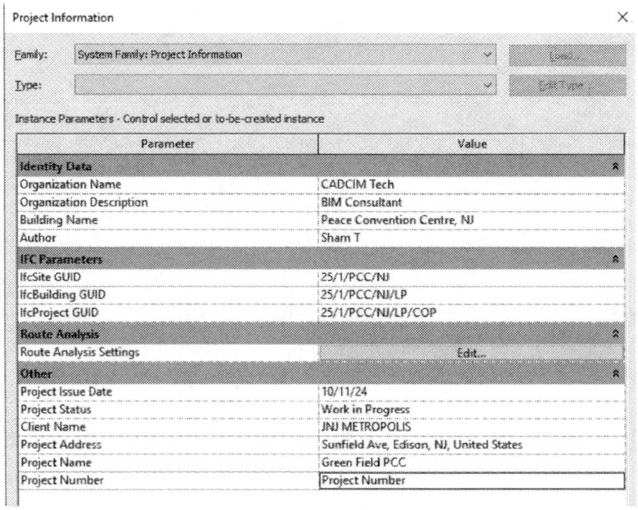

***Figure 2-15** The partial view of the **Project Information** dialog box*

6. In the **Project Information** dialog box, choose the **OK** button; the dialog box is closed.

Creating the Project Parameter

In this section, you will create a shared project parameter to be added to the project information.

1. Choose the **Shared Parameters** tool from the **Settings** panel of the **Manage** tab; the **Edit Shared Parameters** dialog box is displayed, refer to Figure 2-16.

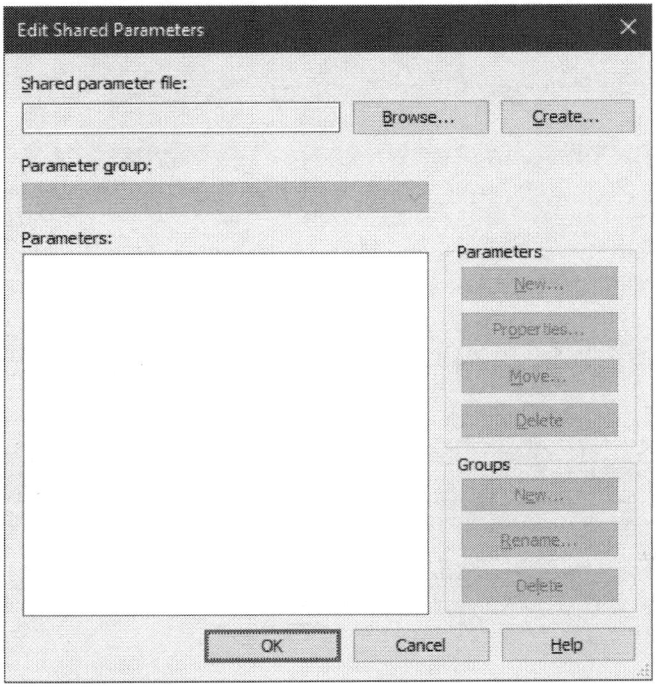

Figure 2-16 The ***Edit Shared Parameters*** *dialog box*

2. In the **Edit Shared Parameters** dialog box, choose the **Create** button; the **Create Shared Parameter File** dialog box is displayed.
3. In the **Create Shared Parameter File** dialog box, create a folder as per the following path *C:\Office-Space\MEP.*
4. Next, open the **MEP** folder and then enter the **Carbon Footprint** text in the **File name** edit box and then choose the **Save** button; the dialog box closes and the **Edit Shared Parameters** dialog box is displayed again.
5. In this dialog box, choose the **New** button in the **Groups** area; the **New Parameter Group** dialog box is displayed.
6. In the displayed dialog box, enter the **Green House Gas** text in the **Name** edit box and then choose the **OK** button; the **New Parameter Group** dialog box is closed and the new group is displayed in the **Parameter group** drop-down list of the **Edit Shared Parameters** dialog box.

7. Now, choose the **New** button in the **Parameters** area; the **Parameter Properties** dialog box is displayed.

8. In this dialog box, enter **Carbon Footprint Factor** in the **Name** edit box and then select the **Energy** option from the **Discipline** and **Type of Parameter** drop-down lists.

9. Now, in the **Parameter Properties** dialog box, choose the **OK** button; the dialog box is closed and the **Edit Shared Parameters** dialog box is displayed with the new parameter added.

10. In the **Edit Shared Parameters** dialog box, choose the **OK** button; the dialog box is closed.

Adding the Project Parameter to Project Information

In this section, you will add the project parameter to the project information.

1. Choose the **Project Parameters** tool from the **Settings** panel of the **Manage** tab; the **Project Parameters** dialog box is displayed.

2. In the **Project Parameters** dialog box, choose the **New Parameter** button; the **Parameter Properties** dialog box is displayed.

3. In the **Parameter Type** area of the **Parameter Properties** dialog box, select the **Shared parameter** radio button and then choose the **Select** button; the **Shared Parameters** dialog box is displayed.

4. In this dialog box, ensure that the **Carbon Footprint Factor** parameter is selected in the **Parameters** area and then choose the **OK** button; the **Shared Parameters** dialog box is closed.

5. In the **Parameter Data** area of the **Parameter Properties** dialog box, ensure that the **Instance** and **Values are aligned per group type** radio buttons are selected. Also, ensure that the **Energy Analysis** option is selected in the **Group parameter under** drop-down list.

6. In the **Categories** area, select the **Project Information** check box from the list box displaying all categories.

7. Now, choose the **OK** button; the **Project Parameters** dialog box is displayed. Choose the **OK** button in the displayed dialog box to close it.
   After assigning the shared parameter to the category of project information, now you need to specify a value to the added parameter in the **Project Information** dialog box.

8. Choose the **Project Information** tool from the **Settings** panel of the **Manage** tab; the **Project Information** dialog box is displayed.

9. In the **Project Information** dialog box, click in the value field corresponding to the **Carbon Footprint Factor** parameter and enter **12000** in it.

10. Next, choose the **OK** button to close the **Project Information** dialog box.

Setting the Project Location

In this section, you will set the location of the project.

1. Choose the **Location** tool from the **Project Location** panel of the **Manage** tab; the **Location and Site** dialog box is displayed.

2. In this dialog box, ensure that the **Location** tab is chosen by default. Now, select the **Internet Mapping Service** option from the **Define Location by** drop-down list, if it is not selected by default.

> **Note**
>
> *On selecting the* ***Internet Mapping Service*** *option from the* ***Define Location by*** *drop-down list, the* ***Google Map*** *browser is activated for browsing the desired location. You should ensure that the internet connection is active at this stage.*

3. In the **Location** tab of the **Location and Site** dialog box, enter **Schererville, IN** in the **Project Address** edit box and then choose the **Search** button; the **Bing Map** browser is displayed under the **Project Address** edit box showing the desired location in the map, as shown in Figure 2-17.

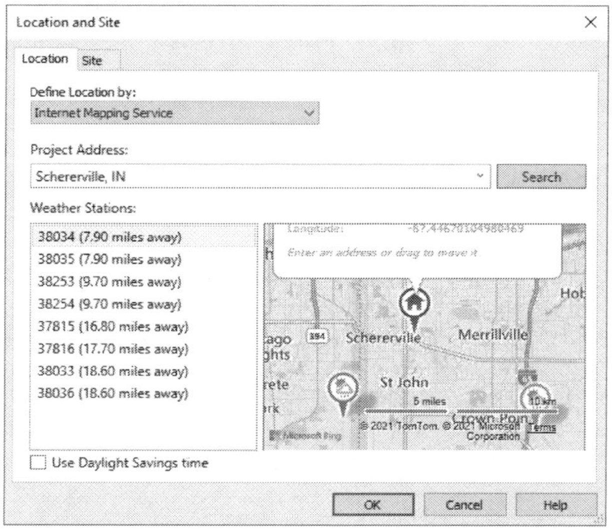

Figure 2-17 The ***Location and Site*** *dialog box*

4. Now, choose the **OK** button to close the **Location Weather and Site** dialog box.

Setting the Browser Organization

In this section, you will set the **Project Browser** to display various views and information related to the project in the order of their discipline.

1. Choose the **Browser Organization** tool from **View > Windows > User Interface** drop-down; the **Browser Organization** dialog box is displayed.

2. In the displayed dialog box, ensure that the **Views** tab is chosen by default and then choose the **New** button; the **Create New Browser Organization** dialog box is displayed.

3. In the **Name** edit box of the displayed dialog box, enter **Office-Space-MEP** and then choose the **OK** button; the dialog box closes and the **Browser Organization Properties** dialog box is displayed.

4. In this dialog box, choose the **Grouping and Sorting** tab; various options in this tab are displayed.

5. Select the **Discipline** option from the **Group by** drop-down list.

6. Ensure that the **All characters** radio button located under the **Group by** drop-down list is selected. Now, select the **Family and Type** option from the **Then by** drop-down list.

7. Next, select the **View Name** option from the **Sort by** drop-down list located at the bottom in the **Browser Organization Properties** dialog box. Ensure that the **Ascending** radio button located below the **Sort by** drop-down list is selected.

8. Choose the **OK** button; the **Browser Organization Properties** dialog box is closed.

9. In the **View** tab of the **Browser Organization** dialog box, select the **Office-Space-MEP** check box. Now, choose the **Apply** button and then the **OK** button; the dialog box is closed and the specified settings are applied to the **Project Browser**, refer to Figure 2-18.

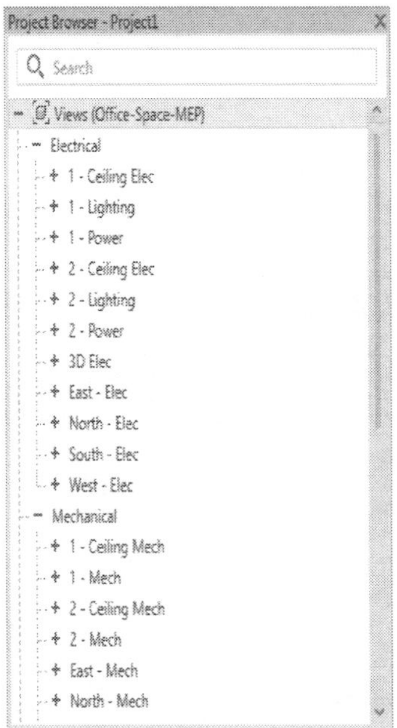

*Figure 2-18 The **Project Browser** displaying the settings specified in the **Office-Space-MEP** browser organization*

Saving the Project

In this section, you need to save the project and the settings using the **Save As** tool.

1. Choose **Save As > Project** from the **File** menu; the **Save As** dialog box is displayed.

2. In this dialog box, browse to the *C* drive and then create a folder with the name **rmp_2025.** Next, create a sub-folder with the name **c02_rmp_2025_tut** in it.

3. In the **File name** edit box, enter **c02_Office-Space_tut1** for Imperial or **M_c02_Office-Space_tut1** for Metric and then choose the **Options** button; the **File Save Options** dialog box is displayed.

4. In this dialog box, enter **5** in the **Maximum backups** edit box, and then in the **Thumbnail Preview** area, select the **3D View: {3D}** option from the **Source** drop-down list.

5. Select the **Regenerate if view/sheet is not up-to-date** check box in this dialog box.

6 Now, choose the **OK** button; the **File Save Options** dialog box is closed and the **Save As** dialog box is displayed.

7. In this dialog box, choose the **Save** button to save the current project file with the specified name and to close the **Save As** dialog box.

Closing the Project

1. To close the project, choose the **Close** option from the **File** menu.

EXERCISE

Exercise 1 Power Plant

In this exercise, you will create a new project file for the *Power Plant* project with the following parameters. **(Expected time: 15 min)**

1. Template file:

For Imperial	**English-Imperial > Systems-Default**
For Metric	**English > Systems-Default_Metric**

2. Project information to be added:

Project Issue Date	**10/05/2025**
Project Status	**Started**
Client Name	**Cadsoft Technologies**
MEP Consultant	**Cadsoft-Technologies**
Mechanical Consultant	**Sham Tickoo**
Project Name	**Power Plant**
Commencement Date	**20/09/2025**
Documented Contract Completion Date	**31/07/2025**
Provisional Period Allowed in Contract	**30 Days**

3. Project Units- Set the various units in the *Power Plant* project as mentioned in the table given next.

Discipline	**Units**	**Format (Imperial)**	**Format (Metric)**
HVAC	Power	Kilowatts	Kilowatts
HVAC	Heat Gain	Kilowatts	Kilowatts
HVAC	Duct Size	Decimal inches	Millimeters
HVAC	Factor	Fixed	Fixed
HVAC	Pressure	Pounds per square inches	Pascals
Electrical	Frequency	Cycles per second	Cycles per second
Electrical	Current	Kiloamperes	Kiloamperes
Electrical	Illuminance	Footcandles	Lux
Electrical	Electrical Potential	Kilovolts	Kilovolts
Piping	Velocity	Feet per second	Meters per second
Energy	Coefficient of Heat Transfer	BTU per hour	Joules

4. File name to be assigned:

For Imperial *c02_Power_Plant_exer1.rvt*
For Metric *M_c02_Power_Plant_exer1.rvt*

Chapter 3

Creating Building Envelopes

Learning Objectives

After completing this chapter, you will be able to:

- *Understand the concept of Levels*
- *Understand the concept of Grids*
- *Work with Project Views*
- *Understand walls*
- *Add doors and windows*
- *Create openings in walls*
- *Create ceilings*
- *Create floors*
- *Create rooms*

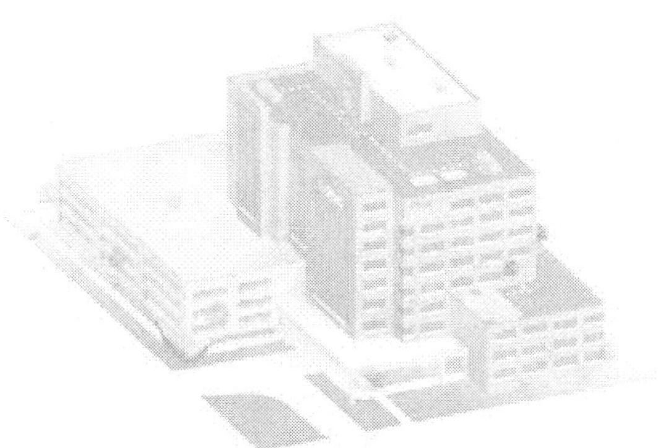

INTRODUCTION

The building envelope behaves as a physical barrier between the interior and exterior of a building. It comprises of walls, floors, roofs, fenestrations and doors. In a project, fenestration are any opening in the structure such as windows, skylights, clerestories, etc. In a Revit project, the building envelope also comprises of beams, columns, datum elements, and standard views.

In an HVAC system, the building envelope is considered as an important aspect of an HVAC design. The building envelope is the outer shell that maintain a dry, heated, or cooled indoor environment and facilitate its climate control. Building envelope design is a specialized area of architectural and engineering practice that draws from all areas of building science and indoor climate control. While working in an MEP project you can either create a building envelope in the current MEP project or link the architectural and structural models created in Revit to the current MEP project.

LEVELS

Levels, in a multistory building, refer to the infinite horizontal planes that define each story of the structure. Autodesk Revit uses levels as references for level-hosted elements such as walls, ducts, AHUs, air terminals, pipes and pipe fixtures, sanitary fixtures, floor, roof, ceiling, and so on. The distance between levels can be used to define the story height of a building model, as shown in Figure 3-1. Autodesk Revit also provides flexibility to create a non-story level or a reference level such as sill level, parapet level, and so on.

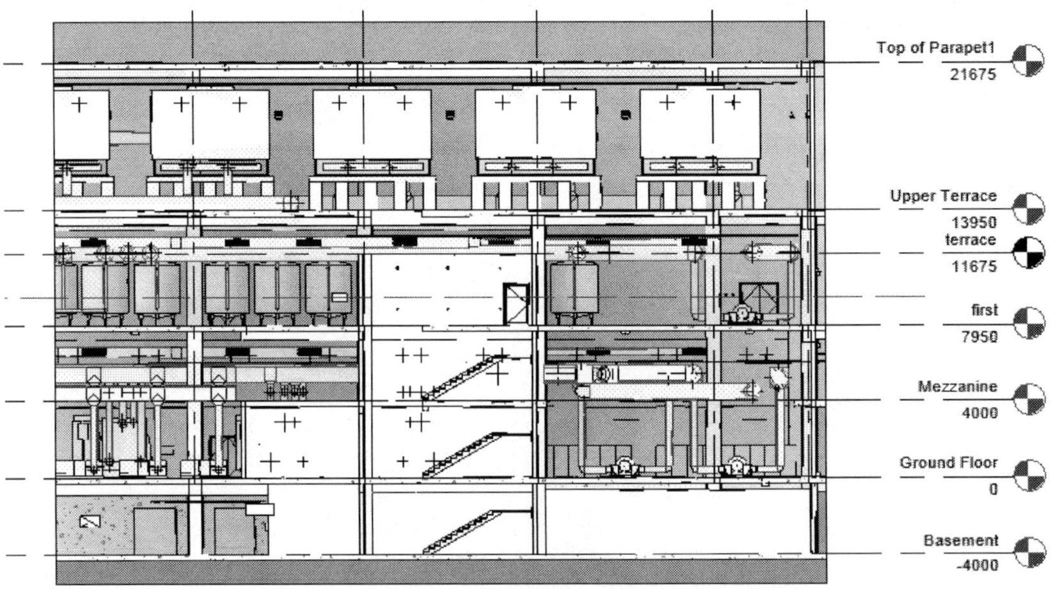

Figure 3-1 *The elevation view of a building model displaying different levels*

Understanding Level Properties

A typical level is represented by a level line, level bubble, level name, level elevation, and so on, refer to Figure 3-2. Using these parameters and controls, you can modify the app earance of a level. The level name is a modifiable parameter that is used to refer to each level. The level elevation is the distance of the level from the base level. The visibility of the level bubble on either side of the level line can be controlled by using the bubble display control. The length alignment control can be used to align level lines. Autodesk Revit provides the 2D or 3D extents control for datums when they are selected. This enables you to change their extents in one or multiple views in which they are visible. When a datum is in 3D mode, any modification made in the 3D view reflects in all views of the building model. Therefore, you cannot modify a specific view in the 3D mode. However, 2D mode can be used to modify a datum in a specific view, thereby making it view-specific.

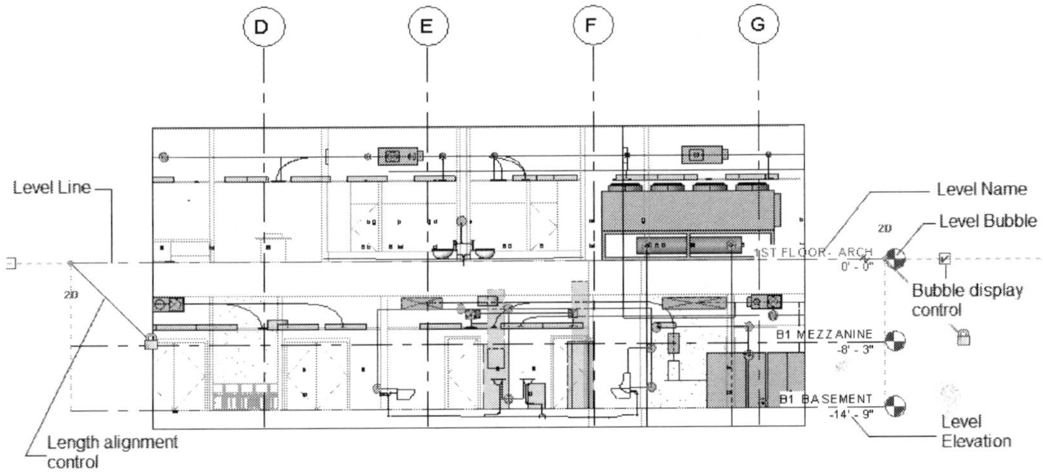

***Figure 3-2** A section view displaying levels with its various components*

Adding Levels

In Autodesk Revit, levels can only be created in elevation or section views, as the **Level** tool is inactive in plan views. To add a level, open a section or elevation view, then use the **Level** tool from the Datum panel under the **Architecture** tab or press **LL**. The **Modify | Place Level** tab will appear, allowing you to choose from sketching options in the Draw panel. You can also select a level type from the Type Selector drop-down. In the Imperial system, available types include **Level: 1/4" Head**, **No Head**, and **Plenum**; in the Metric system, options include **Level: 8mm Head** and **Plenum**. To display the level head, choose **1/4" Head** (Imperial) or **8mm Head** (Metric).

In the **Modify | Place Level** tab, you can use tools like **Line** or **Pick Lines** from the Draw panel to sketch a level. To create associated plan views automatically, check the **Make Plan View** box in the Plan Views panel. You can then click the **Plan View Types** button to choose between **Ceiling Plan** and **Floor Plan** in the dialog box. If unchecked, no associated views will be created. You can also use the **Offset** box in the Options Bar to place a level at a specific distance. When placing a new level in an elevation or section view, moving the cursor near existing levels will show temporary dimensions to help set the desired distance from them.

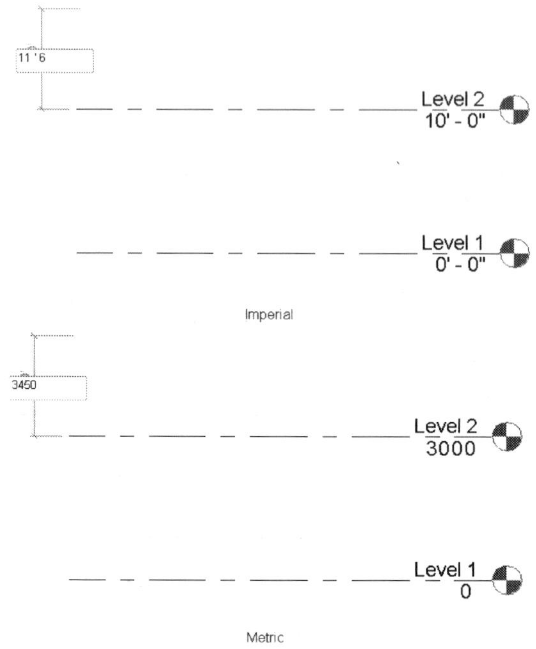

***Figure 3-3** Specifying the distance from the existing level*

Modifying Level Parameters

To change a level type in Revit, select the level and choose a new type from the Type Selector in the Properties palette. You can also adjust properties before placing a level. Levels can be renamed by clicking the name and typing a new one, prompting Revit to optionally rename associated views like floor and ceiling plans.

Controlling the Visibility of Levels

You can control the visibility of levels in any of the project views. To do so, select a level and right-click; a shortcut menu will be displayed, as shown in Figure 3-4. Choose **Hide in View > Category** from the shortcut menu; the level category will be hidden in the current view. To hide a particular level, choose **Hide in View > Elements** from the shortcut menu; only the selected level will be hidden in the current view and will be displayed in all other views.

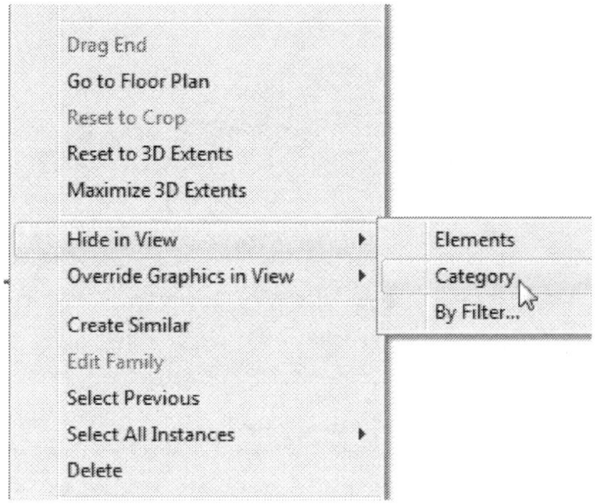

Figure 3-4 *Choosing the* ***Category*** *option from the shortcut menu*

WORKING WITH GRIDS

Autodesk Revit provides you the option of creating rectangular or circular grids for your projects. Using these grids, you can create building envelopes easily and also place the MEP elements at desired locations and intersections.

Modifying Grids

Grids in Revit can be easily modified after creation. To rename a grid, select it and change its name in the Properties palette. You can adjust spacing between grids by editing the temporary dimension or simply drag grids to a new location—select one or multiple grids (using **CTRL**, and drag them. Hold **SHIFT** to restrict movement to straight lines. Right-clicking a highlighted grid opens a shortcut menu with options like **Select Previous**, **Create Similar**, or **View Properties**.

Grid Properties

Like levels, grids too have associated properties. A typical grid consists of a grid line, grid bubble, grid name, and other controls. The usage of the controls such as grid bubble visibility, 2D or 3D extents, and grid bubble break is similar to the usage of controls described for levels.

REFERENCE PLANES

Reference planes in Revit are useful as guidelines for sketching and placing building elements, and they're also essential for creating new family elements. To create one, use the **Ref Plane** tool from the Work Plane panel under the **Architecture** tab or simply type **RP**. Once activated, use the tools in the Draw panel to sketch the reference plane by clicking a start point in the drawing, moving the cursor, and clicking again to define the endpoint. To name the reference plane, select it and enter a name in the **Name** field within the Properties palette.

WORK PLANES

As the name suggests, the work plane is a plane that can be used for sketching elements. In Autodesk Revit, you can create and edit only those elements that lie in the current work plane. The work plane can be horizontal, vertical, or inclined at any specified angle. Each generated view has an associated work plane. This workplane is automatically defined for some standard views such as floor plans.

Setting a Work Plane

In Revit, you can set a work plane based on your project needs by selecting the **Set** tool from the Work Plane panel in the **Architecture** tab, which opens the Work Plane dialog box. This dialog shows the current work plane and allows you to choose a new one from a drop-down list of levels, grids, and reference planes using the Name option.

Controlling the Visibility of Work Planes

You can control the visibility of the current work plane by using the **Show** button from the **Architecture** tab. This button is used to toggle the display of grids in the workplane.

WORKING WITH PROJECT VIEWS

While working on the building envelope and MEP models, you may need to view its different exterior and interior portions in order to add or edit the elements in the design. Revit provides various features and techniques that can be used to view the building model.

Viewing a Building Envelope

In Revit, the default Mechanical, Electrical, Plumbing, or Systems template includes standard views like floor plans, ceiling plans, elevations, sections, and 3D views, all accessible from the Project Browser. You can open any view by double-clicking its name and manage the Project Browser's visibility through View > Windows > User Interface. Floor plan views also display elevation markers for North, East, South, and West, which are clickable.

Overriding the Visibility/Graphic of an Element

You can override the visibility and graphics of any element in a view. To do so, select an element and right-click; a shortcut menu will be displayed. Choose **Override Graphics in View > By Element** from the shortcut menu; the **View Specific Element Graphics** dialog box will be displayed, as shown in Figure 3-5. The options in this dialog box are discussed next.

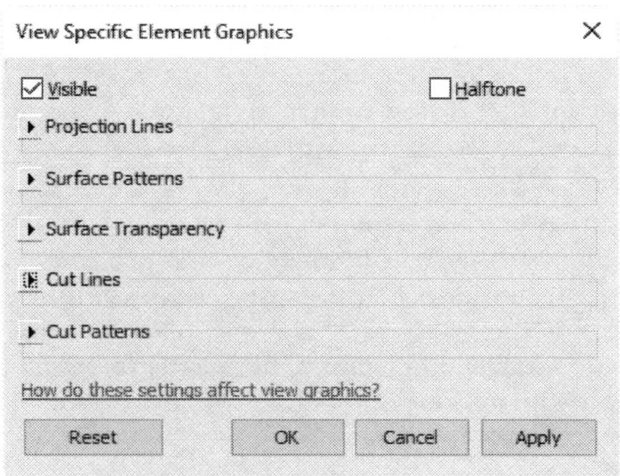

***Figure 3-5** The **View Specific Element Graphics** dialog box*

Overriding the Visibility/Graphic of Element Category

To edit the visibility and graphics of an element category in a project view in Revit, open the view and select an element from the desired category. Right-click and choose **Override Graphics in View > By Category** to open the View-Specific Category Graphics dialog box, then click the button to access the **Visibility/Graphic Overrides** dialog. From here, you can adjust visibility by checking or unchecking categories, and apply graphic overrides such as fill patterns, colors, transparency, halftone, and detail levels. Use the appropriate override buttons to make changes, then click **Apply** to preview and **OK** to confirm and close the dialog box.

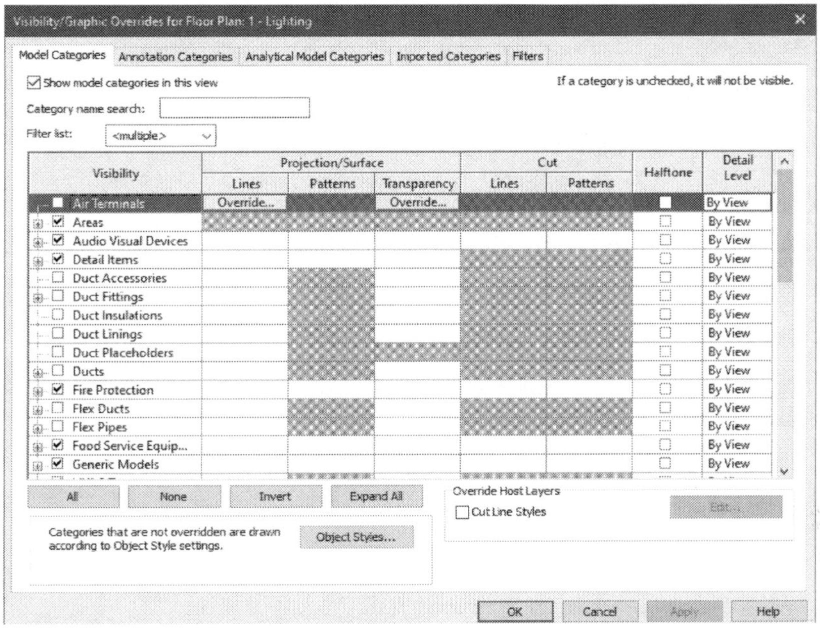

*Figure 3-6 The **Visibility/Graphics Overrides for Floor Plan** dialog box*

Making Elements Transparent

Revit provides a tool that allows you to make elements transparent, enabling you to see through them—useful for viewing a building's interior from the top even after adding roofs or ceilings. To do this, select the element, right-click, and from the shortcut menu, choose **Override Graphics in View > By Element**, which opens the **View Specific Element Graphics** dialog box. In this dialog, expand the **Surface Transparency** section to access the **Transparency slider**, which you can move to the right to increase transparency or input a specific value in the adjacent edit box. Once set, click **OK**, and the selected element will become transparent, allowing visibility through it while still displaying its edges and surface pattern.

Using the Temporary Hide/Isolate Tool

The Temporary **Hide/Isolate** tool allows users to temporarily hide or isolate elements in a project view and becomes available only after selecting elements. Once selected, users can access the tool from the **View Control Bar**, where a cascading menu with six options appears. These options enable hiding or isolating elements or their categories, and once activated, a cyan border appears around the drawing area to indicate that **Temporary Hide/Isolate** mode is active. The

Hide Element tool hides the selected element, while the **Isolate Element** tool displays only the selected elements, hiding the rest.

Plan Views

In Autodesk Revit, you can use the floor and ceiling views to view the building plan of the respective discipline. If you are using the **Systems-Default.rte** template file to start a new project, by default Autodesk Revit opens the **1-Mech** floor plan view. The **Systems-Default.rte** template file contains floor plans for the Electrical, Mechanical, and Plumbing disciplines and ceiling plans for Electrical and Mechanical discipline.

Elevation Views

An elevation view refers to the view of the building model from the four sides, North, East, South, and West. If you are using the default template file, the four sided elevation views are created automatically by Autodesk Revit when the default template file is used.

Section Views

Section views are generated by cutting sections through the building model. These views are created to display various wall elevations, floor heights, and special vertical features of the project. They are also useful in creating and editing elements added to the interior spaces of the building model.

For example, in an office building, to emphasize the salient features of the central atrium, you may need to show a section through the central atrium. Autodesk Revit enables you to create it with relative ease. You can also modify the sectional view to create a section that displays the interior spaces.

Using the Scope Box Tool

Autodesk Revit provides you the option of controlling the visibility of the datum elements in the project views using the **Scope Box** tool. As described earlier in this chapter, the datum planes have an infinite scope and extend throughout the project. Using the **Scope Box** tool, you can limit the boundary for the visibility of the datum planes. You can also specify the views in which these datum planes become visible.

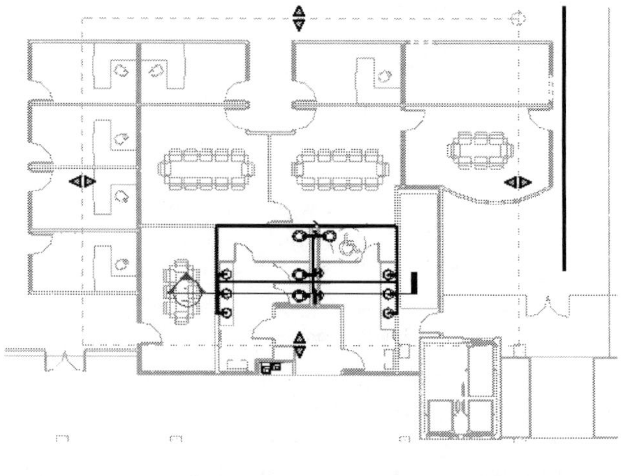

Figure 3-7 *The selected scope box in a plumbing layout*

UNDERSTANDING WALL TYPES

Autodesk Revit provides you with several predefined wall types based on their usage. These wall types are discussed next

Exterior Wall Type

This is the wall type that is primarily used for generating the exterior of the building model. It has predefined wall types, such as **Brick on CMU**, **Brick on Mtl. Stud**, **CMU Insulated**, and so on.

Curtain Wall Type

Autodesk Revit also provides you with the flexibility of creating your own wall type. The walls that you will create can have different functions, which can be modified, depending on their functional usage. In Autodesk Revit, you can create both architectural and structural walls. An architectural wall does not contain analytical properties as the structural walls do. In the next section, various techniques to create and modify architectural walls are discussed.

CREATING ARCHITECTURAL WALLS

In Autodesk Revit, each wall type has specific predefined properties such as composition, material, characteristics, finish, height, and so on. You can select the wall type based on its specific usage in the project. Walls, like most other model elements, can be created in a plan view or a 3D view.

To create an exterior architectural wall, first you need to invoke the **Wall: Architectural** tool and then select the appropriate exterior wall type and specify various properties. To do so, invoke the **Wall: Architectural** tool from the **Build** panel, refer to Figure 3-8; the **Modify | Place Wall** tab will be displayed. To select the type of wall, select an exterior wall type from the **Type Selector** drop-down list in the **Properties** palette, as shown in Figure 3-9. Next, from the **Properties** palette, specify and edit various properties of the wall to be added. Various wall properties and the process to specify them are discussed next.

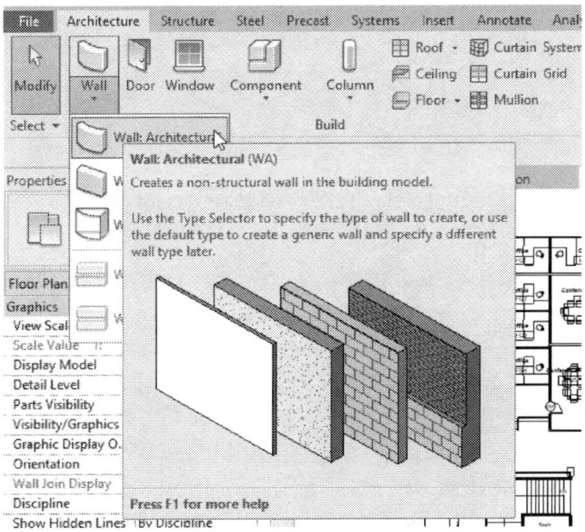

Figure 3-8 *Invoking the* ***Wall: Architectural*** *tool from the* ***Wall*** *drop-down*

Specifying Architectural Wall Properties

In Autodesk Revit, walls like other elements, has two sets of properties, type and instance. These set of properties control the appearance and the behavior of the concerned element.

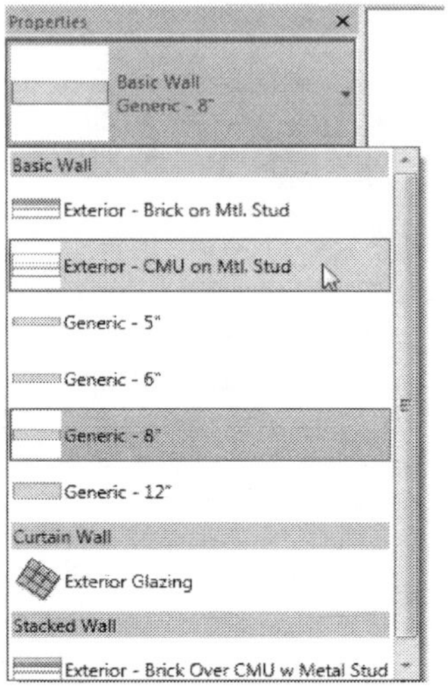

***Figure 3-9** Selecting an exterior wall type from the **Type Selector** drop-down list*

Sketching Walls

The next step after selecting the wall type from the **Properties** palette is to select the sketching tool. Autodesk Revit provides several sketching tools, such as **Line**, **Rectangle**, and so on to sketch the walls of different shapes. These tools, along with the **Options Bar**, can be invoked from the **Draw** panel in the **Modify | Place Wall** tab, as shown in Figure 3-10.

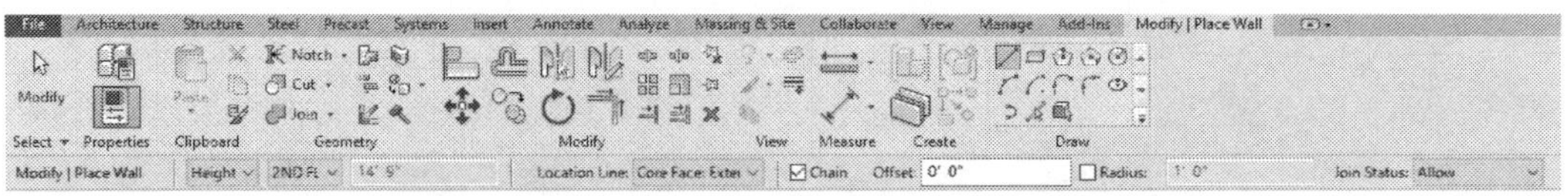

***Figure 3-10** The **Modify / Place Wall** tab*

In the **Modify | Place Wall** tab, you can access different sketching tools from the **Draw** panel and the **Options Bar** in the **Modify | Place Wall** tab. On invoking these tools, you can sketch different wall profiles. The procedures for sketching the straight and circular wall profiles are discussed next.

USING DOORS IN A BUILDING MODEL

A door is a commonly used component in a building model, allowing access between interior and exterior spaces. In Autodesk Revit, various predefined door types are available through the Type Selector drop-down in the Properties palette, and additional types can be loaded from the English-Imperial or Metric folders.

Adding Doors

You can add doors to a building model in Revit using the **Door** tool from the Build panel or by typing **DR**. Since door types are not included in default templates, Revit prompts you to load a door family the first time you use this tool. By choosing "**Yes**," the **Load Family** dialog box appears, where you can browse to the appropriate folder (**English-Imperial** > **Doors** for **Imperial** or **English** > **US** > **Doors** for **Metric**), select the desired family, and click **Open**. In the **Specify Types** dialog box, select one or more door types and click **OK** to load them into the project. Once loaded, select a door type from the **Type Selector** in the Properties palette and click on a wall to place the door.

Understanding Door Properties

The Type Properties dialog box in Revit displays various parameters of a door, organized under categories like Construction, Materials and Finishes, Dimensions, Identity Data, and IFC Parameters. For instance, the Dimensions section includes parameters such as **Height**, **Width**, and **Trim Width**. While you can use the default values, it's recommended to duplicate the original door type before making changes. Clicking the **Duplicate** button allows you to create a new door type with a custom name, inheriting all properties from the original.

Adding a Door to a Wall

In Autodesk Revit, doors can be added in plan, section, elevation, or 3D views by using the **Door** tool and clicking on a host wall. A '+' cursor and temporary dimensions assist with placement, and doors can be repositioned or flipped using control arrows. Door tags are added automatically and can be customized or hidden via the **Visibility/Graphics Overrides** dialog.

ADDING WINDOWS IN A BUILDING MODEL

Windows are key components in any building model, and in Autodesk Revit, they must be hosted on walls. Since default templates don't include window types, you need to load them using the **Load Family** tool from the **Insert** tab. From the **Load Family** dialog box, navigate to the appropriate Windows folder (**Imperial** or **Metric**), select the desired types, and confirm your choices in the **Specify Types** dialog box. Once loaded, the selected window types can be placed into the model as needed.

Adding a Window to a Wall

In Autodesk Revit, windows can be added to walls using the Window tool from the **Architecture** tab by selecting a type from the **Type Selector** in the Properties palette and clicking on a wall. The window symbol appears as the cursor moves over the wall, and temporary dimensions show its distance from nearby elements. You can adjust the window's position using these dimensions or flip its orientation with the twin arrow symbol. Window tags are automatically generated and share the same tag for identical types; these can be edited in the

DOORS AND WINDOWS AS WALL OPENINGS

Autodesk Revit also provides in-built opening types for door and windows. These types are available in the **Openings** subfolder of the **Libraries > English-Imperial** folder for Imperial or the **Openings** subfolder of the **Libraries > English** folder for Metric that can be accessed from the **Load Family** dialog box. To invoke this dialog box, choose the **Load Family** tool from the **Load from Library** panel of the **Insert** tab.

OPENINGS IN THE WALL

In Revit, you can create rectangular openings in curved or straight walls using a plan, elevation, or section view—though elevation or section views are typically preferred for easier placement. To create an opening, open the desired view, choose the **Wall: Architectural** tool from the Opening panel of the Architecture tab, and select the host wall. Click to mark the first corner of the opening, then move the cursor and click again to define the opposite corner. Temporary dimensions will appear, which you can adjust to refine placement.

CREATING ARCHITECTURAL FLOORS

You can add a floor to the current level of a building model using the **Floor: Architectural** tool. You can invoke this tool from the **Build** panel of the **Architecture** tab. On invoking this tool, the **Modify | Create Floor Boundary** tab will be displayed. The **Draw** panel of the **Modify | Create Floor Boundary** tab consists of various tools that are used to draw the floor sketches. You can also sketch the boundary in the 3D view, provided that the work plane is set to the plan view.

Sketching the Floor Boundary

To create a floor, you need to sketch its boundary. There are two methods to sketch a boundary. The first method is to pick the already created walls using the **Pick Walls** tool to define the floor boundary. The other method is to draw the floor profile using the draw tools such as line, rectangle, polygon, arc, and others from the **Draw** panel of the **Modify | Create Floor Boundary** tab.

PLACING CEILINGS

You can add a ceiling to a building model by using the **Ceiling** tool. To do so, invoke this tool from the **Build** panel in the **Architecture** tab; the **Modify | Place Ceiling** tab with various tools for creating the ceiling will be displayed. Since the ceilings are not visible in the floor plan, they are created in the ceiling plan head of each discipline. You can add a ceiling to a project using three different methods, by adding automatic ceiling, by sketching the ceiling, and by using the pick walls method.

Creating an Automatic Ceiling

To add a ceiling in Revit using the automatic method, go to the **Architecture** tab and select the **Ceiling** tool from the **Build** panel. This opens the **Modify | Place Ceiling** tab. Choose **Automatic Ceiling** from the **Ceiling** panel, then select a ceiling type from the **Type Selector** in the **Properties** palette. Open the ceiling plan view, move your cursor inside the room, and when the room boundary is highlighted (usually in red), click to place the ceiling.

Sketching a Ceiling

You can sketch a ceiling boundary to create it. To do so, choose the **Ceiling** tool from the **Build** panel; the **Modify | Place Ceiling** tab will be displayed. In this tab, choose the **Sketch Ceiling** tool from the **Ceiling** panel; the **Modify | Create Ceiling Boundary** tab will be displayed. This tab contains various tools for sketching and modifying the ceiling boundary. To start the sketch of the ceiling boundary, choose the **Boundary Line** tool from the **Draw** panel; a list box containing various drawing tools will be displayed on the right in this panel, as shown in Figure 3-11.

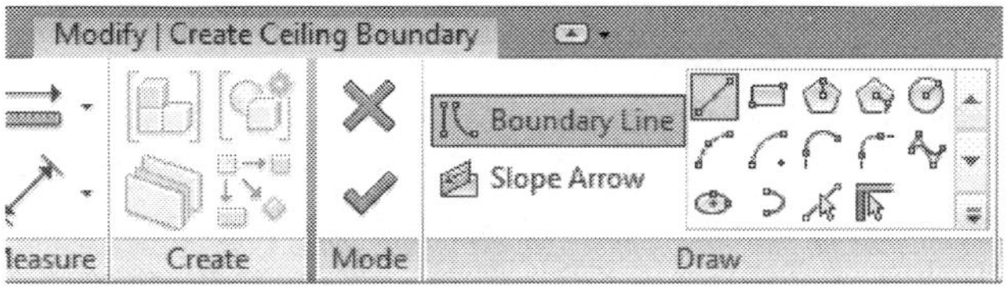

Figure 3-11 *Different sketching tools for ceiling*

The **Line** tool is chosen by default in the **Draw** panel. You can use this tool and the other drawing and modification tools displayed in the **Modify | Create Ceiling Boundary** tab to complete the sketch. Figure 3-12 shows a ceiling boundary that is sketched using various tools displayed in the **Modify | Create Ceiling Boundary** tab.

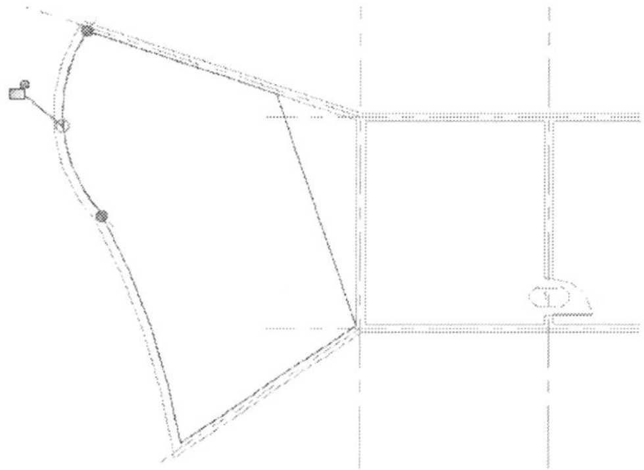

Figure 3-12 *The sketched ceiling boundary*

After completing the sketch, choose the **Finish Edit Mode** button from the **Mode** panel of the **Modify | Create Ceiling Boundary** tab; the ceiling will be created within the sketched profile. Display the section view to view the ceiling in it. The ceiling will appear at a certain height from the floor level, as shown in Figure 3-13.

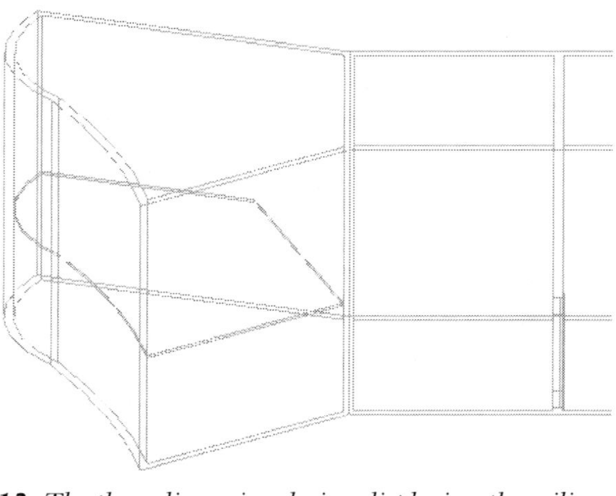

Figure 3-13 *The three dimensional view displaying the ceiling at a height*

Using the Pick Walls Method

The third method to create the ceiling is by picking the wall faces. This is a common method for creating ceilings. To do so, choose the **Ceiling** tool from the **Architecture** tab; the **Modify | Place Ceiling** tab will be displayed. In this tab, choose the **Sketch Ceiling** tool from the **Ceiling** panel; the **Modify | Create Ceiling Boundary** tab will be displayed. In this tab, choose the **Pick Walls** tool from the list box of the **Draw** panel; you will be prompted to pick the walls to define the boundary of the ceiling. Specify the offset value for the ceiling in the **Offset** edit box in the **Options Bar**. Select the **Extend into Wall (to core)** check box, if you want the offset to be measured from the core layer of the walls. Next, pick the walls to create the sketch for the ceiling. In the **Properties** palette, select the ceiling type from the **Type Selector** drop-down list. Next, choose the **Finish Edit Mode** button from the **Mode** panel of the **Modify | Create Ceiling Boundary** tab; the ceiling will be created.

CREATING ROOMS

Room is a part of Revit building elements. Revit provides you the flexibility of creating rooms independent of room tags. Rooms can be created only in the plan view. You can also add rooms from the room schedules. Rooms and areas have the same graphical representation. Also, Revit forms the basis for creating spaces.

Adding Rooms

You can add rooms in the plan view of a specified discipline by using the **Room** tool. To do so, invoke this tool from the **Room & Area** panel, as shown in Figure 3-14; the **Modify | Place Room** tab will be displayed. Now, move the cursor inside the closed boundary that you need to define as room in the drawing; a symbol with a cross hair graphics attached to the cursor will appear. Notice that the size of this cross hair graphics will change according to the area of the room space of the closed boundary. Before you insert the room, you can select the type of room from the **Type Selector** drop-down list in the **Properties** palette. You can select any of the three options: **Room Tag**, **Room Tag With Area**, and **Room Tag With Volume** from the **Type-Selector** drop-down list. You can select the **Room Tag** option to display the inserted room with the tag only. Similarly, you can select the **Room Tag With Area** or **Room Tag With Volume** option from the **Type Selector** drop-down list to attach the room tag to the room to display the information regarding the area or volume of the room. After selecting the required option from the **Type Selector** drop-down list, choose the **Highlight Boundaries** tool from the **Room** panel of the **Modify | Place Room** tab; the **Autodesk Revit 2025** message box will be displayed informing that the room bounding elements are highlighted in the drawing. Choose the **Close** button to close the message box.

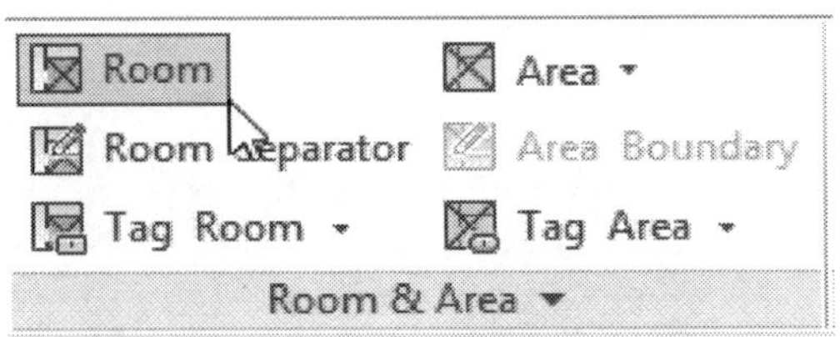

*Figure 3-14 Invoking the **Room** tool*

To add a room with a room tag, choose the **Tag on Placement** tool from the **Tag** panel. You can specify the level up to which the boundary of a room will extend vertically or in upward direction. To do so, select an option from the **Upper Limit** drop-down list in the **Options Bar**. For example, if you have added a room to the floor plan of **Level 1** and you desire to extend the room to **Level 2**, select the **Level 2** option from the **Upper Limit** drop-down list. You can extend the room boundary above the level specified in the **Upper Limit** drop-down list. To do so, click in the **Offset** edit box in the **Options Bar**. The default value displayed in this edit box is **10'** for Imperial or **3000mm** for Metric. You can enter a positive value to extend the room boundary above the level that you have selected from the **Upper Limit** drop-down list. Similarly, you can enter a negative value to extend the room boundary below the selected level. You can orient the room tag horizontally or vertically with reference to the current view, or align it with the walls and boundary lines present in the building model. To specify the orientation of the room tag, select an option from the drop-down list displayed next to the **Offset** edit box in the **Options Bar**. You can select the **Leader** check box to place the room tag with a leader.

To add a room, click in an area enclosed by the room bounding elements such as walls, as shown in Figure 3-15. The room will be added and the area of the room will be equal to the area enclosed by the walls.

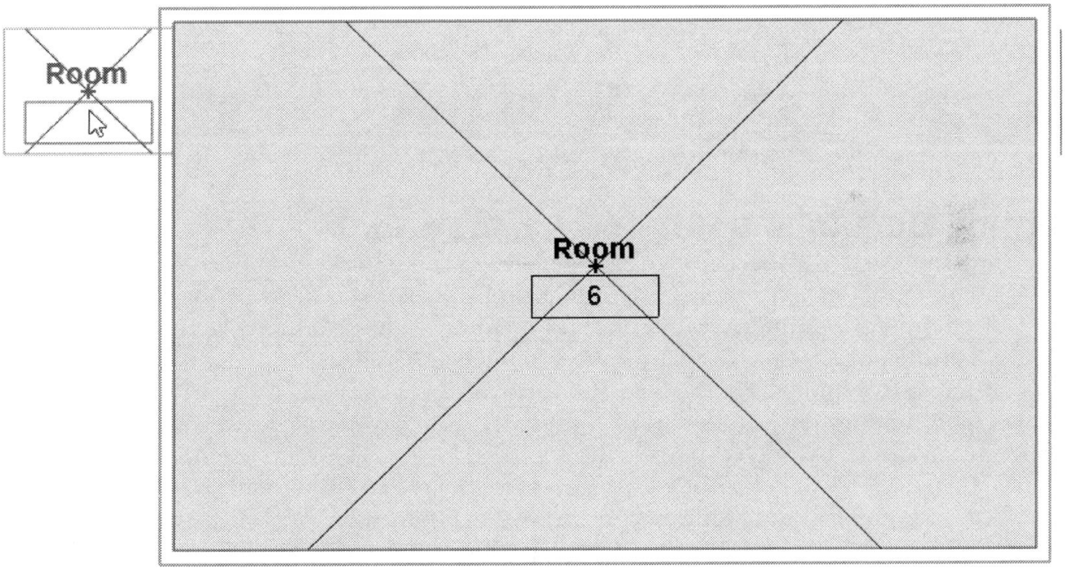

***Figure 3-15** The graphical representation of the room*

Calculating Room Volumes

To enable the room volume computations, choose the **Area and Volume Computations** tool from the **Room & Area** panel; the **Area and Volume Computations** dialog box will be displayed. In the **Volume Computations** area of the **Computations** tab, select the **Areas and Volumes** radio button. By default, the **Areas only** radio button is selected in the **Volume Computations** area. Next, choose the **OK** button; the volume computation of the room will be enabled and displayed in its instance properties. To view the computed volume of a room, select it from the drawing; the **Properties** palette will be displayed. In the **Properties** palette, the value in the value column of the **Volume** parameter shows the computed volume of the room.

While computing the room volume, the room volume calculation engine of Autodesk Revit looks up and down from the measurement height to access the vertical limit of the room-bounding element. You can set the lower and upper limits of the room bounding elements using the **Properties** palette. The **Base Offset** parameter in the value column in this palette defines the limit of the lower boundary. It is used to limit the extent of the room with no floors or prevent floors to leak from floor openings while computing the volume of the room. The volume is also displayed in the drawing view if the rooms are added in the drawing view with the **Room Tag With Volume** option selected from the **Type Selector** drop-down list.

CUTTING OPENINGS IN A WALL, FLOOR, AND CEILING

In Autodesk Revit, you can create an opening in the wall, floor, structural floor, ceiling, and structural elements such as beams and braces. To do so, invoke any of the tools from the **Opening** panel of the **Architecture** tab, as shown in Figure 3-16. From this panel, you can choose any of these five options, **By Face**, **Wall**, **Vertical**, **Shaft**, or **Dormer** to create an opening. In Revit, to create openings, you need to change the **Discipline** parameter of view to **Architecture**.

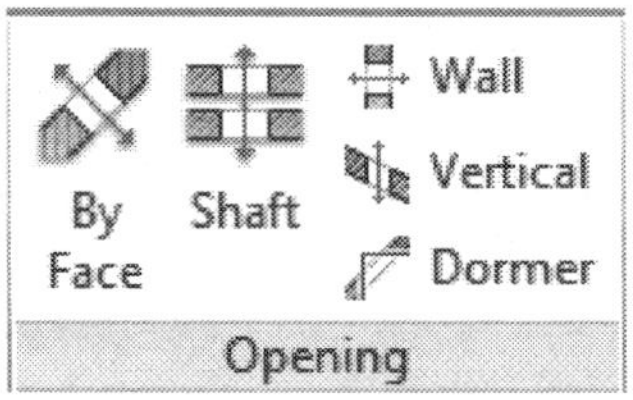

Figure 3-16 *The tools in the **Opening** panel*

Creating Openings Using the By Face Tool

You can use the **By Face** tool to create openings on the faces of floors, or ceilings in the building model. This tool is useful in a project when you need to cut an opening in a floor for stairway and in the ceiling for a chimney. To do so, invoke this tool from the **Opening** panel; you will be prompted to select a planar face of required floor, ceiling, beam, or column. Select the face on which you want to create the opening; the **Modify | Create Opening Boundary** tab will be displayed. In this tab, you can use various tools to sketch the opening in desired view. Now, open the required view and sketch the opening using the reference planes, as shown in Figure 3-17. After sketching the opening boundary on the selected face, choose the **Finish Edit Mode** button from the **Mode** panel to finish the sketch of the opening and then exit the **Modify | Create Opening Boundary** tab. Figure 3-18 shows the opening created perpendicular to the selected face.

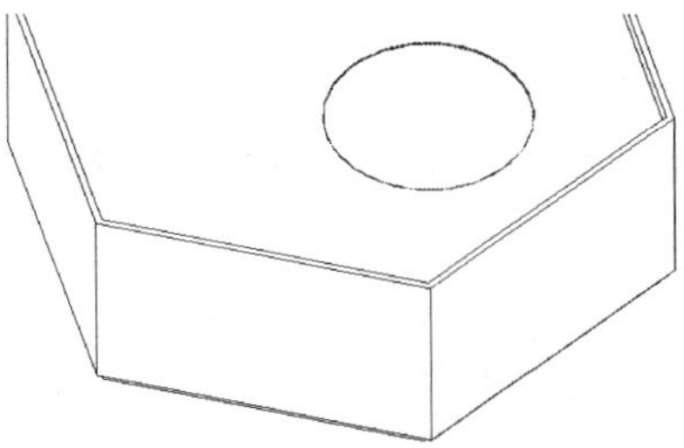

Figure 3-17 *Sketching the opening in the ceiling plan using the reference planes*

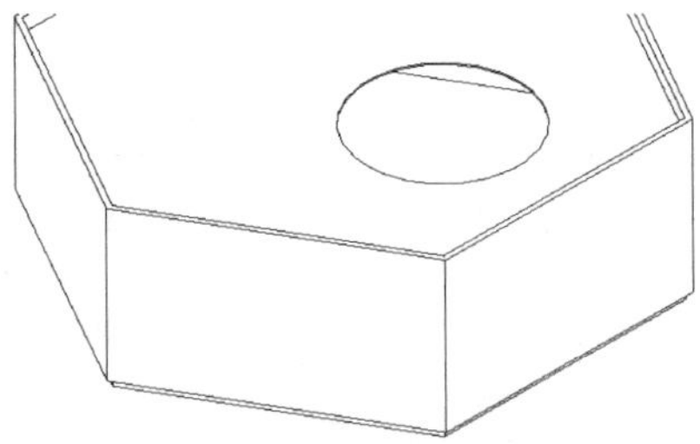

Figure 3-18 *Opening created perpendicular to the selected face*

Creating Openings Using the Vertical Tool

To cut a vertical opening, choose the **Vertical** tool from the **Opening** panel and select the required ceiling, or floor; the **Modify | Create Opening Boundary** tab will be displayed. You can use various tools from this tab to sketch the boundary of the opening. Sketch the opening in the appropriate view using the sketching tools. After sketching the opening boundary, choose the **Finish Edit Mode** button from the **Mode** panel; a vertical opening will be created in the selected element. You can use the sketching tools to draw a sketch of appropriate size, as shown in Figure 3-19. The opening can also be viewed in the 3D view, as shown in Figure 3-20.

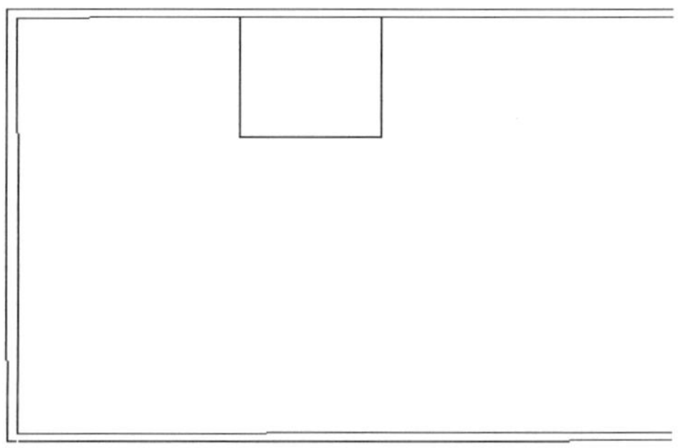

Figure 3-19 *Sketching the opening in the ceiling plan view*

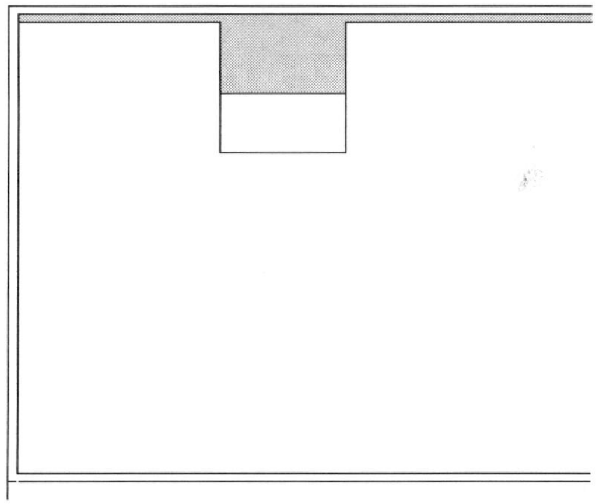

Figure 3-20 *3D view of the opening in ceiling*

Creating Openings Using the Wall Tool

To create a rectangular opening in a wall, choose the **Wall** tool from the **Opening** panel in the **MEP** tab; you will be prompted to select a wall. Select the wall in which you want to cut an opening and then sketch a rectangular opening of the required size by clicking and dragging the cursor.

Creating Openings Using the Dormer Tool

To create a dormer opening in the dormer roof, choose the **Dormer** tool from the **Opening** panel; you will be prompted to select the roof in which you want to create a dormer opening. Select the roof; the screen will enter the sketch mode. Choose the **Pick Roof/Wall Edges** tool from the **Pick** panel of the **Modify | Edit Sketch** tab and pick the boundary of the dormer to create the dormer opening. Next, choose the **Finish Edit Mode** button from the **Mode** panel; a dormer opening will be created.

Note

The boundary that you will pick for the dormer opening in a roof should be an edge of the selected roof, wall, or both, and should form a closed loop.

Creating Openings Using the Shaft Tool

To cut an opening up to the entire height of a building, choose the **Shaft** tool from the **Opening** panel; the **Modify | Create Shaft Opening Sketch** tab will be displayed. Choose a suitable sketching tool from this tab to create an opening of the required shape. Next, choose the **Finish Edit Mode** button from the **Mode** panel; the opening will be created passing through the entire height of the building. Make sure that before sketching the opening, you select the required work plane and the view to sketch the opening.

You can also specify the levels that will be cut by the opening. It will help you to restrict the opening to a particular level. To specify the levels, select the opening; the **Modify | Shaft Openings** tab will be displayed. Select a level for the **Base Constraint** parameter in the **Properties** palette to start the opening. Next, select a level for the **Top Constraint** parameter to end the opening. The opening will be cut through the selected levels.

TUTORIAL

Tutorial 1 Office Building I

In this tutorial, you will create the exterior and interior walls, add grids, and modify levels in office building based on the sketch plan shown in Figure 3-21. The dimensions have been given only for reference and are not to be used in this tutorial. The project file and the parameters to be used for creating the walls and for adding grids and modifying levels are given next.

(Expected time: 30 min)

1. Project file-
   For Imperial *Systems Default*
   For Metric *Systems-Default_Metric*
2. Exterior Wall type-
   For Imperial **Generic- 9"**.
   For Metric **Generic- 230mm**.
3. Interior Wall type-
   For Imperial **Generic - 5"**.
   For Metric **Generic - 90mm**.
4. Location line parameter- **Wall Centerline**; Top Constraint- **Up to Level 2**.
5. Rename Level 1 as the Ground Floor
6. Grids to be created in the plan view.
7. File name to be assigned:
   For Imperial *c03_Office_BuildingI_tut1.rvt*
   For Metric *M_c03_Office_BuildingI_tut1.rvt*

The following steps are required to complete this tutorial:

a. Open the required template file.
   For Imperial *Systems Default*
   For Metric *Systems-Default_Metric*
b. Invoke the **Wall: Architectural** tool from the ribbon.
c. Select the required exterior wall type from the **Properties** palette.
   For Imperial **Generic- 9"**
   For Metric **Generic- 230mm**

d. Select the required interior wall type from the **Properties** palette.
   For Imperial **Generic- 5"**
   For Metric **Generic- 90mm**
e. Modify Top Constraint- **Up to Level: Level 2** and Location Line - **Wall Centerline** as wall properties using the **Properties** palette, refer to Figure 3-22.
f. Invoke the **Rectangle** tool and then sketch the exterior walls based on the given parameters, refer to Figures 3-23 through 3-20.
g. Sketch the interior walls based on the given parameters, refer to Figures 3-63 through 3-69.
h. Modify levels by renaming them, refer to Figure 3-21.
i. Add grids using the **Grid** tool, refer to Figures 3-22 through 3-24.
j. Save and close the project.

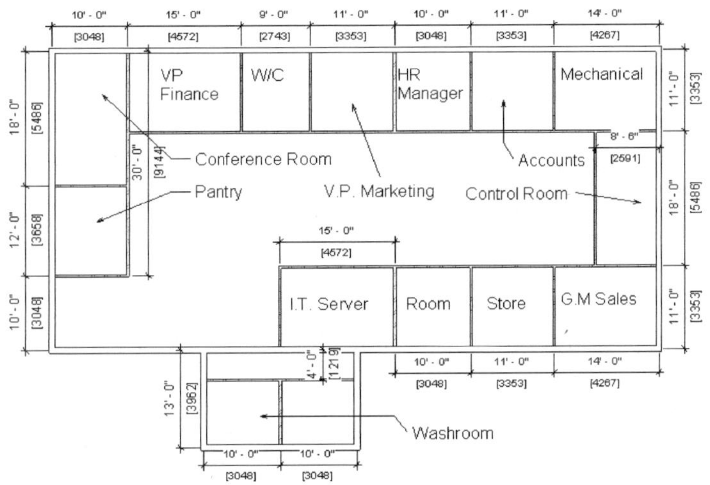

Figure 3-21 Sketch plan for creating exterior walls for the Office Building

Opening a New Project and Using the Template File

1. Choose **New > Project** from the **File** menu; the **New Project** dialog box is displayed.

2. In the **New Project** dialog box, choose the **Browse** button and then select the following template:

   For Imperial **Systems-Default**
   For Metric **Systems-Default_Metric**

   Next, choose the **Open** button and then the **OK** button; the desired template file is loaded. Notice that the **Project Browser** now shows several views that are preloaded in the template file.

3. In the **Project Browser**, select **1-Mech** under the **Mechanical** head, if it is not selected by default and right-click to display a flyout. From this flyout, choose **Duplicate View > Duplicate**; **1 - Mech Copy 1** is displayed under the **Mechanical** head.

4. Double click on **1 - Mech Copy 1** to display the corresponding view. Now, in the **Properties** palette, click on the value field corresponding to the **Discipline** parameter. Select the **Architectural** discipline from the drop-down list and choose the **Apply** button to apply the changes.

5. Now, the **Architectural** head is added under the **Views (Discipline)** in the **Project Browser** and under that head, **1 - Mech Copy 1** is displayed.

6. Select the **1 - Mech Copy 1** under the **Architectural Plan** head and right-click to display a flyout. From the flyout, choose the **Rename** option; the edit box is displayed. Enter **Level 1** in the edit box and choose the **OK** button; the dialog box is closed and the view is renamed to **Level 1**.

7. Repeat the procedure followed in steps 3 through 6 to duplicate another level as **Level 2** from **Mechanical** head into **Architectural** head, as shown in Figure 3-22.

8. In the **Project Browser**, select **East-Mech** under the **Mechanical** head and **Elevations** sub-head of the **Mechanical** head and right-click to display a flyout. From this flyout, choose **Duplicate View > Duplicate**; **East - Mech Copy 1** is displayed under the **Elevations** sub-head in the **Mechanical** head.

9. Double-click on **East - Mech Copy 1** to display the corresponding view. Now, in the **Properties** palette, click on the value field corresponding to the **Discipline** parameter. Select the **Architectural** discipline from the drop-down list and choose the **Apply** button to apply the change.

10. Now, in the **Project Browser** window, **East - Mech Copy 1** is displayed under the **Views (Discipline) > Architectural > Elevations (Building Elevation)**.

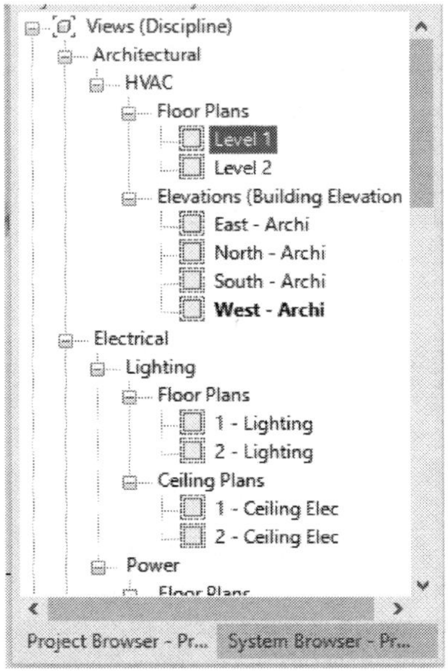

*Figure 3-22 The **Properties** palette displaying other views under the **Elevations** sub-head*

11. Select **East - Mech Copy 1** under the **Architectural Plan** head and right-click to display a flyout. From the flyout, choose the **Rename** option; the edit box is displayed. Enter **East - Archi** in the edit box and choose the **OK** button; the dialog box is closed and the view is renamed to **East - Archi**.

12. Repeat the procedure followed in steps 8 through 10 to add other views under the **Elevations** sub-head, refer to Figure 3-22.

Invoking the Wall: Architectural Tool and Selecting the Wall Type

In this section, you will sketch an architectural wall using the **Wall** tool.

1. Double-click on the **Level 1** view under the **Views (Discipline) > Architectural > HVAC > Floor Plans**; the corresponding view is displayed.

2. Invoke the **Wall: Architectural** tool from the **Architecture > Build > Wall** drop-down; the **Modify | Place Wall** tab is displayed.

3. In the **Type Selector** drop-down list of the **Properties** palette, select the **Generic - 8"** wall type for Imperial or **Generic - 200mm** wall type for Metric unit system.

4. In the **Properties** palette, choose the **Edit Type** button; the **Type Properties** dialog box is displayed.

5. Choose the **Duplicate** button from the upper right corner of this dialog box; the **Name** dialog box is displayed. Enter **Generic- 9"** for Imperial (**Generic- 230mm** for Metric system**)** in the **Name** edit box and then choose the **OK** button; the dialog box is closed and the **Generic - 9"** for Imperial system (**Generic - 230mm** for Metric system) is selected in the **Type** drop-down list.

6. Now, choose the **Edit** button from the value field corresponding to the **Structure** parameter; the **Edit Assembly** dialog box is displayed.

7. In the **Edit Assembly** dialog box, click on the value field of the **Thickness** column corresponding to the **Structure[1]** function parameter. Enter **9"** for Imperial or enter **230mm** for Metric and then choose the **OK** button; the **Edit Assembly** dialog box is closed.

8. Now, choose **OK** button; the **Type Properties** dialog box is closed and the **Generic - 9"** wall type is selected in Imperial system or **Generic - 230mm** is selected in the **Type Selector** drop-down list.

Modifying the Properties of the Exterior Wall

After sketching the wall type, you need to modify the instance properties of the wall type using the **Properties** palette.

1. In the **Properties** palette, ensure that the **Location Line** parameter has **Wall Centerline** as the default value. Click in the value field of the **Top Constraint** instance parameter; a drop-down list is displayed. Select **Up to Level: Level 2** from the drop-down list displayed, as shown in Figure 3-23 and choose the **Apply** button.

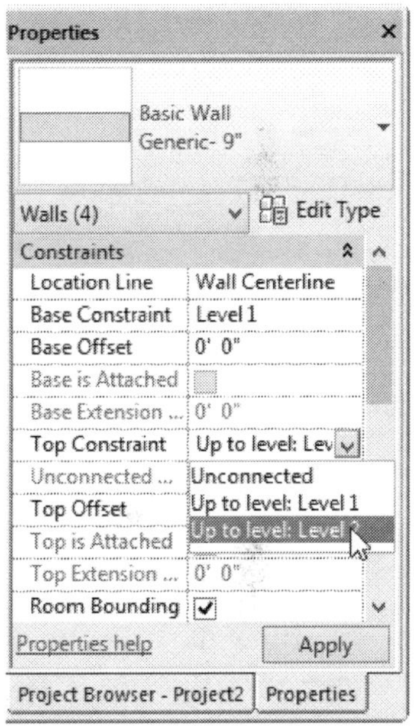

Figure 3-23 *Setting the* ***Top Constraint*** *parameter using the* ***Properties*** *palette*

Sketching the Exterior Wall Segment

In this section, you will sketch the exterior wall segment.

1. Invoke the **Rectangle** tool from the **Draw** panel of the **Modify | Place Wall** tab.

2. Click between the four elevation symbols in the drawing area to specify the first point. Next, move the cursor toward bottom right to draw a rectangle. On doing so, a rectangle starts creating from a specified point. Now, click to specify the endpoint, as shown in Figure 3-24.

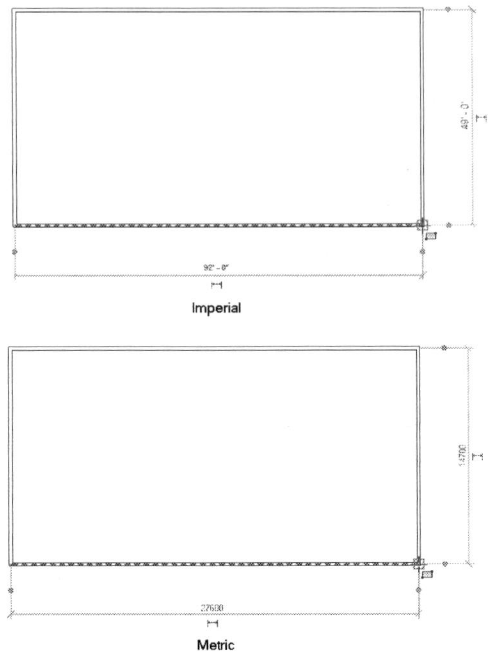

Figure 3-24 The temporary dimensions being displayed on the sketched wall

3. Click on the temporary dimension of the horizontal wall displayed; an edit box appears showing the current dimension of the wall segment.

4. Enter **80'** for Imperial or **24000 mm** for Metric in the edit box and then press ENTER; the length of the horizontal wall is modified to the entered value.

5. Similarly, click on the temporary dimension of vertical wall; an edit box appears. Enter **40'** for Imperial or **12000 mm** for Metric in the edit box and then press ENTER; the length of the vertical wall is modified to the entered value. Press ESC twice to exit the **Modify | Walls** tab.

   The external wall profile is drawn with the specified dimensions, as shown in Figure 3-25.

Figure 3-25 The completed exterior wall segment of Office Building

Sketching Other Exterior Wall Segments

In this section, you need to create other exterior wall segments.

1. Choose the **Wall: Architectural** tool from the **Architecture > Build > Wall** drop-down; the **Modify | Place Wall** tab is displayed.
2. Choose the **Line** tool from the **Draw** panel of the **Modify | Place Wall** tab. Ensure that the **Chain** check box is selected in the **Options Bar**. Now, place the cursor to the lower left corner of the wall and move toward right. When temporary dimension appears, enter **20'** for Imperial or enter **6000 mm** for Metric and then press ENTER.
3. Draw **13'0"** line in Imperial system or **3900 mm** line in Metric system from that point in downward direction, as shown in Figure 3-26.

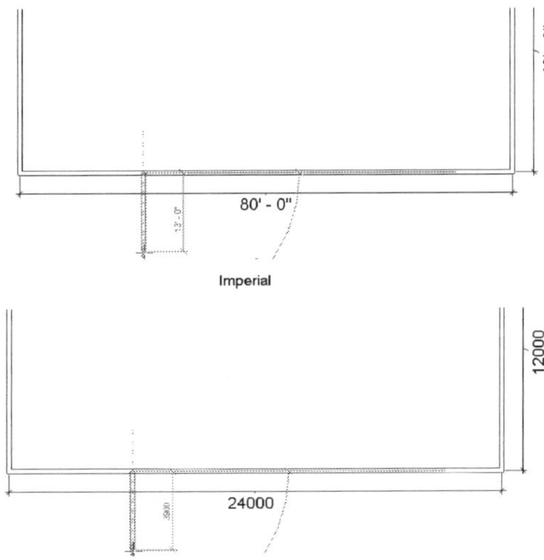

***Figure 3-26** The other exterior wall segment created outside the Office Building*

4. After drawing the line, the wall starts creating dynamically with one end attached to the specified point and the other end attached to the cursor. Move the cursor horizontally toward the right so that you see a dashed horizontal line inside the wall segment. Now, enter **20'0"** for Imperial or **6000 mm** for Metric as the value of the length; an edit box is displayed with the dimension you have entered, as shown in Figure 3-27.

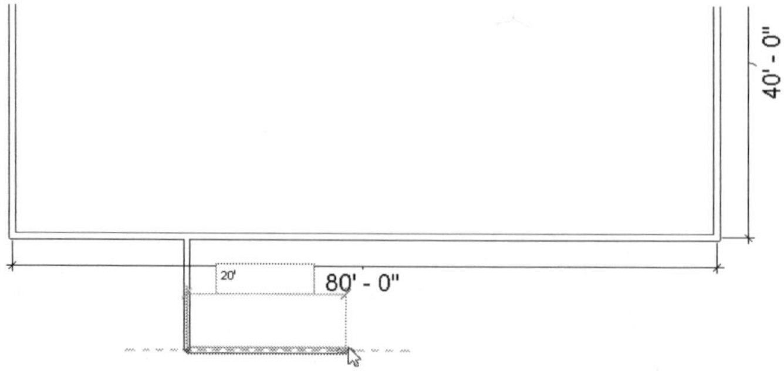

Imperial

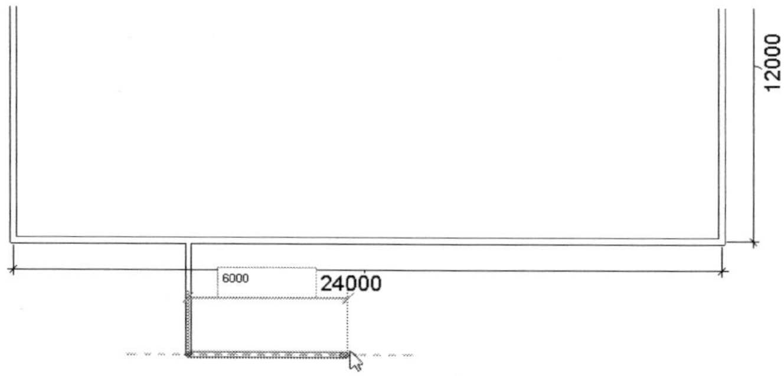

Metric

Figure 3-27 *Creating the second exterior wall segment*

5. Now you need to draw the third wall segment. To do so, move the cursor upward and enter **13'** for Imperial or enter **3900 mm** for Metric as the value of the length and press ENTER. The other exterior wall segments are created, as shown in Figure 3-28.

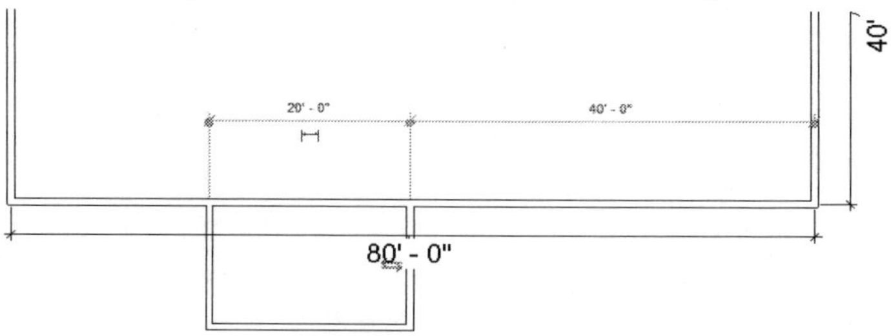

Imperial

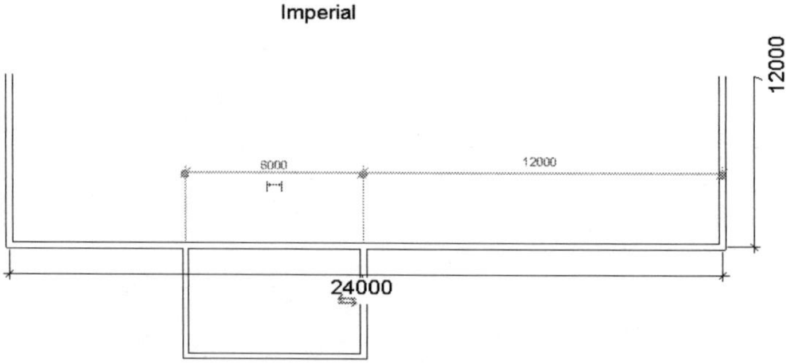

Metric

Figure 3-28 *The completed exterior profile of Office Building*

6. Now, press ESC or choose the **Modify** button to clear the selection.

Selecting the Interior Wall Type

In this section, you will select the interior wall type.

1. Invoke the **Wall: Architectural** tool; the wall instance parameters are displayed in the **Properties** palette. In this palette, select the required wall type from the **Type Selector** drop-down list, as shown in Figure 3-29.

   For Imperial **BasicWall Generic - 5"**
   For Metric **BasicWall Generic - 90mm**

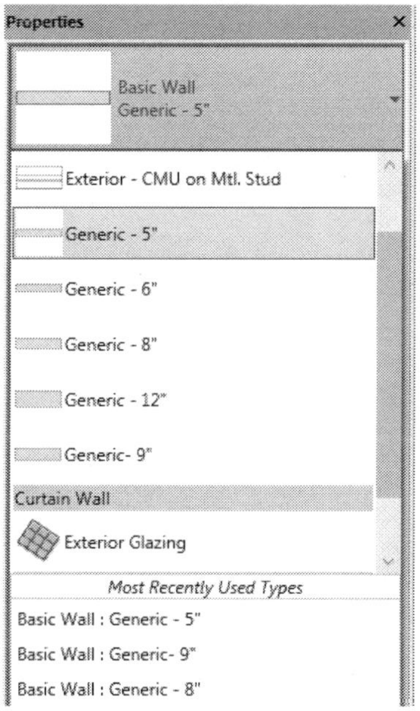

***Figure 3-29** Selecting the interior wall type from the **Type Selector** drop-down list*

2. In the **Options Bar**, select the **Wall Centerline** option from the **Location Line** drop-down list, if it is not selected by default.

Sketching Other Interior Walls

Next, you will sketch other horizontal or vertical interior walls by specifying their start point and end point using different object snap options.

1. Move the cursor near the top left end of the wall and then start moving the cursor horizontally toward the right. Enter **10'0"** for Imperial or **3000 mm** for Metric when the temporary dimensions and the intersection object snap appear, as shown in Figure 3-30. Now, press ENTER; the starting point of the first interior wall is specified.

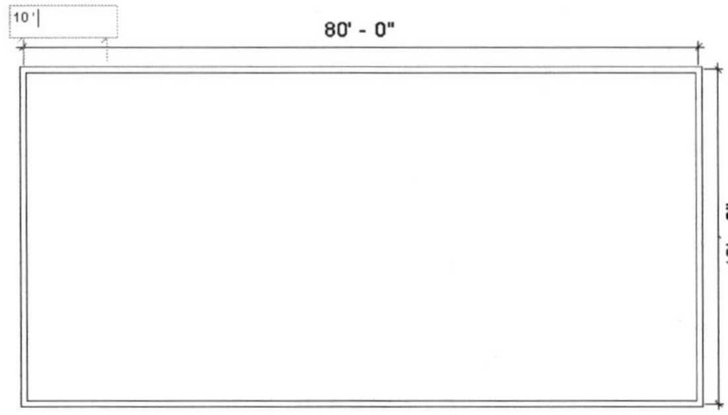

Imperial

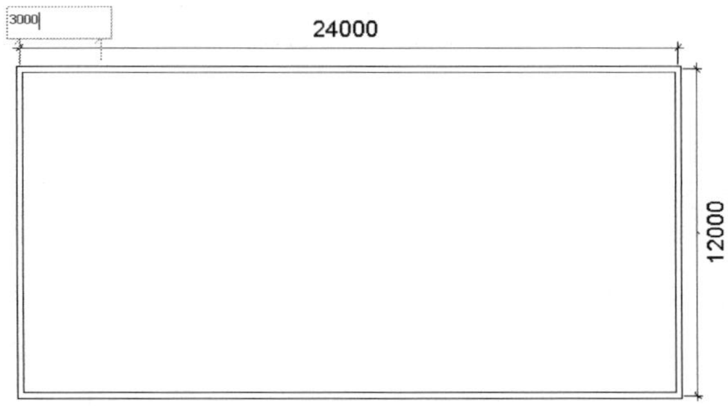

Metric

Figure 3-30 *Specifying the distance for starting the first interior wall segment*

2. Next, move the cursor vertically downward and enter **30'** for Imperial (**9000 mm** for Metric) and close the wall segment, as shown in Figure 3-31; the interior wall segment is sketched.

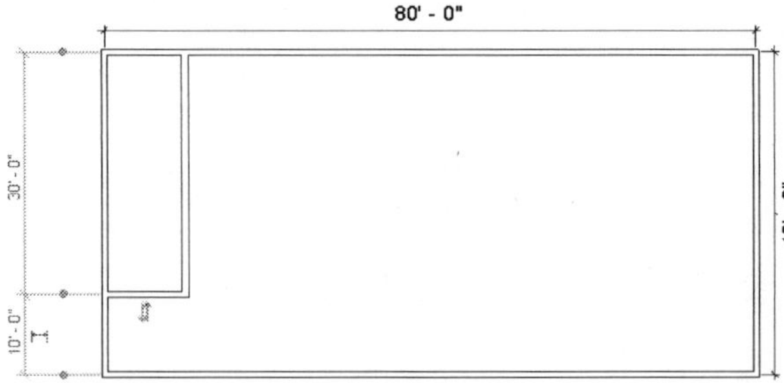

Imperial

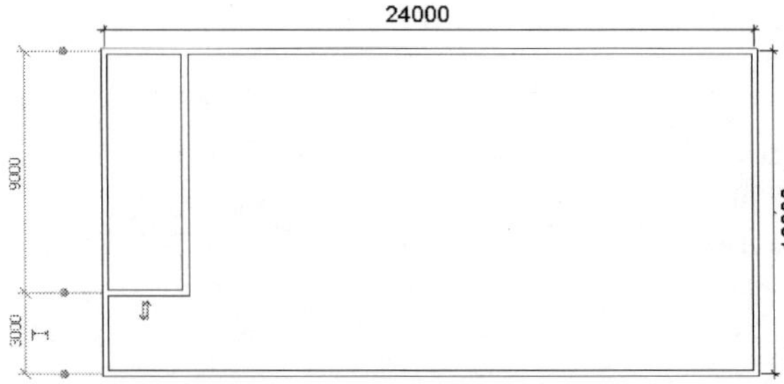

Metric

Figure 3-31 *Specifying the first interior wall segment*

3. To sketch the second interior wall, move the cursor to the upper endpoint of the interior wall you just created and then move the cursor vertically downward. Enter 11'0" for Imperial or 3300 mm for Metric when the temporary dimension and the intersection object snap appears as shown in Figure 3-32.

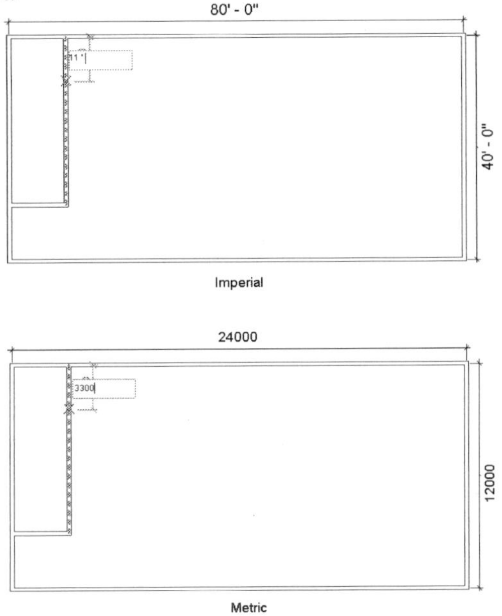

Figure 3-32 *Specifying the starting point of second interior wall segment*

4. Press SHIFT and move the cursor toward right near the right exterior wall segment. Now, click to specify the location of the endpoint of the wall segment; the second interior wall segment is sketched, as shown in Figure 3-33.

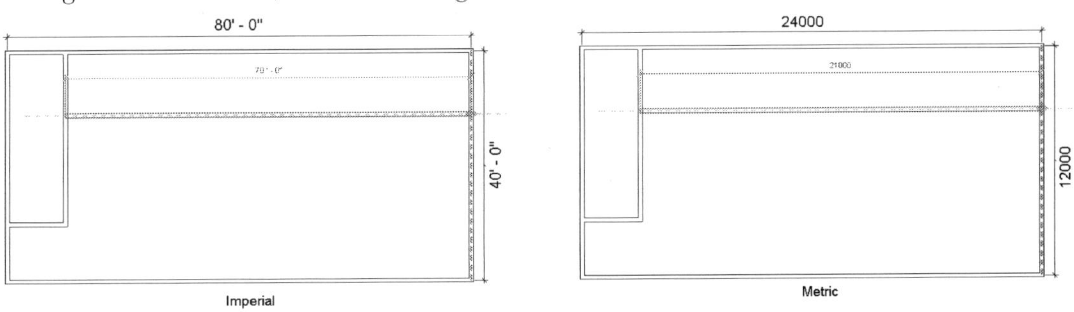

Figure 3-33 *Sketching the second interior wall*

5. To sketch the third interior wall, place the cursor toward the lower right corner of the wall segment and move the cursor upward. Enter **11'0"** for Imperial or **3300** mm for Metric and then press ENTER, as shown in Figure 3-34; the starting point of the interior wall is specified.

6. Next, move the cursor in a horizontal direction toward the left. Enter **50'0"** for Imperial and **15000** for Metric and then press ENTER, as shown in Figure 3-35. Ensure that the **Chain** option is selected. Now, move the cursor vertically downward near the lower exterior wall segment; the interior wall segment is sketched, as shown in Figure 3-36.

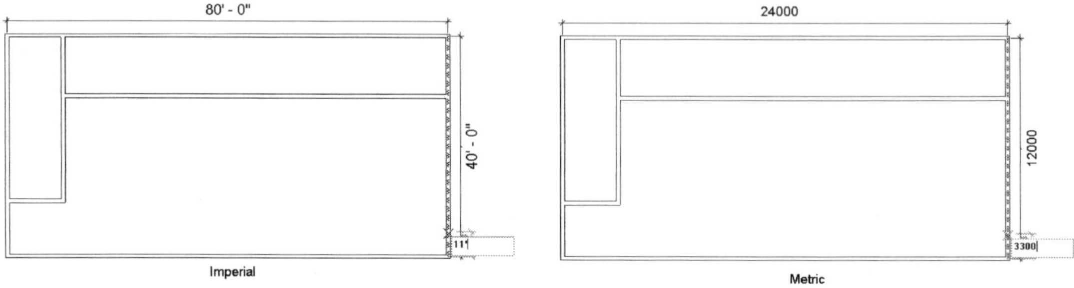

***Figure 3-34** Specifying the distance for starting the first interior wall segment*

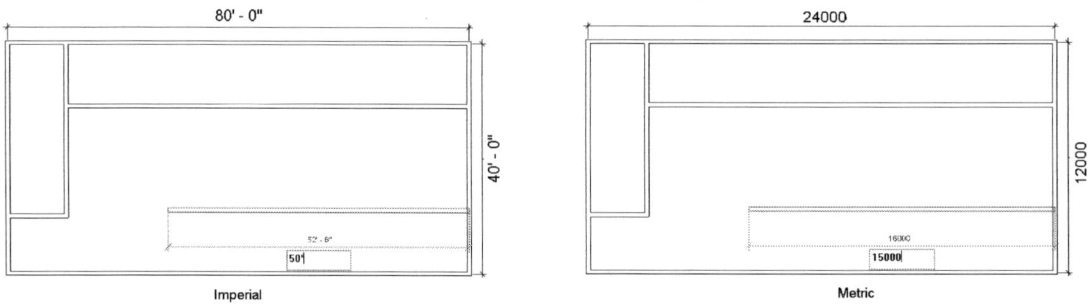

***Figure 3-35** Specifying the distance for starting the interior wall segment*

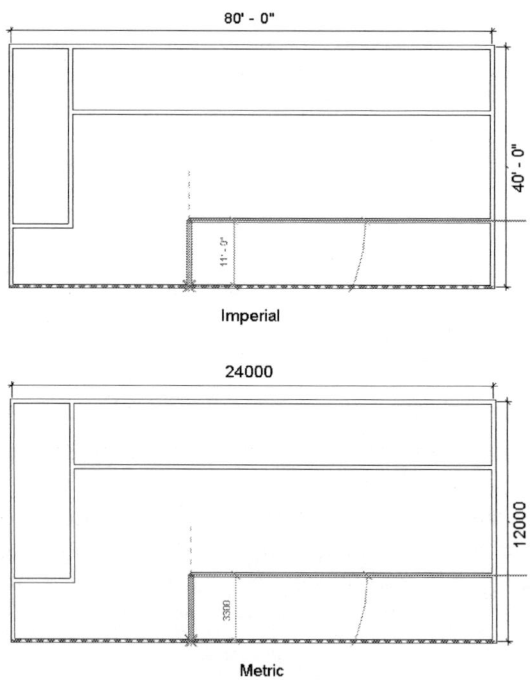

***Figure 3-36** Sketched wall segment*

7. Similarly, you can draw other interior walls by using the **Line** tool, refer to Figure 3-37.

8. Choose the **Modify** button to exit the tool.

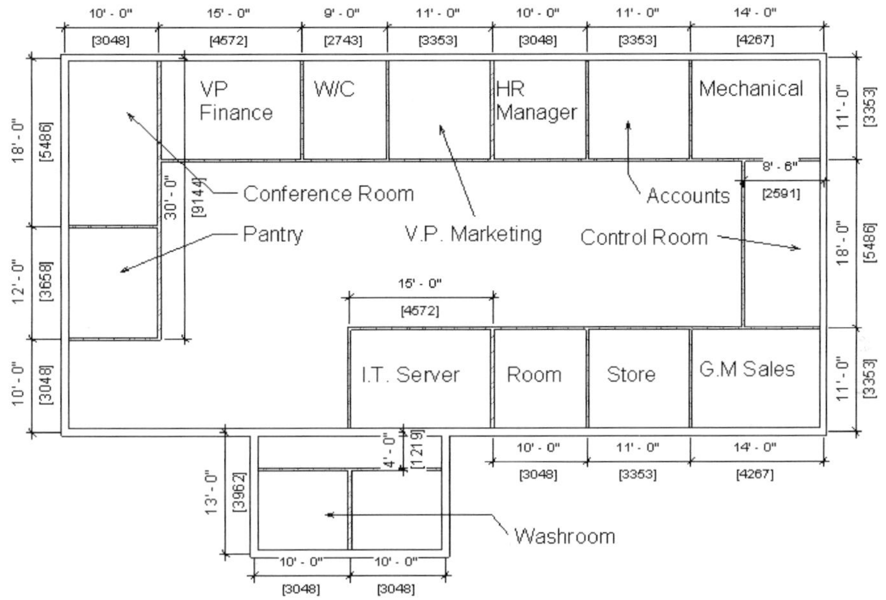

Figure 3-37 Layout of internal walls for Office Building

Modifying Levels

In this section, you will modify levels in the elevation view.

1. In the **Project Browser**, double-click on the **North - Archi** under the **Elevations (Building Elevation)** head; the north elevation is displayed within the existing levels in the drawing window.

2. Choose the **Zoom in Region** tool from the Navigation Bar to enlarge the right portion of the elevation showing the levels, refer to Figure 3-73.

3. To rename the levels, move the cursor over **Level 1** in the **Project Browser** and right-click; a shortcut menu is displayed.

4. Choose the **Rename** option from the shortcut menu; the edit box is displayed.

5. In the displayed edit box, enter **Ground Floor** and choose the **OK** button; you are prompted to verify whether you want to rename the corresponding levels and views.

6. Choose the **Yes** button to rename the level and views. The level is immediately renamed in the elevation view.

7. Similarly, rename the **Level 2** as **Roof** , as shown in Figure 3-38.

Figure 3-38 Renamed levels and views for the Office Building

Creating Grid Lines

You can use the plan view to add grids to the project using the **Grid** tool. Grids are automatically numbered as they are created. Now, you will create Grids in the sequence as shown in the sketch plan.

1. Double-click on **Ground Floor** from the **Floor Plans** head in the **Project Browser** to display the ground floor plan in the drawing window.

2. Next, choose the **Grid** tool from the **Datum** panel of the **Architecture** tab; the **Modify | Place Grid** tab is displayed.

3. Now, ensure that the **Line** tool is chosen in the **Draw** panel.

4. Move the cursor near the top left corner of the exterior wall profile till a vertical extension line is displayed. Click to specify the start point of the grid line when the temporary dimension of **3'0"** for Imperial or **900 mm** for metric is displayed from the centerline of the exterior wall, as shown in Figure 3-38.

5. Move the cursor vertically downward and click outside the south wall to specify the endpoint of the grid line, as shown in Figure 3-39.

Note

*If grid lines are not visible in the drawing, choose the **Visibility/Graphics** tool from the **Graphics** panel of the **View** tab to display the **Visibility/Graphic Overrides for Floor Plan** dialog box. In the **Annotation Categories** tab of this dialog box, select the **Grids** check box; the grid lines will become visible.*

The same procedure can be followed to draw grid lines for the interior walls. As the thickness of the interior wall is 5" or 90 mm, you can specify 2 1/2" or 64 mm as the offset distance to draw grid lines for the interior walls.

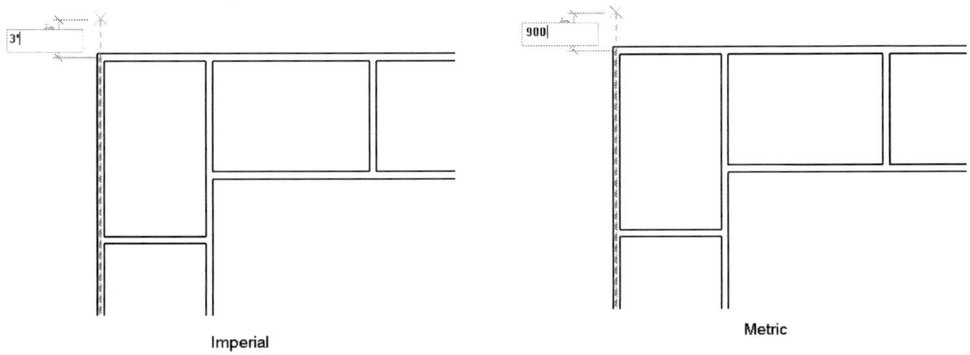

Figure 3-39 *Specifying the start point of the grid line*

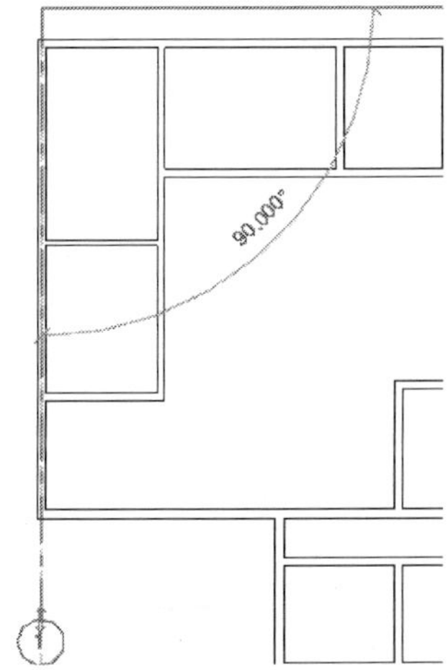

Figure 3-40 *Specifying the grid line drawn at first point*

6. Repeat steps 3, 4, and 5 to create other vertical and horizontal grid lines in the sequence of their numbers using the **Pick Line** tool. After adding grid lines, press ESC twice to exit. Figure 3-41 shows the floor plan after adding grid lines.

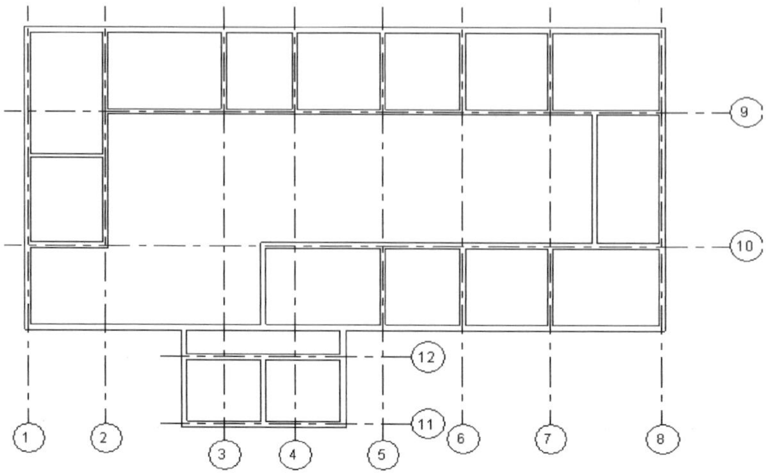

Figure 3-41 *The horizontal and vertical grid lines created for wall centerline*

Note
You can rename a grid by selecting it from the drawing and then entering a value corresponding to the value column for the ***Name*** *parameter in the* ***Properties*** *palette. The entered value will be the new name of the grid.*

This completes the tutorial for creating walls, adding grids, and modifying levels of the building envelope project.

Saving the Project

In this section, you need to save the project and the settings using the **Save As** tool.

1. Choose **Save As > Project** from the **File** menu; the **Save As** dialog box is displayed.
2. In this dialog box, browse to *C:\rmp_2025\c03_rmp_2025_tut* and then enter **c03_Office_BuildingI_tut1** for Imperial or **M_c03_Office_BuildingI_tut1** for Metric in the **File Name** edit box.
3. Now, choose the **Save** button; the **Save As** dialog box closes and the project file is saved.

Closing the Project

1. Choose the **Close** option from **File** menu.

EXERCISE

Exercise 1 Residential Building

Create the exterior and interior walls of a residential building and then add doors and windows to them. Create floor and ceiling and then add grids to the walls of the Residential building, refer to Figure 3-42. Do not add dimensions or texts as they are given only for reference. Figure 3-43 shows the three dimensional view of the residential building with the added floors and ceilings. The project parameters for this exercise are given next. **(Expected time: 1 Hour)**

1. Project File -
   For Imperial **Systems-Default**.
   For Metric **Systems-Default_Metric**.

2. Discipline - **Architectural**.

3. Rename Level 1 - First Floor, Level 2 - Second Floor.

4. Exterior wall type -
   For Imperial **Basic Wall** - **Exterior Brick on Mtl. Stud**.
   For Metric **Basic Wall** - **Exterior Brick on Mtl. Stud**.

5. Interior wall type -
   For Imperial **Basic Wall: Generic- 5"**.
   For Metric **Basic Wall: Generic- 90 mm**.

6. Height of wall - **Top Constraint** - **Upto Level 2**.

7. Door type to be used:
   For Imperial Main door - **Double - Glass 2 - 36" x 84"**
   Bedroom doors - **Single - Flush 30" x 84"**
   Washroom door - **Single Flush Vision - 36" x 84"**
   For Metric Main door - **Double - Glass 2 - 1830 x 1981mm**
   Bedroom doors - **Single - Flush 0762 x 2134mm**
   Washroom door - **Single Flush Vision - 0915 x 2134 mm**

8. Window type to be used:
   For Imperial **Fixed with Trim - 36" x 24"**
   For Metric **Fixed with Trim - 0915 x 0610 mm**

9. Floor type -
   For Imperial **Floor: Generic- 12"**.
   For Metric **Floor: Generic- 300mm**

10. Ceiling type - **Generic**.

11. File name to be assigned:
    For Imperial *c03_Residential-Building_exer1.rvt*

For Metric *M_c03_Residential-Building_exer1.rvt*

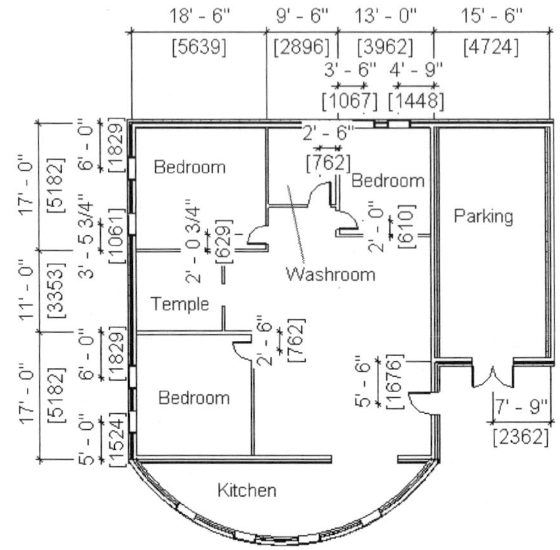

Figure 3-42 *Sketch plan of walls with doors and windows, and floor and ceiling added to the building*

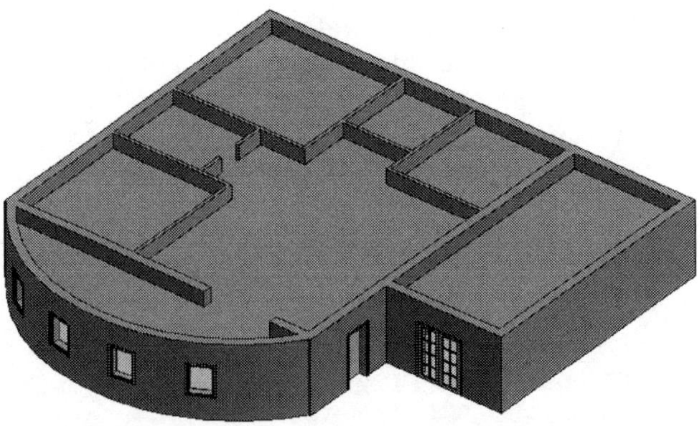

Figure 3-43 *Three dimensional view of residential building with added floors and ceilings*

Chapter 4

Creating Spaces and Zones and Performing System Analysis

Learning Objectives

After completing this chapter, you will be able to:

- *Create Spaces*
- *Modify Spaces*
- *Create Zones*
- *Modify Zones*
- *Perform System Analysis*

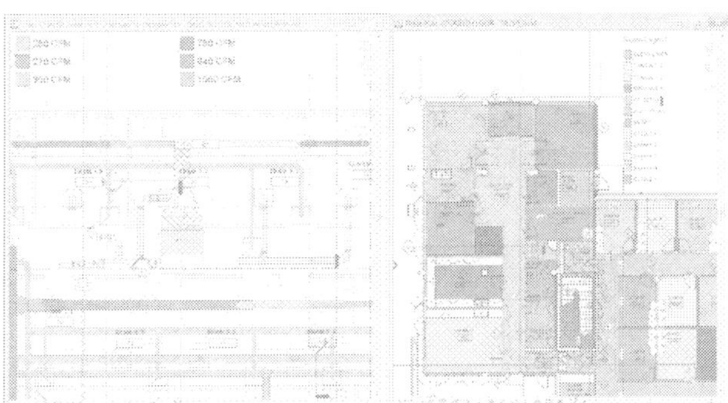

INTRODUCTION

In Autodesk Revit, you can model spaces and create zones to track building design changes and assign HVAC loads for analysis. These loads can be exported via gbXML files to simulation tools like Trane TRACE 700 or HAP. Accurate space modeling is crucial for **HVAC** design but often time-consuming, especially when changes occur. Revit simplifies this by allowing load analysis using **EnergyPlus**, providing annual and peak heating/cooling loads to guide energy-efficient design decisions. Since Revit 2020, the systems analysis feature allows grouping spaces into thermal zones and connecting them to various HVAC systems. In Revit 2025, this has been enhanced to simulate detailed HVAC systems, including radiant and central plant systems, using default high-performance settings that can be customized. This chapter covers how to prepare space models, create zones, and perform system analysis using Revit 2025.

BIM CONCEPTS *In a Revit MEP project aligned with the ISO 19650 framework, the effective use of spaces and zones is essential for accurate analysis, coordination, and documentation. Spaces in Revit represent the volumes within architectural boundaries, enabling detailed energy analysis, heating/cooling load calculations, and airflow simulations. Zones, on the other hand, group multiple spaces for system-based analysis, such as HVAC zoning. To streamline the BIM workflow, define spaces early, ensuring they match the architectural model precisely. Use shared parameters for naming and numbering to maintain consistency.*

SPACE MODELING FOR BUILDING ANALYSIS

Accurate and efficient space modeling in Revit is essential for successful load analysis. Each space must include key components such as building construction elements, internal loads, and external load factors, as all of these influence the overall building performance. These components have specific inputs that significantly impact the heating and cooling loads within each space, making detailed modeling crucial for reliable results.

Creating Spaces

You need to create spaces in a building model to perform a load analysis. In Revit, spaces are created from room bounding elements such as walls, floors, ceilings, and roofs, and from room separation lines. After creating the initial set-up for the MEP project and linking the desired architectural model in the project or creating the building envelope, the architectural elements that make up a room define your space accurately.

Figure 4-1 shows a linked architectural model. Select the linked model from the drawing area; the properties of the linked model will be displayed in the **Properties** palette. In the palette, choose the **Edit Type** button; the **Type Properties** dialog box will be displayed. In this dialog box, select the check box corresponding to the **Room Bounding** parameter; the spaces in the linked model will be bounded by room. Now, choose **OK** and **Apply** to close the **Type Properties** dialog box.

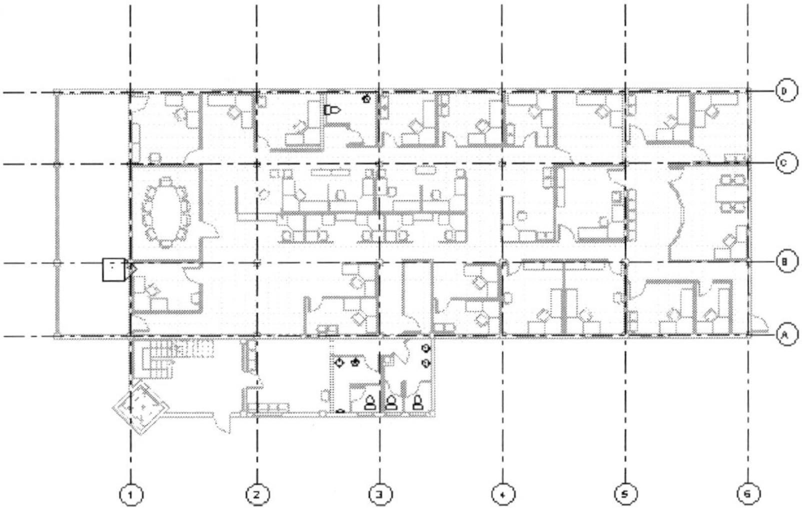

Figure 4-1 *The linked architectural model*

Modifying Spaces

After adding spaces to the model, you may need to modify the design specification or adjust the spaces depending on the result of the heating and cooling load analysis. In some cases, you may need to increase or decrease the vertical extents of the space. You may also need to split, combine, delete, or move a space. These modifications are discussed next.

COLOR SCHEMES

Color schemes, as shown in Figure 4-2, are used to visually categorize and represent different spaces in a project by applying specific color codes. These schemes can be tailored to various plan views—such as different floors—and can be based on categories like gross area, rentable area, or specific space properties (e.g., room cavity ratio, cooling load, space type, etc.). For instance, different colors can be used to distinguish areas like offices, storage, or accounts. A legend is typically added to identify these areas by color. To implement a color scheme, you must first create and modify it using the **Edit Color Scheme** dialog box, then apply it to the desired project view.

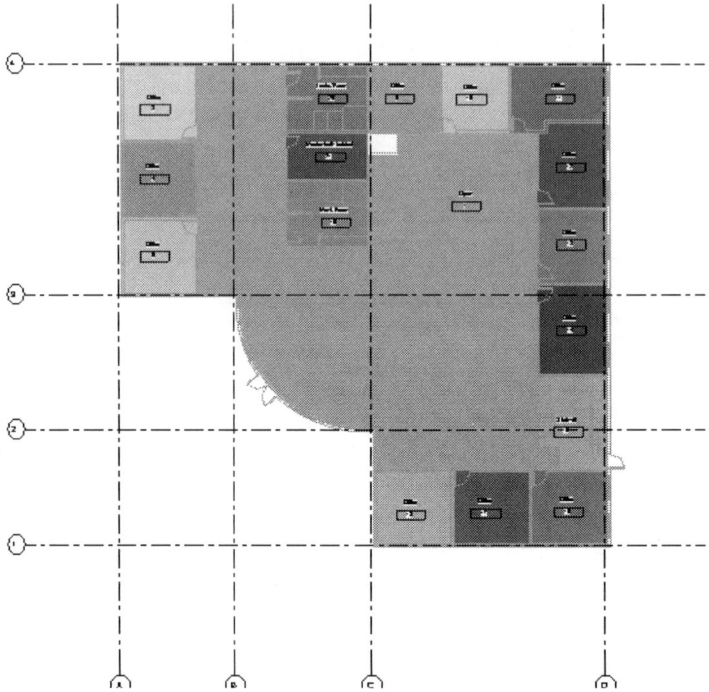

Figure 4-2 *The color scheme of spaces categorized by its assigned number*

Creating and Editing Color Schemes

Before creating a color scheme, the areas and spaces must be defined in the plan view. A color scheme can be based on any space parameter, such as number, name, or area—so these details should be added beforehand. To create a color scheme, open the **Edit Color Scheme** dialog box by selecting the **Color Schemes** tool from the **Spaces & Zones** panel in the **Analyze** tab, as shown in Figure 4-3. Alternatively, you can access it by clicking the `<none>` button in the Properties palette under the **Color Scheme** parameter. In the dialog box, select a category from the **Category** drop-down, click **Duplicate** to create a new scheme, and enter a name in the **New Color Scheme** dialog. Add a title in the **Scheme Definition** area, then choose a parameter from the **Color** drop-down to define how colors will be applied. Note that the available options in this list depend on the selected category.

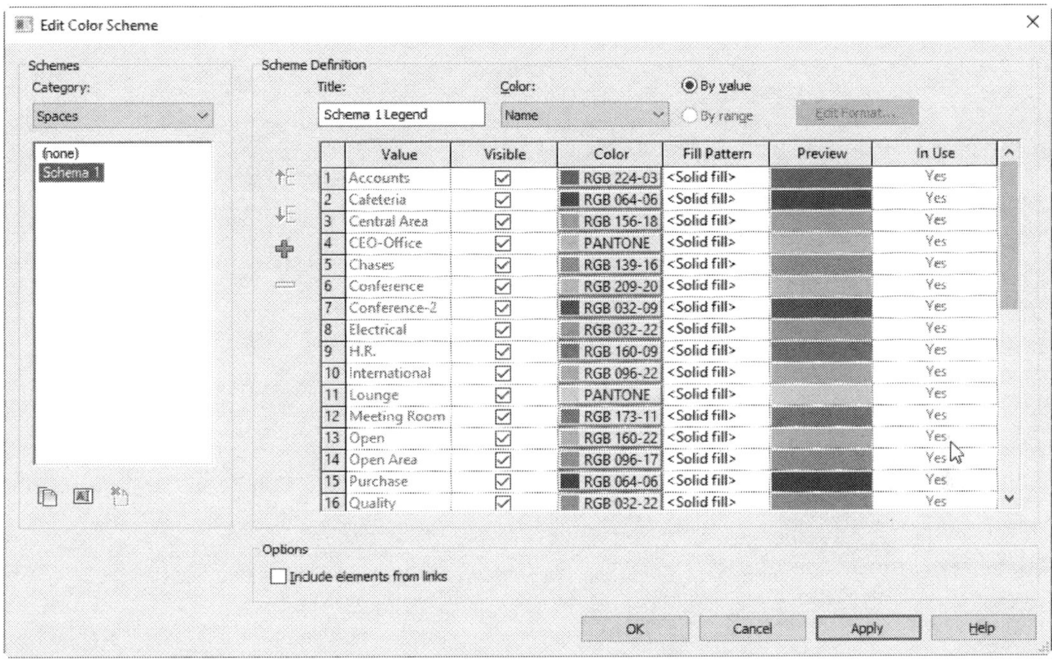

*Figure 4-3 The **Edit Color Scheme** dialog box*

Note

*Make sure you select the correct parameter from the **Color** drop-down list to create a color scheme. The values for the parameters should be predefined. For example, if you select the **Number of People** parameter to create a color scheme for the plan view, then the **People** parameter in the **Properties** palette for the spaces must be predefined.*

Applying a Color Scheme to the Spaces

To apply a color scheme in the required plan view, you can use the **Properties** palette of the current view. In the **Properties** palette of the current view, choose the **<none>** button in the **Value** column of the **Color Scheme** parameter; the **Edit Color Scheme** dialog box will be displayed. Select the required color scheme and choose the **OK** button; colors will be applied to the areas or rooms based on the color scheme selected.

Adding Color Scheme Legends

Color scheme legends help identify different rooms and areas using assigned colors and are part of the Annotation Tag category. These legends can be placed multiple times within a plan view and are influenced by the Annotation crop region—excluding them from a view is possible by adjusting this region. To add a color fill legend, select the **Color Fill Legend** tool from the **Color Fill** panel under the **Annotate** tab. In the **Properties** palette, choose a color scheme from the **Type Selector** drop-down, then click in the drawing area to place it. You can move or resize the legend using the drag symbol. **Figure 4-4** illustrates a floor plan with a color scheme legend showing room areas based on their occupied space.

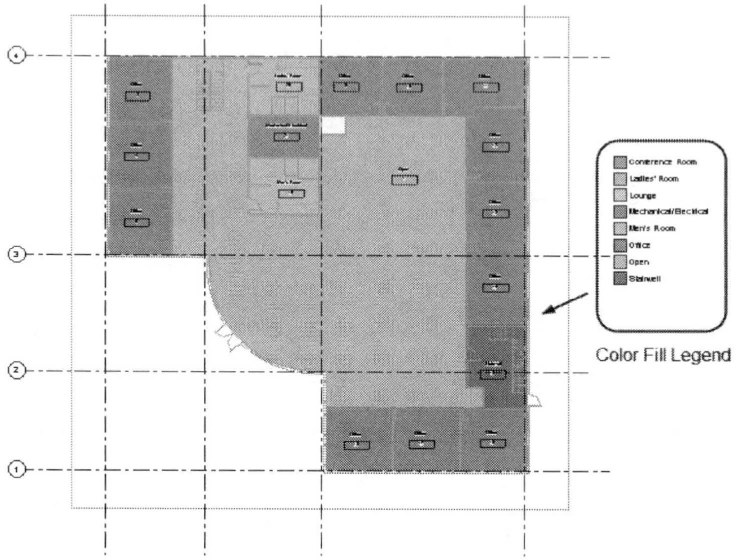

Figure 4-4 *The floor plan displaying the color scheme*

Modifying a Color Scheme

To modify a color scheme, select the existing color scheme legend from the drawing area and choose **Scheme > Edit Scheme** from the **Modify Color Fill Legends** tab; the **Edit Color Scheme** dialog box will be displayed. In this dialog box, modify the colors, fill patterns, title, parameter, and other values as required and choose the **Apply** button to view the changes. Choose the **OK** button to retain the changes.

Modifying a Color Scheme Legend

To modify a color scheme legend, select the existing color scheme legend and drag the blue circular drag control to move the color swatches into new columns. Also, drag the cursor down to move the color swatches into one column. Next, choose the **Edit Type** button in the **Properties** palette; the **Type Properties** dialog box will be displayed. In this dialog box, you can modify various parameters related to text, swatches, and other elements in the **Color Fill** legend. After modifying the parameters, choose the **OK** button to close the dialog box. You can also display the title of the color scheme legend by selecting the **Show Title** check box.

CREATING ZONES FROM SPACES

In an MEP project, spaces with similar environmental and design needs are grouped into zones to control conditions like temperature and humidity. Zoning allows better airflow management, such as increasing airflow in high-load areas or turning it off in unoccupied zones. Zones can include both occupied and unoccupied areas, even across different levels. Revit uses zone and space data during heating and cooling load analysis to estimate energy demands. Every project starts with a Default zone, where all spaces are initially placed. Unbounded zones, which contain no spaces, can be used for design, fire protection, or retaining zone information. These can be moved within views, unlike bounded zones that become fixed once spaces are assigned.

Adding and Modifying HVAC Zones

To add a new zone, first open the project view and enable the System Browser window by selecting the System Browser checkbox from **View** > **Windows** > **User Interface** (Figure 4-5). In the System Browser, choose **Zones** from the View drop-down to display a list with a Default node, which contains all project spaces by default. Next, select the **Zone** tool from the **Spaces & Zones** panel under the **Analyze** tab; the **Edit Zone** tab appears, and the view fades with a "+" zone icon on the cursor (Figure 4-6). Ensure Add Space Zone is selected in the Mode panel, then hover over a space and click to add it to the zone—its color will fade. Repeat this to add more spaces. To remove a space, select **Remove Space** from the Mode panel; a "–" icon appears with the cursor. Click the highlighted space to remove it—it will fade, indicating removal.

Figure 4-5 The ***Edit Zone*** *contextual tab*

PERFORMING SYSTEM ANALYSIS

In Revit 2025, the heating and cooling loads analysis has changed to System Analysis. System analysis helps mechanical design engineers to define all the main components and a set of relationships that comprise a mechanical HVAC system, without physically modeling it. The workflow for performing the system analysis is discussed next.

Setting the Model

Setting the model for system analysis is the first step toward performing the System Analysis. There are three different ways to set up the model. You can use the model that contains walls, floors, roofs, and so on for system analysis. Alternatively, you can use the massing model or a combination of the massing and building model to perform the analysis.

Adding the Analytical System and Defining the System Zone

After selecting the type of model and specifying its location, the next step in the workflow for system analysis will be to add the analytical system to be used in the model and then define the system zone required to be analyzed.

Creating the Analytical Model

After you have created system-zones and assigned zone equipment, generate the analytical spaces and revise the analytical model as needed.

A separate energy model consisting of surfaces and volumes is created directly from your design model. When you create the energy model, analytical spaces are automatically assigned to zone equipment. Energy settings allows you control different factors to consider as you create the energy model.

To create an energy model, choose the **Create Energy Model** tool from the **Energy Optimization** tool of the **Analyze** tab; the **Create Energy Model** window will be displayed. Choose the **Create the Energy Analytical Model** option; the **Analytical Spaces** tab will be displayed showing the analytical spaces of the model, as shown in Figure 4-6. Also, separate schedules for analytical spaces and analytical surfaces will be generated and displayed under the **Schedules/Quantities** head in the **Project Browser**. Note that when you create the energy model, analytical spaces are automatically assigned to zone equipment.

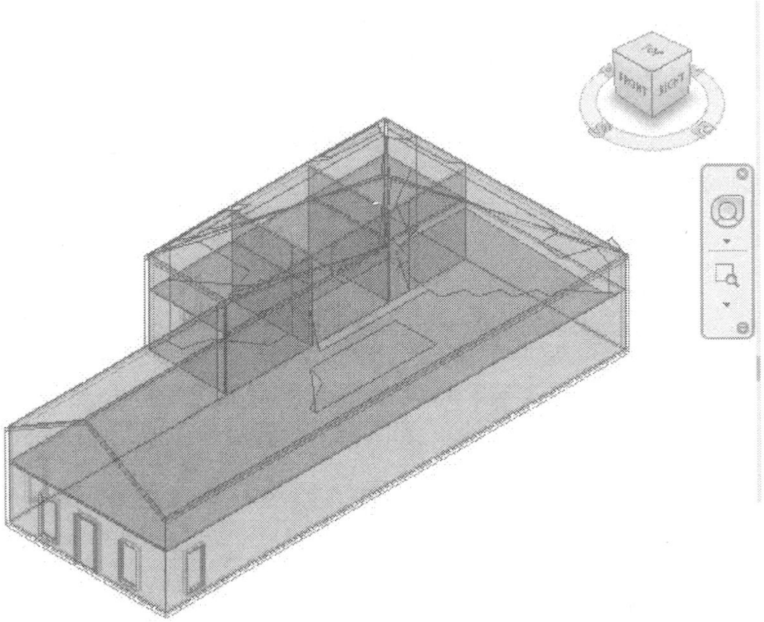

Figure 4-6 *The created analytical model*

Running the System Analysis

In Revit 2025, system analysis uses EnergyPlus simulation software through two default workflows: HVAC Systems Loads and Sizing, and Annual Building Energy Simulation. To start, go to the Energy Optimization panel in the Analysis tab and open the **Systems Analysis** dialog. Select a workflow, enter a report name, and set the report folder path in Energy Settings. After confirming the folder, click **Run Analysis**. The results will appear under **Reports** > **Analysis Reports** in the **Project Browser** and be saved in the chosen folder.

TUTORIALS

General instructions for downloading tutorial files:

1. Download the *c04_rmp_2025_tut.zip* file for this tutorial from *https://www.cadsofttech.com*. The path of the file is as follows: *Books > Civil/GIS BOOKS > Revit MEP > Revit MEP 2025 For Novices*

2. Now, save and extract the downloaded folder at the following location:
C:\rmp_2025

Tutorial 1 Placing Spaces- Office Space

In this tutorial, you will open a linked architectural model and then add spaces to it.

(Expected time: 1 hr 15 min)

1. Template file:
For Imperial: **English Imperial > Systems-Default**.
For Metric: **English > Systems-Default_Metric**

2. File name to be used (Linked model):
For Imperial: *c04_archi_spaces_rmp2025.rvt*
For Metric: *M_c04_archi_spaces_rmp2025.rvt*

3. File name to be assigned:
For Imperial: *c04_Office-Space_tut1.rvt*
For Metric: *M_c04_Office-Space_tut1.rvt*

The following steps are required to complete this tutorial:

a. Start a Revit 2025 session.
b. Open a new project and link an architectural model.
c. Add a Plenum level.
d. Rename and modify the Plenum level.
e. Place a spaces in the 1-Mech view.
f. Rename and renumber the spaces.
g. Enable the space visibility.
h. Place space in the separated area.
i. Place a space in the Chase area.
j. Save the project using the **Save As** tool.
k. Close the project using the **Close** tool.

Starting Autodesk Revit 2025

1. Choose **Start > All Programs > Autodesk > Revit 2025 (for Windows 10)**; the Revit interface window is displayed.

Opening a New Project and Linking the Architectural Model

In this section, you will open a new project and link the downloaded *.rvt* file to it.

1. Choose **New > Project** from the **File** menu; the **New Project** dialog box is displayed.
2. In the **New Project** dialog box, choose the **Browse** button; the **Choose Template** file dialog box is displayed. In this dialog box, select the *Systems-Default.rte* (for Metric *Systems-Default_Metric.rte*) from the **English Imperial** folder (for Metric **English** folder) and choose **OK**; the **Choose Template** dialog box is closed and the selected template is loaded in the file.

3. Now, in the **New Project** dialog box, select the **Systems-Default** (for Metric **Systems-Default_Metric**) from the **Template** file drop-down list and then choose the **OK** button; the **New Project** dialog box is closed and the new project is opened. Make sure that the **Project** radio button is selected in the **Create new** area.

   In the new project file, you will download the architectural model in the current project view and then link the architectural model in the **1-Mech** view under the **Mechanical > HVAC > Floor Plans** view displayed in the **Project Browser**.

4. In the **Project Browser**, ensure that the **1-Mech** node is selected under **Mechanical > HVAC > Floor Plans** and then choose the **Link Revit** tool from the **Link** panel of the **Insert** tab; the **Import/Link RVT** dialog box is displayed. In this dialog box, browse to the *C:\rmp_2025\c04_rmp_2025_tut* folder and then select the *c04_archi_spaces_rmp2025.rvt* file. For Metric select the *M_c04_archi_spaces_rmp2025.rvt* file.

5. Now, choose the **Open** button; the selected file is linked to the current project and is displayed in the project view, as shown in Figure 4-7.

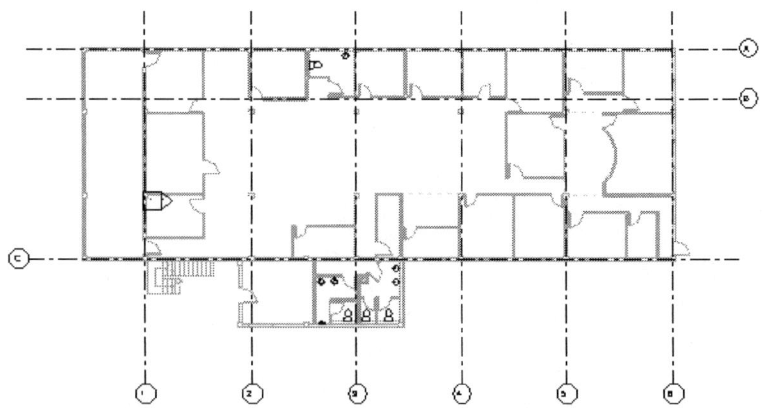

Figure 4-7 *The drawing view with the linked model*

6. Now, move the cursor over the linked model and click when a blue border is displayed around it; the **Modify | RVT Links** contextual tab is displayed.

7. In this tab, choose the **Type Properties** tool from the **Properties** panel; the **Type Properties** dialog box is displayed.

8. In this dialog box, select the check box displayed in the **Value** field corresponding to the **Room Bounding** parameter. The architectural components (such as walls and floors) are closed so that they are recognized as boundaries for spaces.

Note

While working with a linked file, make sure that the roof is defined as room bounding element and the ceiling is defined as a non-room bounding element. These components are defined in the architectural model file and not in the MEP model file.

9. In the **Type Properties** dialog box, choose the **OK** button to close it. Now, in the **Modify | RVT Links** contextual tab, choose the **Modify** button in the **Select** panel; the linking process of the architectural model is now complete.

Adding a Plenum Level

In this section, you will create a plenum level for the spaces. The plenum levels are created to place spaces in the plenum areas (between the ceiling and the floor) of the building. You must place spaces in all areas (occupied and unoccupied) of the building to achieve an accurate heating and cooling load analysis.

1. In the **Project Browser**, expand **Mechanical > HVAC> Elevations (Building Elevation)**, and double-click in the **East - Mech** node; the east elevation view is opened.

2. Choose the **Level** tool from the **Datum** panel of the **Architecture** tab; the **Modify |Place Level** tab is displayed along with various options in the **Plan Views** and the **Draw** panels.

3. In the **Properties** palette, select the **Level Plenum** option from the **Type Selector** drop-down list.

4. In the **Plan Views** panel, ensure that the **Create Plan View** check box is selected and then choose the **Plan View Types** button; the **Plan View Types** dialog box is displayed, as shown in Figure 4-8. In this dialog box, you will notice that the **Ceiling Plan**, **Structural Plan,** and **Floor Plan** options are selected and highlighted in the **Select view types to create** list box.

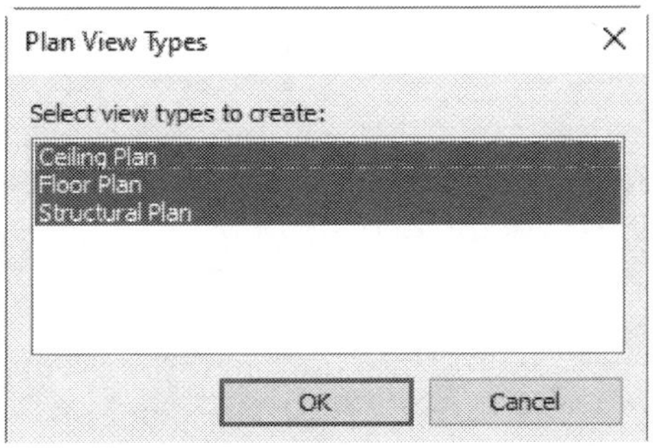

Figure 4-8 *The* ***Ceiling Plan, Structural Plan*** *and the* ***Floor Plan*** *options selected in the* ***Plan View Types*** *dialog box*

5. In the list box, click on the **Structural Plan** and **Ceiling Plan** options to clear the selection and then choose the **OK** button; the **Plan View Types** dialog box is closed.

6. Now, in the **Draw** panel of the **Modify | Place Level** contextual tab, ensure that the **Line**

tool is chosen and then move and place the cursor near the endpoint of **Level 1** level. Move the cursor up. On doing so, you will notice a temporary dimension emerging from the left endpoint of the **Level 1**, as shown in Figure 4-9.

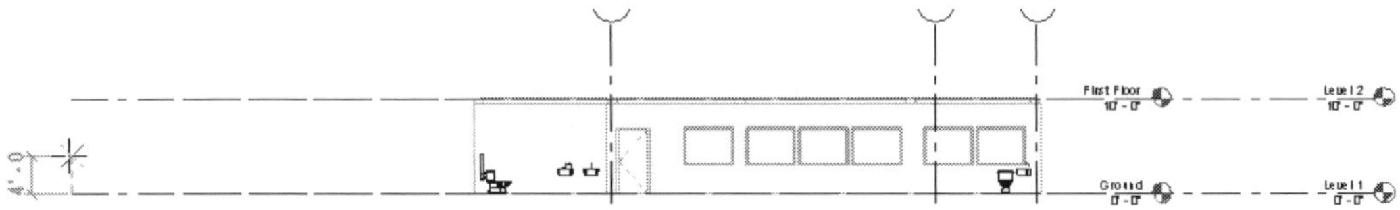

Figure 4-9 *The temporary dimension emerging from the **Level 1** line*

7. Now, type **8'6" (2591mm)** and press ENTER; the level starts at a distance of 8'6" (2591mm) from the **Level 1** level.

8. Now, move the cursor toward right and place it over the endpoint of **Level 1**. Click when the extension snap appears along with an alignment line emerging from **Level 1** line; the plenum level is created with a default name of **Mechanical 3**.

9. In the **Modify | Place Level** tab, choose the **Modify** button from the **Select** panel to exit the **Level** tool.

10. In the **Project Browser**, double-click on the **Mechanical 3** (for Metric **Level 3**) node in **Mechanical > HVAC > Floor Plans** node; the floor plan view of the **Mechanical 3** level is displayed in the drawing window.

Renaming and Modifying the Plenum Level

1. In the **Properties** palette, click on the value field corresponding to the **View Name** parameter and enter **1-Mech-Plenum** in the edit box to replace the existing name. Press ENTER; the **Confirm Plan View Rename** message box is displayed. Choose the **Yes** button in the displayed message box; the level is renamed.

2. In the **Properties** palette, choose the **Mechanical Plan** button displayed in the value field corresponding to the **View Template** parameter; the **Apply View Template** dialog box is displayed.

3. In the dialog box, select the **<None>** option from the **Names** list box and then choose the **OK** button; the **Apply View Template** dialog box is closed. On doing so, the default view template that was assigned to the view is removed.

4. In the **Properties** palette, choose the **Edit** button displayed in the value field corresponding to the **View Range** parameter; the **View Range** dialog box is displayed.

5. In the **Primary Range** area of the dialog box, select the **Level Above (Level 2)** option from the **Top** drop-down list. Also, select the **Associated Level (1-Mech-Plenum)** option from the **Bottom** drop-down list. Now, click in the **Offset** edit box displayed next to the **Cut plane** drop-down list and enter **1' (304mm)** to replace the existing value.

6. In the **View Depth** area of the **View Range** dialog box, select the **Associated Level (1-Mech-**

Plenum) option from the **Level** drop-down list. Choose **Apply** and then **OK**; the **View Range** dialog box is closed.

7. In the **Project Browser**, under **HVAC > Floor Plans**, double-click on **1- Mech** and maximize the window. Now, choose the **Close Inactive** tool from the **Windows** panel of the **View** tab to close the displayed elevation view. Alternatively, you can choose the **1-Mech** tab in the user interface to activate the view.

Note
*If the active window is in the tile mode, then the **Close Inactive** tool is disabled. If more than one window is opened in the current session, choose the **Tile Views** tool to tile the windows displayed.*

Placing Spaces in the 1-Mech View

In this section, you will place the spaces in the **1-Mech** floor plan view.

1. Choose the **Space** tool from the **Spaces & Zones** panel of the **Analyze** tab; the **Modify | Place Space** tab is displayed.

2. In the **Tag** panel of this tab, make sure that the **Tag on Placement** button is chosen by default. In the **Options Bar**, select the **1-Mech-Plenum** option from the **Upper Limit** drop-down list.

   In the **Options Bar**, ensure that the **0'0"** (**0.0**) value is specified in the **Offset** edit box, the **Horizontal** option is selected in the drop-down list displayed next to the **Offset** edit box, the **Leader** check box is cleared and the **New** option is selected in the **Space** drop-down list.

3. In the drawing area, zoom into the office area located in the upper-left corner of the building. Now, place the cursor in it, until the space snaps to the room-bounding elements, as shown in Figure 4-10. Click when the room bounding is highlighted; the new space is placed with the default name **Space** and number **1**.

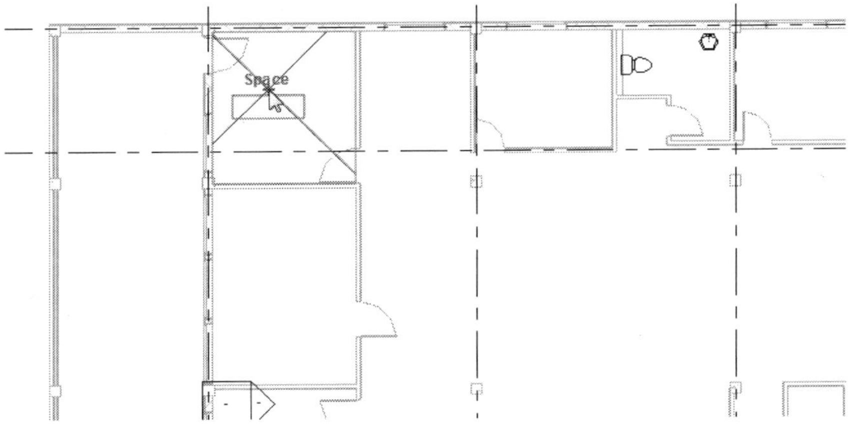

Figure 4-10 *Placing the space inside the room-bounding area*

4. Ensure that the **Space** tool is invoked. Then, press ZF; the project view is zoomed to fit the drawing area. This displays the entire floor plan placed at the center of the drawing area.

> **Tip**
> *You can right-click in the drawing window and select the **Zoom to Fit** option from the shortcut menu displayed*

5. Now, move the cursor to the large open area at the center of the floor plan, and after the space snaps to the room bounding elements, click to place a space, as shown in Figure 4-11.

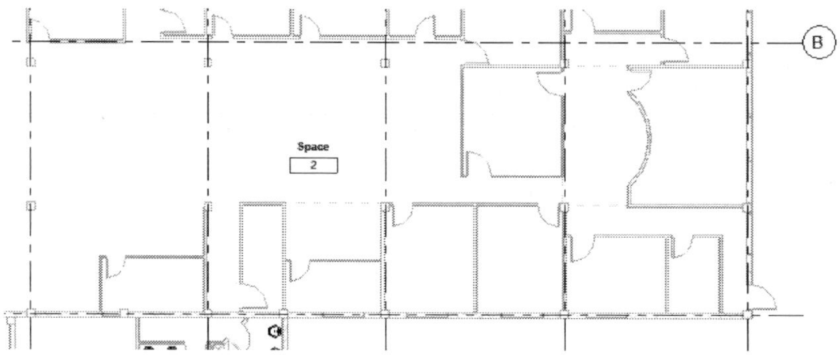

Figure 4-11 *The space added to the large open area*

Make sure that you have placed the space tag toward right in the open space. Later in the tutorial, you will separate the open space near the entrance and place a space there.

6. In the **Modify | Place Space** contextual tab, choose the **Modify** button from the **Select** panel; the **Space** tool is exited.

Renaming and Renumbering the Spaces

In this section, you will rename and renumber the spaces added.

1. Zoom in the space tag in the office area in the upper-left corner and then click on the **Space** text; an edit box is displayed. Enter **CEO-Office** in the edit box and press ENTER; the space name is changed.

> **Tip**
> *In a project, you can also rename and renumber the spaces by using the **Properties** palette. Naming and number of a space should be done as per the standard practices followed and should be similar to the names and numbers of the rooms specified by the Architect.*

2. Now, in the same space, double- click on the text **1**; an edit box is displayed. Enter **101** to replace the previous number and press ENTER; the space number is changed.

3. Repeat steps 1 and 2 and rename and renumber the space at the central area as **Central Area** and **110**, respectively, refer to Figure 4-12.

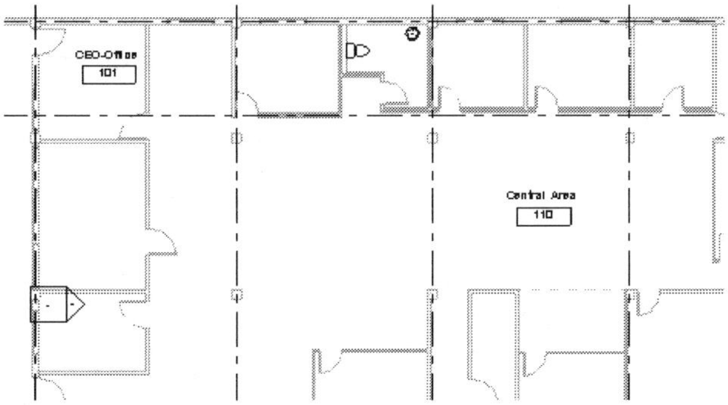

Figure 4-12 *The central area space renamed and renumbered*

Enabling the Space Visibility

In this section, you will enable the visibility of the space interior fills and markers.

1. Ensure that the **1-Mech** floor plan view is active and then choose the **Visibility/Graphics** tool from the **Graphics** panel of the **View** tab; the **Visibility/Graphics Override for Floor Plan: 1-Mech** dialog box is displayed.

2. In the dialog box, ensure that the **Model Categories** tab is chosen by default, and then in the **Visibility** column, expand the **Spaces** node. Ensure that the check box corresponding to the **Spaces** node is selected.

3. Now, under the **Spaces** node, ensure that the **Color Fill** check box is selected and then select the **Interior** and **Reference** check boxes.

4. Choose the **OK** button; the the **Visibility/Graphics Override for Floor Plan1-Mech** dialog box is closed and the added spaces will be displayed with the specified settings, as shown in Figure 4-13.

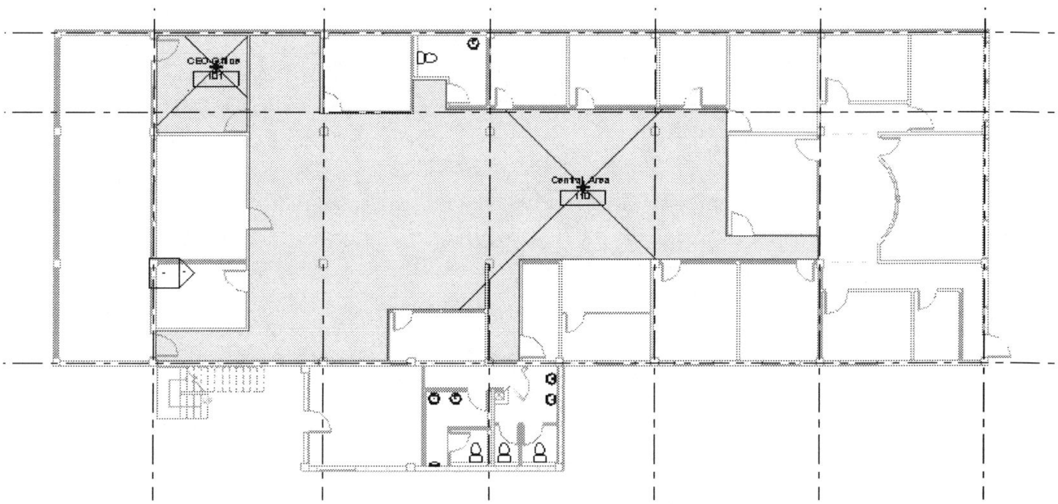

Figure 4-13 *The markers and the interior fills displayed in the added spaces*

Splitting the Space

You need to place a space in the area next to the building entrance because this area will be heated and cooled more often than the rest of the open spaces. The entrance area is considered semi-bounded. To place a space in this area, you need to make it fully-bounded by drawing space separation lines. In this section, you will split the **Central Area** space.

1. Ensure that the **1 - Mech** view is active and enter ZR; the cursor changes into a zoom region icon. Click at the point above the **CEO-Office** space area and drag the mouse downward right below the **Central Area** space, refer to Figure 4-14. and click; the central region of the office area is zoomed.

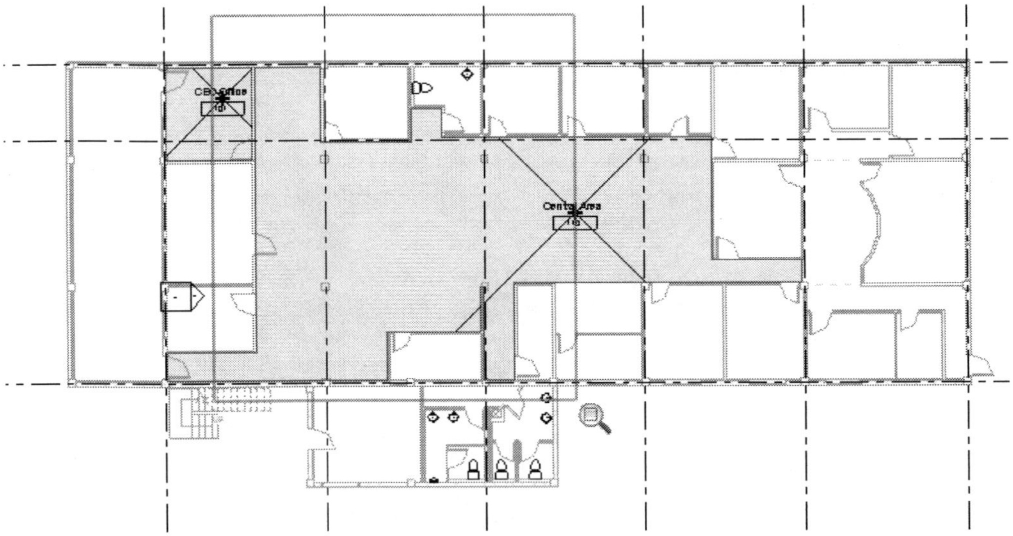

Figure 4-14 *Specifying the zoom area*

2. Choose the **Space Separator** tool from the **Spaces & Zones** panel of the **Analyze** tab; the **Modify | Place Space Separation** contextual tab is displayed. In the **Draw** panel of the tab, ensure that the **Line** tool is chosen by default. Also, in the **Options Bar**, ensure that the **Chain** check box is selected and the value specified in the **Offset** edit box is **0'0"** (**0.0**).

3. Now, place the cursor at the endpoint of the space boundary of the **Central Area** near the **Toilet** area, as shown in Figure 4-15. Click when the **Endpoint** snap marker appears; the start point of the separation line is specified.

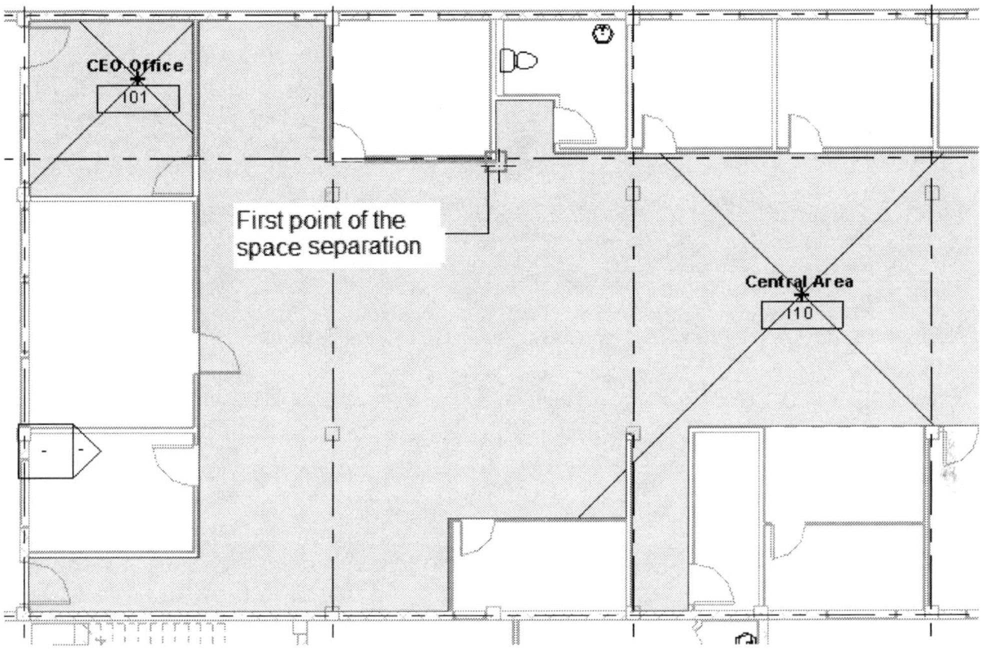

Figure 4-15 *Specifying the start point of the separator line*

4. Now, move the cursor vertically downward; a dashed line along with temporary dimension appears. Enter **26'5" (8052mm)** and press ENTER; the separation line is created and the **Central Area** is adjusted to the new separation, as shown in Figure 4-16. Now, choose the **Modify** button in the **Select** panel of the **Modify | Place Space Separation** contextual tab to exit the **Space Separator** tool.

Note

Space separation lines are MEP-specific room bounding lines that separate areas where a wall is not required or is not possible to create one. After the areas are separated, spaces can be placed in them. Although room separation lines are recognized in Revit, space separation lines are not recognized in Architecture discipline of Revit.

Tip

Once you create a space, you can inspect its various properties such as its area and volumes. The area and volume of the newly created spaces are 1015.39SF and 8377.05CF, respectively. Also, you can check the various parameters for energy analysis before performing an energy analysis.

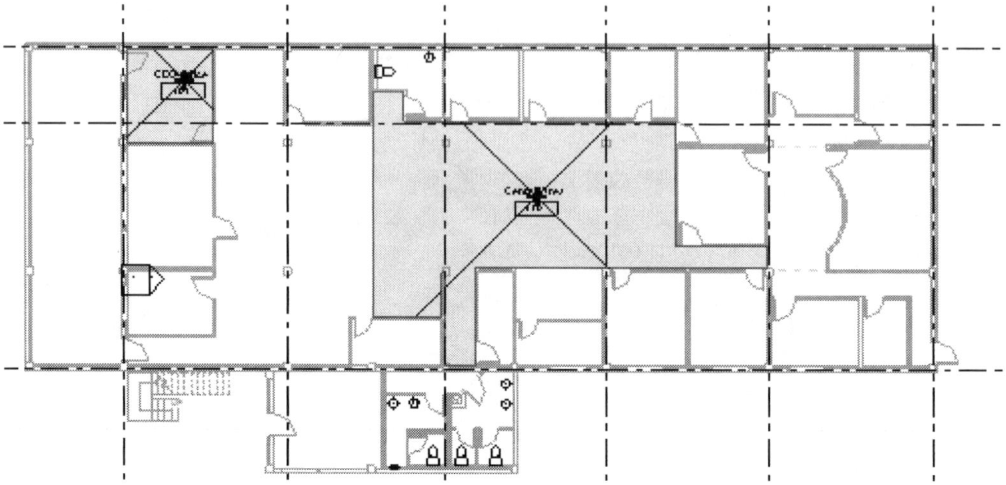

Figure 4-16 The ***Central Area*** *space in the plan*

Placing a Space in the Separated Area

In this section, you will add a space to the area separated from the **Central Area** space.

1. Choose the **Space** tool from the **Spaces & Zones** panel of the **Analyze** tab; the **Modify | Place Space** tab is displayed.

2. In this tab, ensure that the **Tag on Placement** button is chosen. Also, in the **Options Bar**, select the **1-Mech Plenum** from the **Upper Limit** drop-down list and then enter **0"(0.0)** in the **Offset** edit box. In the **Options Bar**, ensure that the **Horizontal** option is selected from the drop-down list displayed next to the **Offset** edit box, the **Leader** check box is cleared, and the **New** option is selected from the **Space** drop-down list.

3. Now, move and place the cursor in the separated area; the room bounding space snap is displayed. Click inside the area, the space is placed in the area.

4. Now, choose the **Modify** button and then select the newly created space; the **Properties** palette displays the instance property of the selected space.

5. In the **Properties** palette, click in the value field corresponding to the **Number** parameter. Next, type **115** and press ENTER.

6. Now, click in the value field of the **Name** parameter. Then, type **Lounge** and press ENTER. Choose the **Modify** button from the **Select** panel of the **Modify| Spaces** tab to exit the modification of the space.

The space number and the name is changed, as shown in Figure 4-17.

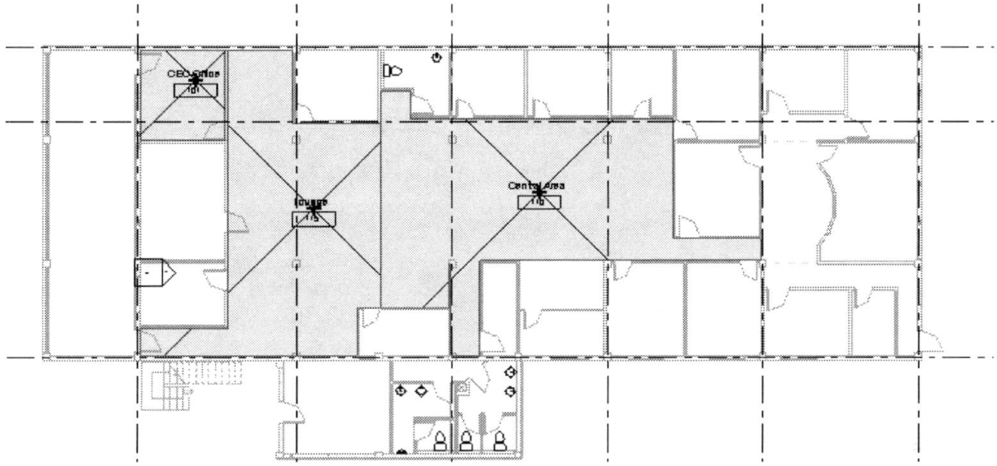

Figure 4-17 *The added space renamed and renumbered*

Placing a Space in the Chases Area

In this section, you will add a space to the chases area. A chase is a space in a building to accommodate services such as pipes or cables. Note that the space components are placed in chases to permit a reliable heating and cooling load analysis.

1. Zoom in the chases area that is located below the central area, as shown in Figure 4-18.

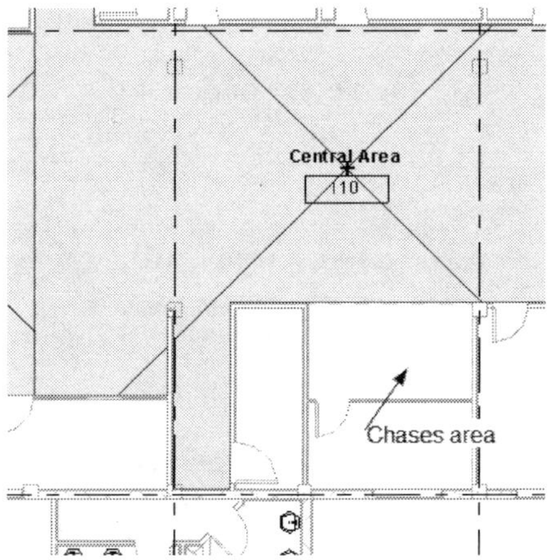

Figure 4-18 *Zooming in the chases area*

2. Invoke the **Space** tool from the **Spaces & Zones** panel of the **Analyze** tab and then in the **Options Bar**, select **Level 2** from the **Upper Limit** drop-down list.

3. Click in the **Offset** edit box, enter **1'6" (457mm)**, and press ENTER.

4. Next, click in the chases area, refer to Figure 4-19, and click when the space snap is highlighted.

5. Now, choose the **Modify** button from the **Select** panel of the **Modify | Place Space** contextual tab and then select the added space.

6. In the **Properties** palette, enter **Chases** and **121** in the value fields of **Name** and **Number** edit boxes, respectively. Figure 4-38 shows the **Chases** space added to the model.

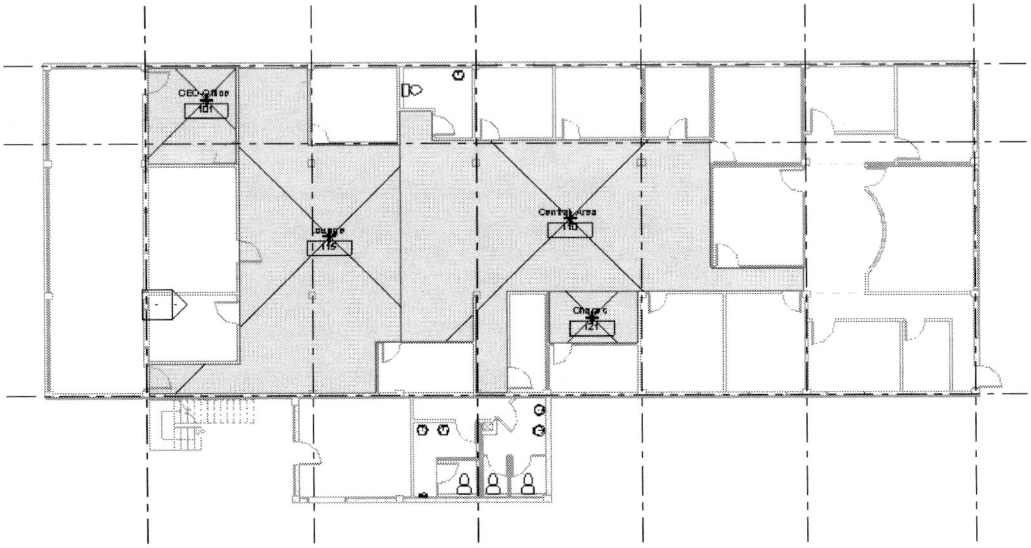

Figure 4-19 *Chases space added to the model*

7. Choose the **Modify** button from the **Select** panel of the **Modify | Space** contextual tab to exit the selection of the space.

Saving the Project

In this section, you need to save project and settings using the **Save As** tool.

1. To save the project with settings, choose **Save As > Project** from the **File** menu; the **Save As** dialog box is displayed, as you are saving the project for the first time.

2. In this dialog box, browse to the *c:\rmp_2025\c04_rmp_2025_tut* folder and then in the **File name** edit box, enter the text **c04_Office-Space_tut1** (for Metric **M_c04_Office-Space_tut1**) and then choose the **Options** button; the **File Save Options** dialog box is displayed. In this dialog box, ensure that the **Active view/sheet** option is selected from the **Source** drop-down list in the **Thumbnail Preview** area.

3. Now, choose the **OK** button; the **File Save Options** dialog box is closed and the **Save As** dialog box is displayed.

4. In this dialog box, choose the **Save** button to save the current project file with the specified name and close the **Save As** dialog box.

Closing the Project

1. To close the project, choose the **Close** option from **File** menu; the file is closed.

EXERCISE

Exercise 1 Creating Spaces

Download the *c04_Conference-Center_exer1.rvt* (for Metric *M_c04_Conference-Center_exer1.rvt*)file from *https://www.cadsofttech.com*. The path of the file is as follows: *Books > Civil/GIS BOOKS > Revit MEP > Revit MEP 2025 For Novices*

In this exercise, you will open an architectural model and then add spaces to it. Name and number the spaces, as shown in Figure 4-20.

Note
You can add properties to the spaces as per your design requirement.

1. Project view to be used :
   Views > Floor Plans > Mechanical > 1ST FLOOR- HVAC Space

2. File name to be assigned: *c04_Conference-Center_exer1a.rvt* (For Metric *M_c04_Conference-Center_exer1a.rvt*).

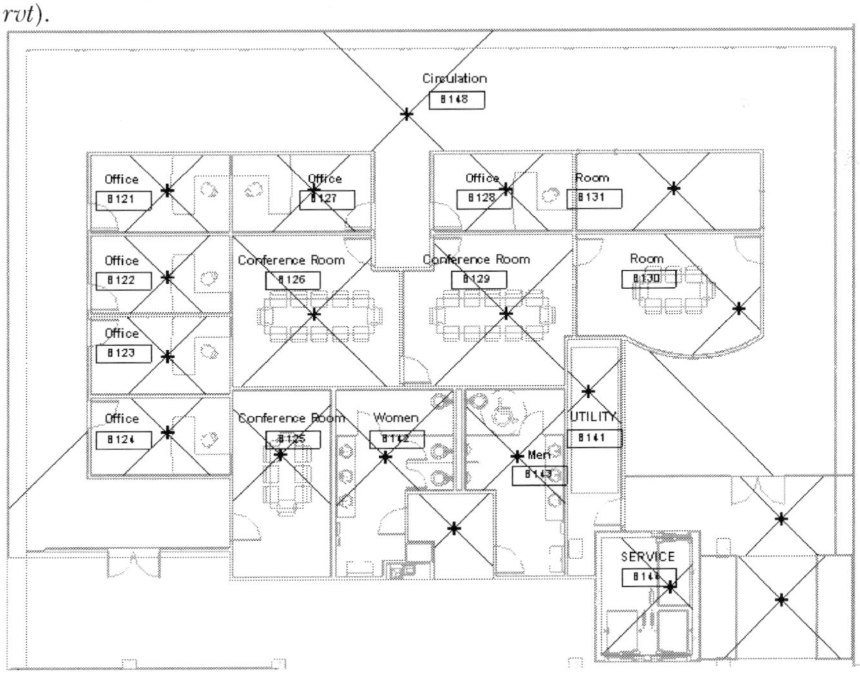

Figure 4-20 *Spaces created for the Conference-Center project*

This page is intentionally left blank

Chapter 5

Creating an HVAC System

Learning Objectives

After completing this chapter, you will be able to:

- *Create an HVAC system*
- *Generate an HVAC system layout*
- *Create ducts and duct fittings*

INTRODUCTION

Heating, Ventilation, and Air Conditioning (HVAC) is the technology of indoor and vehicular environmental comfort. The goal of HVAC is to provide thermal comfort and acceptable indoor air quality. In this chapter, you will learn about various tools and options to create HVAC systems for a project. In Revit, you can use different tools to create, modify, and inspect HVAC systems in a project. These tools allow you to model and analyze systems and check whether your project design values meet the engineering standards fixed by the organization you are working with.

CREATING AN HVAC SYSTEM

In an MEP project, the HVAC system is designed for three important functions, namely heating, ventilating, and air conditioning of a building. These three functions are closely interrelated and the design, installation, and control systems of these functions are integrated into one or more HVAC systems in a project.

In Revit, you create the HVAC systems using different tools available in the **HVAC**, **Fabrication**, and **Mechanical** panels of the **Systems** tab, as shown in Figure 5-1. Apart from these tools, you can use automatic system creation tools to create duct routing layouts to connect the supply and return system components to the HVAC system.

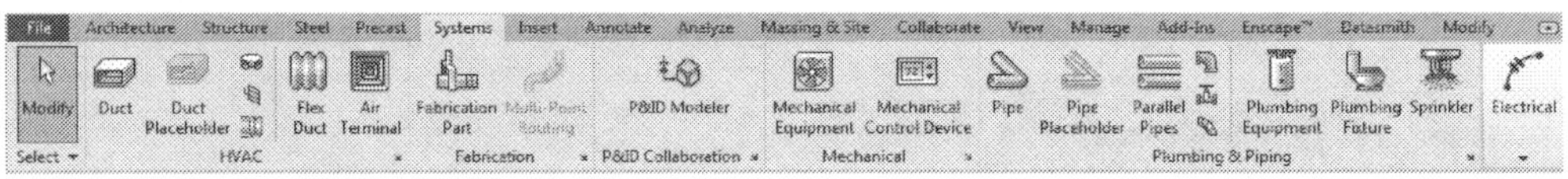

Figure 5-1 *Various tools for HVAC systems in the* ***Systems*** *panel*

Understanding Air Terminals

While working on an HVAC system, it is important that you add sufficient air terminals to the spaces to fulfill air flow requirement. The two ventilation methods that are used for a HVAC system are Natural Ventilation and Mechanical Ventilation.

Natural Ventilation: This method uses natural air circulation through windows, solar chimneys, and trickle vents around the house.

Mechanical Ventilation: This method uses various systems and devices like fans, ducts, intake or exhaust vents, and air filters to replace the air.

Adding Air Terminals

In Revit, adding air terminals starts with loading the appropriate type from the library and setting an offset value (for non-hosted terminals) along with the designed airflow. Use the Load Family tool from the Insert tab to open the **Load Family** dialog box. Navigate to the Air Terminals folder (e.g., **English-Imperial** > **Mechanical** > **MEP** > **Air Side Components** > **Air Terminals**) and select the desired family, as shown in Figure 5-3. Click **Open** to load the selected terminals into the project. You can then add them using the **Air Terminal** tool in the HVAC panel of the Systems tab.

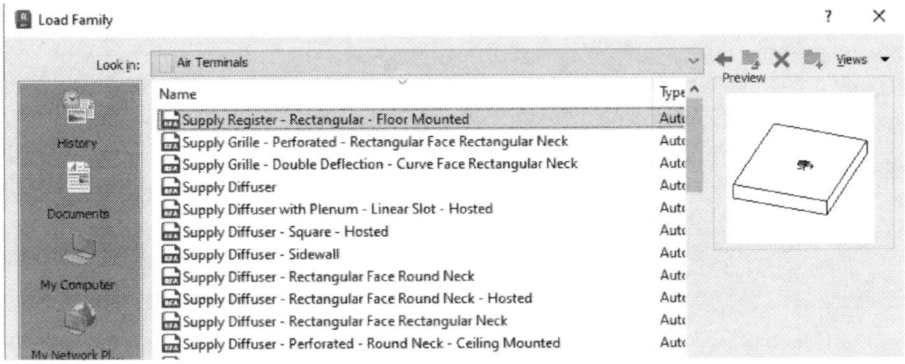

***Figure 5-2** Partial view of the **Load Family** dialog box displaying various families of air terminals*

Tip

Apart from loading the standard air terminals from the Revit library that comes with the software, you can upload various custom based air terminals created by BIM modelers across the team.

Adding Air Equipment

While designing an HVAC system, you need to add sufficient and specified air equipment to control the flow of air from the duct to the air terminals. In Revit, adding air equipment involves loading the appropriate type of air equipment family(ies), placing the offset, and rotating the equipment in the proper direction to connect it to the ducts in the HVAC system.

Adding Mechanical Control Devices

While designing an HVAC system, you need to add specific mechanical control devices such as thermostats, humidistats, and sensors. In Revit, adding mechanical control devices involves loading the appropriate type of family(ies), placing the offset, and selecting the appropriate host.

To load a mechanical equipment into your project, invoke the **Load Family** dialog box and then browse to **English-Imperial > Mechanical > MEP > Information and Communication > Building Controls** folder (for Metric, browse to **English > US > Mechanical > MEP > Information and Communication > Building Controls** folder). In the **Building Controls** folder, as shown in Figure 5-3, you can select the appropriate family required for the HVAC system.

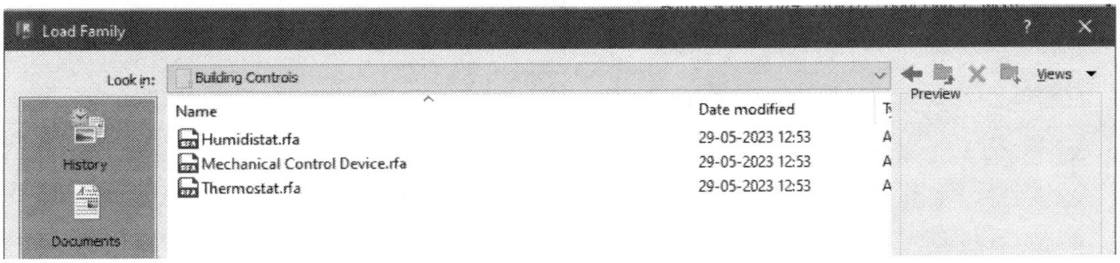

***Figure 5-3** Partial view of the **Load Family** dialog box displaying various mechanical control devices*

Creating Air Supply System

After adding air terminals and equipment, you can create an air supply system by selecting them in the drawing. This activates the **Modify | Multi Select** tab. In the Create System panel, click the **Duct** tool to open the **Create Duct System** dialog box (Figure 5-4). Select "**Supply Air**" as the system type, name the system, and click **OK**. If using a **VAV** box, the Select Connector dialog box will appear—choose a connector and confirm. The system will then appear in the System Browser under Mechanical. The **Modify | Duct Systems** tab also becomes available for further editing and duct layout creation.

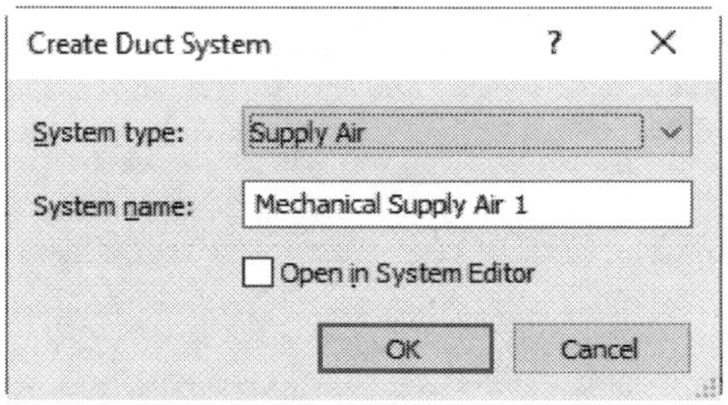

Figure 5-4 *The **Create Duct System** dialog box*

Creating Return Air Terminals, Air System, and Air Ductwork

After creating the air supply system, return air terminals must be added to complete the **HVAC** setup. Start by loading the necessary return air terminal families from the **Air Terminals** folder using the **Load Family** dialog. Then, select the Air Terminal tool from the **HVAC** panel in the Systems tab, set the offset and **CFM** flow values, and place the terminals. Finally, create a return air system to connect the terminals with the **HVAC** equipment, similar to the supply air system.

Inspecting the Duct System

After creating the ductwork in the HVAC system, inspect the airflow to ensure it meets design standards. Select a duct, go to the Analysis panel in the **Modify | Ducts** tab, and click the System Inspector tool. In the floating panel, select Inspect to display red arrows showing airflow direction. Moving the cursor over the duct sections reveals flags with flow information, as shown in Figure 5-5.

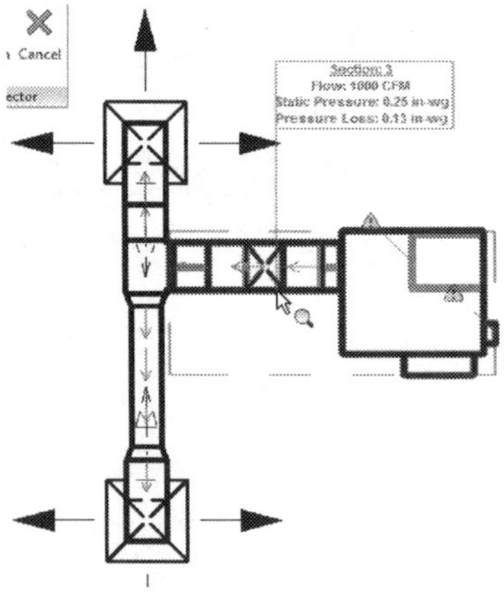

Figure 5-5 *The flag displaying the information of a duct in the system*

Checking the Duct System

After inspecting the air flow in the system, you need to check the connectivity of the ductwork in the system. To do so, choose the **Check Duct Systems** tool from the **Check Systems** panel in the **Analyze** tab; the various disconnects in the ducts, if any, will be shown as warning in the drawing.

Creating Duct Legend

After creating the ducts, you will create duct legend to indicate color fills associated with the ductworks in the **HVAC** system. The duct legend will help you to visualize the air flow requirement in different ducts with the help of color fills. To create a duct legend, choose the **Duct Legend** tool from the **Color Fill** panel in the **Analyze** tab and click in the drawing view; the **Choose Color Scheme** dialog box will be displayed. In the dialog box, select any of the options from the **Color Scheme** drop-down list and choose **OK**; the color fills will be displayed in the ducts and legend will be displayed at the location where you have clicked. Figure 5-6 shows a color fill legend and color fills in ducts in an **HVAC** layout.

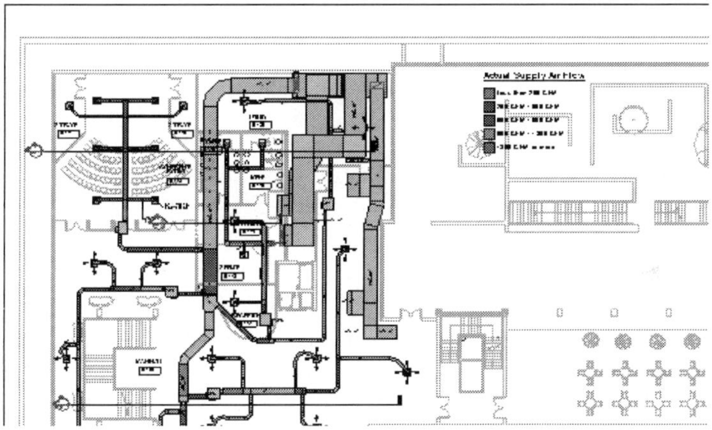

Figure 5-6 *The duct legend and color fills displayed in an HVAC layout*

Different Components of an HVAC System

The components of an HVAC system include various types of mechanical equipment and air terminals that you will add to the project. In an MEP project, you will add these components in the floor plan view or in the ceiling plan view.

The air terminals can be a hosted or non-hosted to a ceiling. To place the hosted air terminals, you need to activate the desired ceiling plan view from **Views > Mechanical > HVAC > Ceiling Plans** node in the **Project Browser**. The equipment are generally non-hosted families. To place the equipment, activate the desired floor plan view from **Views > Mechanical > HVAC > Floor Plans** node of the **Project Browser**.

The various options to add different air terminals and mechanical equipment to an HVAC system are discussed next.

Recommended Practices for Creating HVAC Systems

While working in an HVAC system, there are certain recommended practices that help you to create HVAC systems effectively. These practices are as follows:

1. While placing air terminals and equipment, it is recommended to copy the existing air terminals and equipment to the desired locations. This will help the copied elements in inheriting the properties such as elevation and type from the parent element. This practice will save time because you do not have to specify similar information for multiple components.

2. It is always recommended to schedule information related to spaces in the project, which will be useful during the design process of the HVAC system, irrespective of whether the information will become part of the design documentation or not.

GENERATING HVAC SYSTEM LAYOUTS

After you create an HVAC system, you will add different components to it and use automatic layout tools to review possible routing solutions. The automatic layout tools will provide you with multiple layout solutions that you can explore to suit your design requirement. If you do not find the required set of connections, you can manually modify the layout path solutions.

An HVAC system layout is a representation of the physical connection between different components in an HVAC system. For example, you can create duct layouts to connect air system components. Revit will provide you with tools to generate the layout of the ductwork automatically when components are added to the HVAC systems.

While generating layouts using the automatic layout tool, you will notice that each layout solution consists of blue and green layout lines. Blue lines represent the main duct layout, and green lines represent the branches duct layout. You can generate three types of routing solutions for a layout: Network, Perimeter, and Intersection. The table below describes the solution types.

Solution Type	Description
Network	Provides up to six solutions, each consisting of a main segment with branches at 90 degrees from the main segment.
Perimeter	Provides five solutions that wrap the main segment around the perimeter of the components.
Intersection	Provides up to eight solutions, each consisting of a main segment routed over the connectors of the components. Each solution consists of branches extending perpendicular from the main segment down to the components.

To create an auto layout for the HVAC system in Revit, first select the desired air terminals; this opens the **Modify | Air Terminals** tab. Choose the **Duct** tool from the **Create Systems** panel to open the **Create Duct System** dialog. Select **Supply Air** as the system type, name it "**Supply System-01**," and click **OK**. A blue dotted rectangle will appear around the system, and the **Modify | Duct Systems** tab will open. Click **Edit System**, then select mechanical equipment from the **System Equipment** dropdown. After editing, click **Finish Editing System** to complete setup. To generate the layout, select an air terminal and click **Generate Layout** from the **Layout** panel; the **Generate Layout** tab will appear, as shown in **Figure 5-7**.

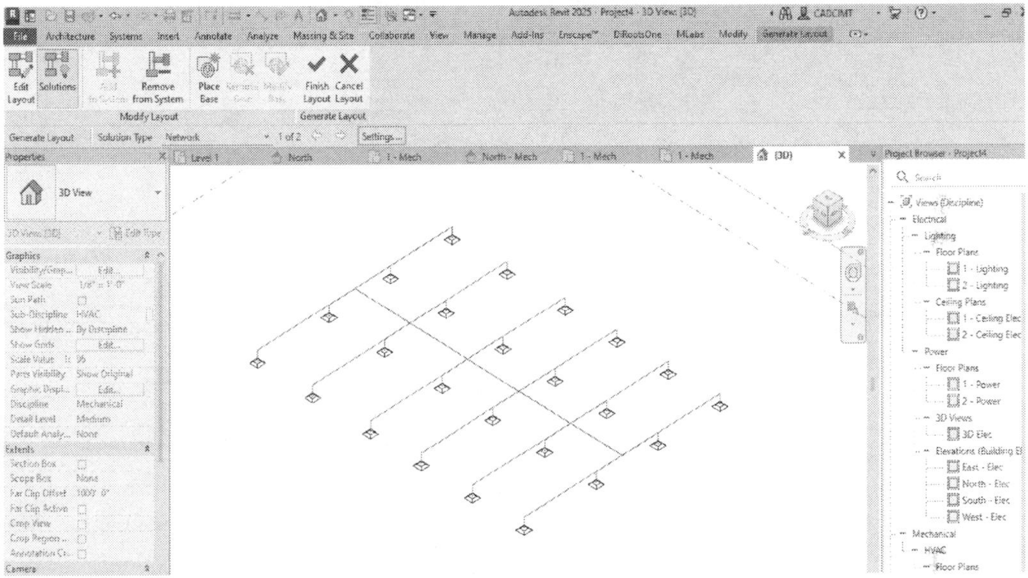

Figure 5-7 *The options for generating duct layout along with the generated layout*

In the **Options Bar** of this tab, choose the **Settings** button; the **Duct Conversion Settings** dialog box will be displayed, as shown in 5-8.

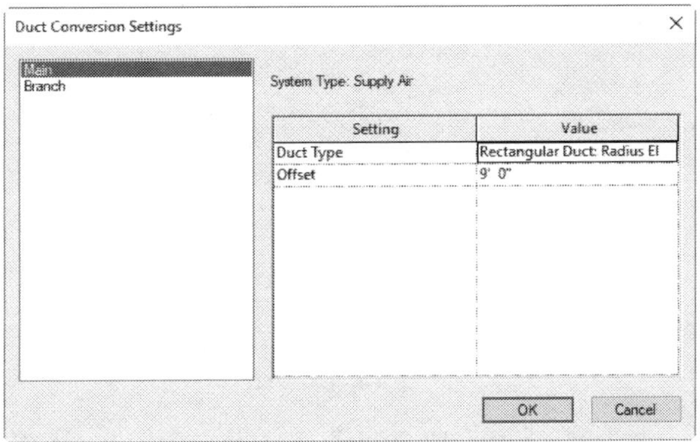

Figure 5-8 *The* ***Duct Conversion Settings*** *dialog box*

In the **Duct Conversion Settings** dialog box, select the **Main** option in the left pane to display related settings in the right pane. Choose a duct type from the drop-down list in the **Duct Type** field and enter a vertical offset value in the **Offset** field. Then, select the **Branch** option in the left pane to view its settings in the right pane, as illustrated in **Figure 5-9**.

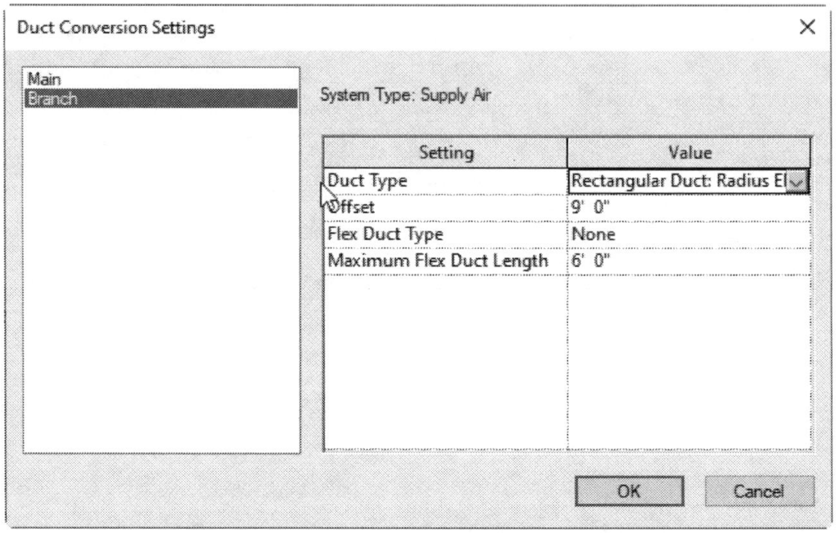

Figure 5-9 *The options for branch ducts in the* ***Duct Conversion Settings*** *dialog box*

In the **Duct Conversion Settings** dialog box, specify the **Duct Type** for the branch in the right pane, then set values for **Offset**, **Flex Duct Type**, and **Maximum Flex Duct Length**. Click **OK** to close the dialog. In the Options Bar, choose a **Solution Type** like **Network** to display a default layout, as shown in **Figure 5-10**. Use the **Next Solution** button to preview alternatives. Finally, in the **Generate Layout** tab, click **Finish Layout** to create the layout, shown in 3D in **Figure 5-11**.

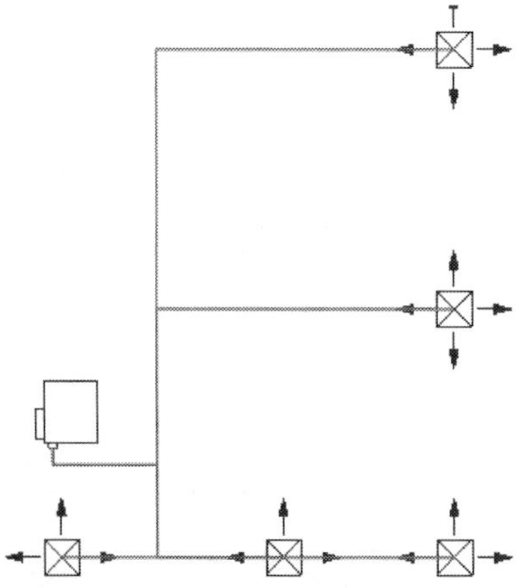

Figure 5-10 *Floor plan displaying a network solution of the HVAC system layout*

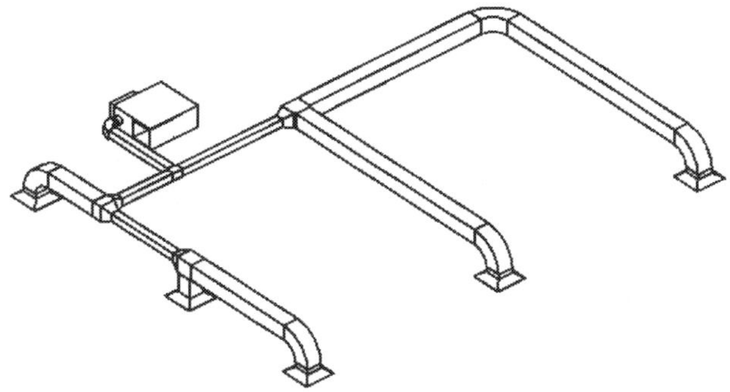

Figure 5-11 *Three-dimensional view of an HVAC system layout*

CREATING DUCTS AND DUCT FITTINGS

In an HVAC system, the ducts are used to supply and remove air from a space. The airflow inside the ducts includes supply air, return air, and exhaust air.

In Revit, the ducts are created using the automatic connection snaps. When you place the duct segments in a project, Revit automatically connects two segments of the ducts with appropriate duct fittings. This method saves your time as there will be no need to join or trim elements.

Creating Ducts

To create a duct segment in a drawing, open the desired view from the **Project Browser** and then choose the **Duct** tool from the **HVAC** panel in the **Systems** tab; the **Modify | Place Duct** contextual tab will be displayed, as shown in Figure 5-12.

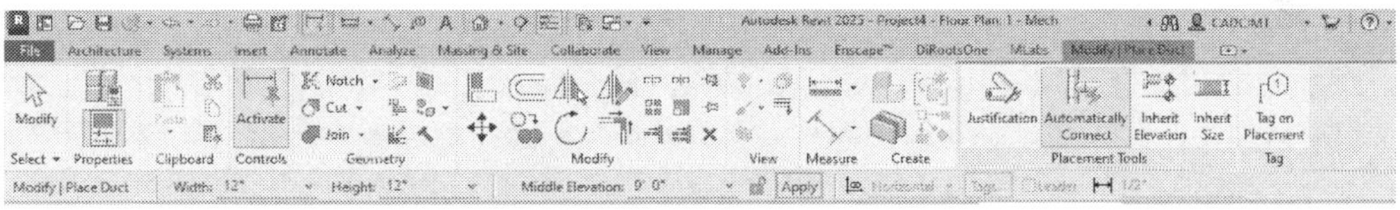

Figure 5-12 *Different options in the **Modify / Place Duct** contextual tab*

In the **Placement Tools** panel, the **Automatically Connect** button is enabled by default, allowing ducts to snap to connectors automatically. Use the **Inherit Elevation** button to match the elevation and size of the connected component. In the **Tag** panel, click **Tag on Placement** to tag ducts during placement. In the **Options Bar**, you can specify **Width**, **Height** (default: 12" or 300 mm), and **Offset** (default: 9' or 2750 mm). The **Lock/Unlock Specified Elevation** button locks the offset, and the **Apply** button saves the settings. In the **Properties** palette (see **Figure 5-13**), select the duct type from the **Type Selector**. Under **Constraints**, set **Horizontal** and **Vertical Justification** (default: Center), **Reference Level**, and **Middle Elevation** to define duct placement and alignment.

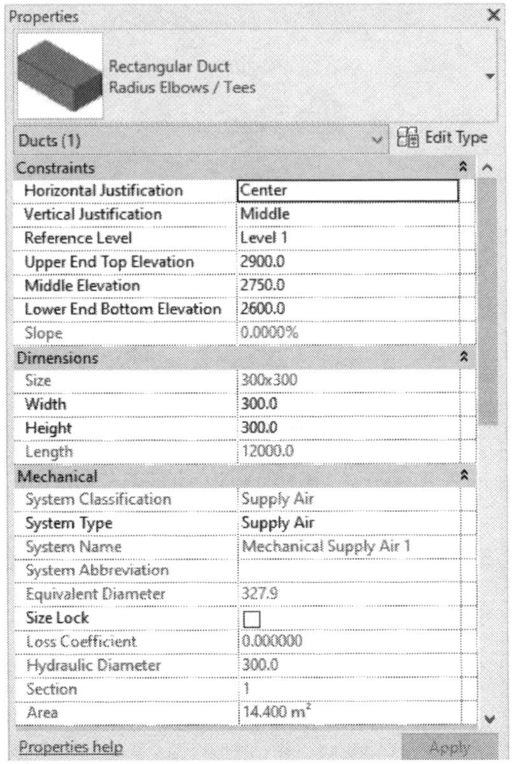

Figure 5-13 *Different instance properties of the duct*

Under the **Mechanical** head of the **Properties** palette, click in the value field corresponding to the **System Type** parameter and select an option from the drop-down list to specify the type of system specified for the duct. By default, the **Supply Air** option is selected. In the drop-down list, you can also select the **Exhaust Air** or **Return Air** option. Under the **Mechanical** head, you can select the check box corresponding to the **Size Lock** parameter to lock the size of the duct for editing.

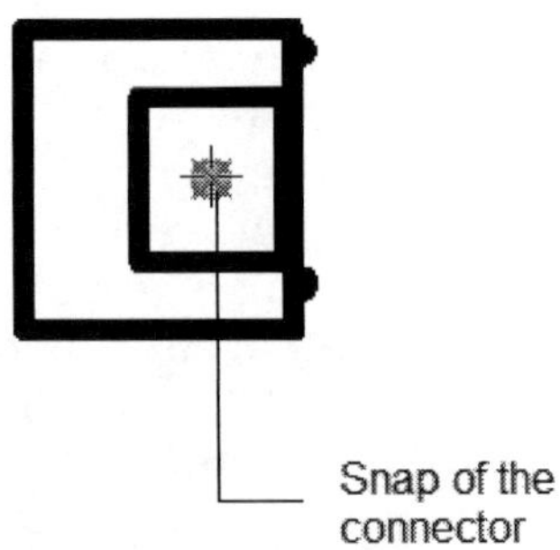

Figure 5-14 *The appearence of the snap point at the connector*

After specifying various options in the **Properties** palette, click in the drawing area to specify the start point of the duct. You can also snap to the connector point of an existing equipment to automatically connect to it. To do so, place the cursor at the connector point of the equipment; a snap will appear, as shown in Figure 5-14. Now, click at the snapped point and drag the cursor toward the desired direction; a preview of the duct with specified cross-section and justification will appear along with the cursor, as shown in Figure 5-14. Now, click ox`n the desired location to specify the end point of the duct segment. On clicking at the desired location, a duct segment will be created. You can now click on more locations to define other duct segments. After defining the duct segments, choose the **Modify** button from the **Select** panel to exit the tool. Figures 5-15 and 5-16 show the elevation and three dimensional views, respectively, of a ductwork in an HVAC system.

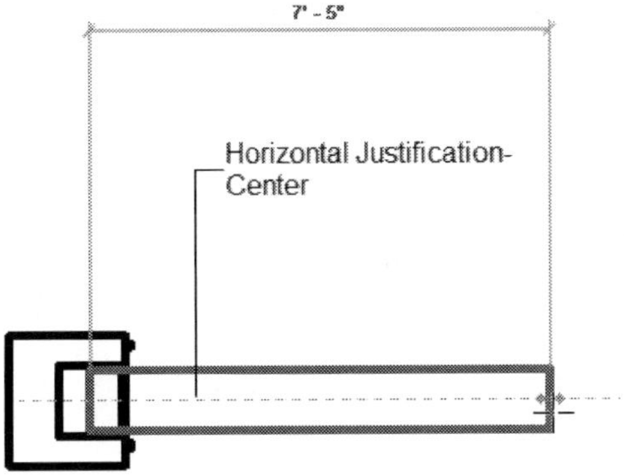

Figure 5-15 *The appearance of a duct segment along with the cursor*

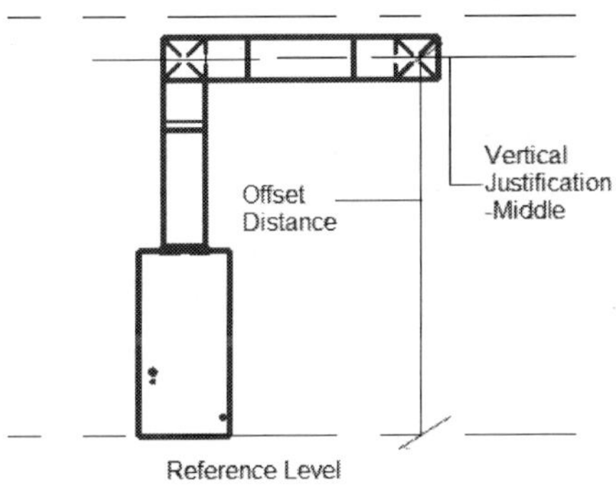

Figure 5-16 *The elevation view of a duct system*

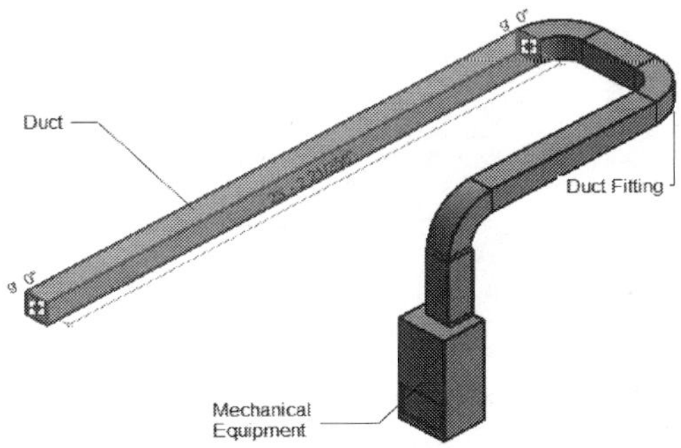

Figure 5-17 *Three dimensional view of a duct system*

FABRICATION DETAILS IN REVIT

In Autodesk Revit, you can create a model with fabrication parts reflecting the fabrication intent of the model. This will enable you to share a more detailed model with the contractors and thereby produce more accurate estimates and bids. In addition, you can quickly generate fabrication-level models, and move directly to fabrication retaining all the information in the model required in BIM.

The creation of fabrication parts can have different workflows. You can place a fabrication part and then detail it or you can convert a Revit model with generic duct, pipe, or electrical containment parts into fabrication parts.

For any of the workflows, you need to first specify a fabrication configuration. A fabrication configuration consists of fabrication content and associated data of the products to support coordinated modeling and detailing for fabrication and construction. The configuration specified must be saved in a specified network or on a local drive. Every Revit model will have a unique configuration of fabrication. Alternatively, you can specify a fabrication profile which is a collection of services specific to a model. If you select the Global profile, all services in the specified configuration are made available.

After configuring the fabrication, you can place Fabrication Parts in a Revit model or convert Generic Parts into Fabrication Parts. In the sections ahead, you will learn to configure the fabrication and add load services in the project, place fabrication parts in the model, and convert generic parts into fabrication parts.

Configuring the Fabrication and Loading Services

To configure fabrication in Revit, you need to invoke the **Fabrication Settings** dialog box. To open the **Fabrication Settings** dialog box, choose the **Fabrication Settings** button from the **Fabrication** panel of the **Systems** tab, refer to Figure 5-18. Alternatively, you can press FS to display the **Fabrication Settings** dialog box shown in Figure 5-19. In this dialog box, the following tabs are displayed: **Fabrication configuration** and **Connection Indicators**. The **Fabrication configuration** tab is chosen by default. The options in the **Fabrication configuration** tab help in specifying a fabrication configuration and profile, and load fabrication services into the model. The options in the **Connection Indicators** tab help to specify connection indicators for the **Route and Fill** tool.

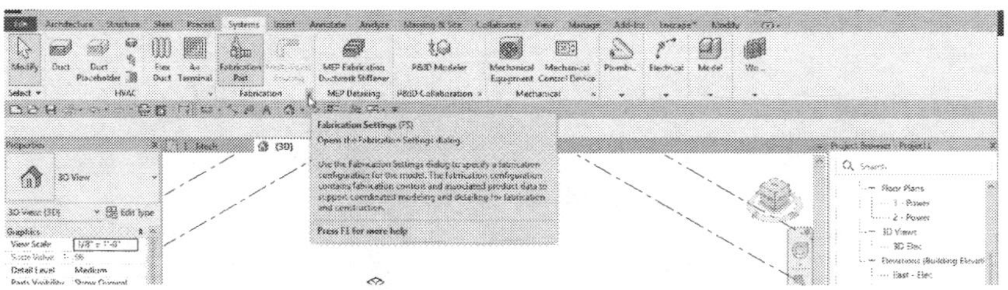

Figure 5-18 *Choosing the **Fabrication Settings** button from the **Fabrication** panel*

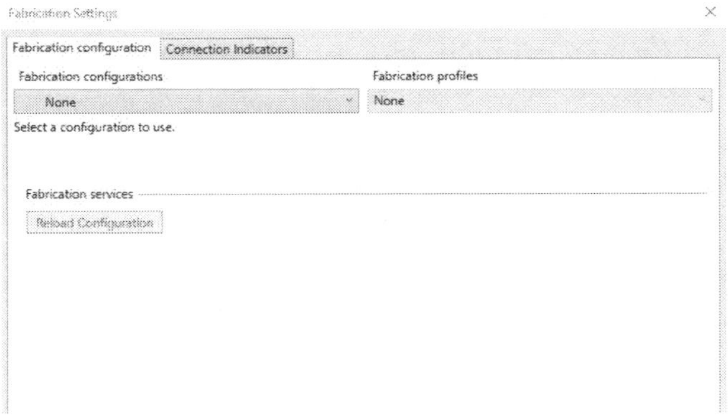

*Figure 5-19 The partial view of the **Fabrication Settings** dialog box*

Fabrication configuration Tab

In the **Fabrication configuration** tab of the **Fabrication Settings** dialog box, you can select a configuration to be used for the fabrication from the **Fabrication configurations** drop-down list. In this drop-down list, the **None** option is selected by default. You can select the **Revit MEP Imperial Content V2.2** or **Revit MEP Metric Content V2.2** option for the Imperial or Metric unit system, respectively, from the drop-down list. After selecting the fabrication configuration, select an option from the **Fabrication profiles** drop-down list to specify the profile for filtering the list of services and additional parts available in the project for the specified configuration. By default, the **Global** option is selected in this drop-down list. As a result, all the services and additional parts of the specified configuration will be available in the **Services** and **Parts** tabs. In the **Services** tab, you can select the desired options for service(s) from the **Unloaded services** area and then choose the **Add** button to add them to the **Loaded services** area. The services that are loaded now can be used in the fabrication detailing. Similarly, in the **Parts** tab, you can select the desired parts from the **Unloaded additional parts** area and then choose the **Add** button to add them to the **Loaded additional parts** area, refer to Figure 5-20.

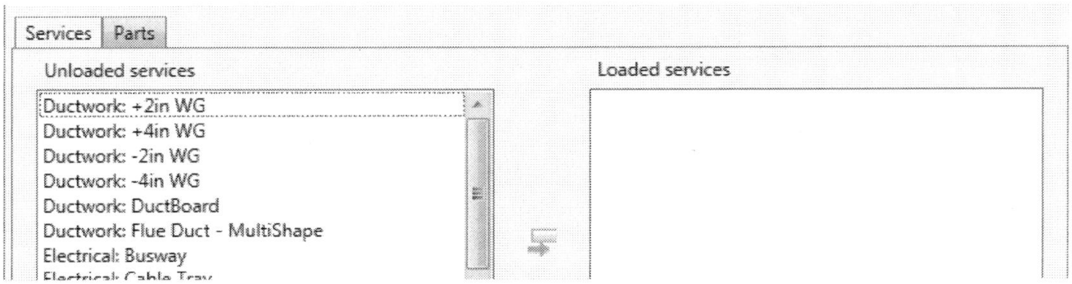

*Figure 5-20 The options in the **Services** tab*

Connection Indicators Tab

In the **Colors** area of the **Connection Indicators** tab, various options are available to change the color of the connection arrow in a fabrication part displayed in the plan and 3d views. You can use the **Plan** option to change the color of the arrows displayed in the plan view. Similarly, you can use the **Towards** and **Away** option to change the color of the arrows displayed in the fabrication part in a 3d view. You can refer to Figure 5-21 for different arrows in the fabrication parts.

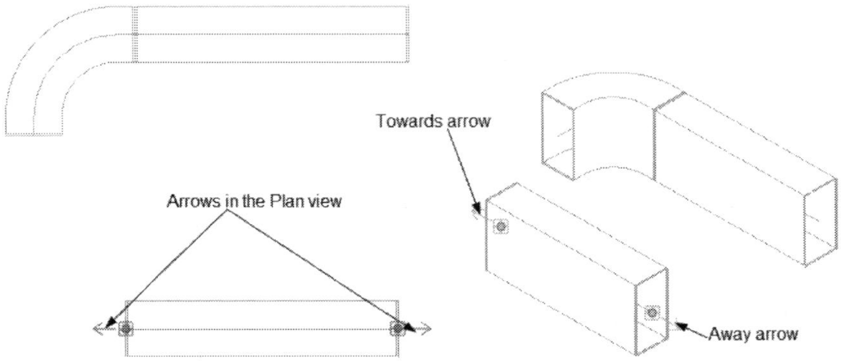

Figure 5-21 *Display of different arrows in the fabrication parts*

Placing the Fabrication Parts

After configuring the Fabrication and loading the services, you can place the fabrication parts in the model. To place the fabrication parts in a model, choose the **Fabrication Part** tool from the **Fabrication** panel of the **Systems** tab; the **MEP Fabrication Parts** palette will be displayed in the screen, as shown in Figure 5-31. Alternatively, you can invoke the **Fabrication Part** tool by pressing PB. In the **Fabrication Parts** palette, select the desired service from the **Service** drop-down list. Note that the services that are loaded in the **Fabrication Settings** dialog box will only be displayed in the **Service** drop-down list. After selecting the desired option from the **Service** drop-down list, select the group to which the part will belong from the **Group** drop-down list. On doing so, the list of services belonging to the specified group will be displayed in the display area of the **Services** tab of the palette. Now, choose a particular service from the display area and add it in the project to place the fabrication.

TUTORIAL

General instructions for downloading tutorial files:

1. Download the *c05_rmp_2025_tut.zip* file for this tutorial from *https://www.cadsofttech.com*. The path of the file is as follows: *Books > Civil/GIS BOOKS > Revit MEP > Revit MEP 2025 For Novices.*

2. Now, save and extract the downloaded zip file at the following location: *C:\rmp_2025\ c05_rmp_2025_tut*

Note
The default unit system used in the tutorials is Imperial.

Tutorial 1 Placing Air Terminals

In this tutorial, you will place air terminals in the ceiling of the rooms. Also, you will create new views, modify air terminal parameters, and place air terminals in the ceiling plan.

(Expected time: 45 min)

1. File to be used: *c05_Office-Space_tut1.rvt* (*M_c05_Office-Space_tut1.rvt*)
2. File name to be assigned: *c05_Office-Space_tut1a.rvt* (*M_c05_Office-Space_tut1a.rvt*)

The following steps are required to complete this tutorial:

a. Open the *05_Office-Space_tut1.rvt* (*M_05_Office-Space_tut1.rvt*) file.
b. Modify the ceiling plan.
c. Add the supply air terminal.
d. Add the return and exhaust air terminal.
e. Save the project by using the **Save As** tool.
f. Close the project using the **Close** tool.

Opening the Project

In this section, you will open the *c05_Office-Space_tut1.rvt* (*M_ c05_Office-Space_tut1.rvt)* file.

1. Choose **Open > Project** from the **File** menu; the **Open** dialog box is displayed.

2. In the **Open** dialog box, browse to *C:\rmp_2025\c05_rmp_2025_tut* and select the *c05_Office-Space_tut1.rvt.* (*M_c05_Office-Space_tut1.rvt*) file. Now, choose the **Open** button; the selected file opens in the Revit.

Note

The architectural model named c04_archi_spaces_rmp_2025.rvt (for Metric, M_c04_archi_spaces_rmp_2025.rvt) linked in this tutorial file is located in the c04_rmp_2025_tut folder.

*While linking a model to the project in Revit, it is recommended to maintain the relative path as the path type for the link. However, if the link is lost, you need to reload the file using the **Manage Links** dialog box which has already been discussed in the previous chapters.*

Modifying the Ceiling Plan View

In this section, you will modify the ceiling plan view using the **View Range** dialog box.

1 In the **Project Browser**, expand **Views (Discipline) > Mechanical > HVAC > Ceiling Plans**, and double-click on **1 - Ceiling Mech** to make it an active view.

2. In the **Properties** palette, choose the **Edit** button corresponding to the **View Range** parameter under the **Extents** head; the **View Range** dialog box is displayed.

3. In the **Primary Range** area of this dialog box, select the **Associated Level (Level1)** option from the **Top** drop-down list and then enter **8' 4"** (**2540mm**) in the **Offset** edit box.

4. Enter **0** in the **Offset** edit box corresponding to the **Cut plane** drop-down list.

5. In the **View Depth** area of the **View Range** dialog box, select the **Associated Level**

(Level 1) option from the **Level** drop-down list and enter **8' 4"** (**2540mm**) in the **Offset** edit box corresponding to the **Level** drop-down list.

6. Choose **Apply** and then **OK**; the specified settings are applied to the current view and the **View Range** dialog box is closed.

Note

On modifying the ranges of the ceiling plan view, the fixtures and fittings that will be placed in the model will be visible.

Adding the Supply Air Terminal

In this section, you will place the supply air terminals in the ceiling of the rooms.

1. Choose the **Air Terminal** tool from the **HVAC** panel of the **Systems** tab; the **Modify | Place Air Terminal** contextual tab is displayed.

2. Choose the **Load Family** tool from the **Mode** panel of the contextual tab; the **Load Family** dialog box is displayed.

3. In the dialog box, browse to **English-Imperial > Mechanical > MEP > Air Side Components > Air Terminals** folder (for Metric, browse to **English > Mechanical > MEP > Air Side Components > Air Terminals** folder) and then select the **Supply Diffuser - Hosted (M_Supply Diffuser-Hosted)** family from the list displayed.

4. Next, choose **Open**; the **Load Family** dialog box is closed and the selected family is loaded in the project file.

5. In the **Properties** palette, select the **Supply Diffuser - Hosted: Workplane-based Supply Diffuser** option (for Metric **M_SupplyDiffuser - Hosted: Workplane-based Supply Diffuser**) from the **Type Selector** drop-down list.

6. In the **Placement** panel of the **Modify | Place Air Terminal** contextual tab, choose the **Place on Face** button.

7. In the drawing area, move the cursor toward the upper left corner of the plan into the **CEO-Office** area and click to place at the location shown in Figure 5-22.

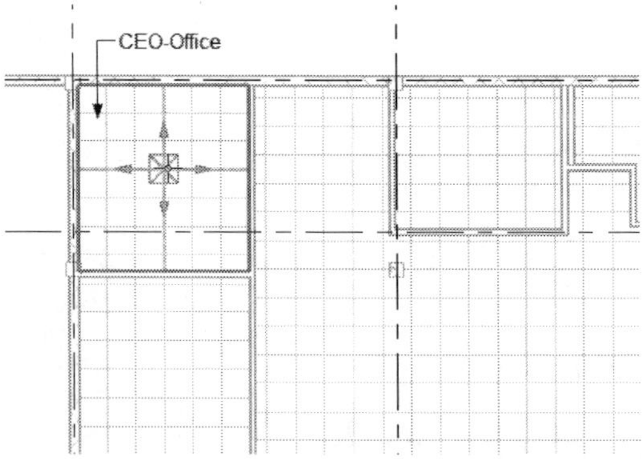

Figure 5-22 *The supply air terminal placed in the CEO-Office area*

Note
The air terminal will be placed on the ceiling.

8. In the **Properties** palette, enter **300 CFM** (**150 L/S**) in the edit box corresponding to the **Flow** parameter under the **Mechanical - Flow** head.

9. Now, choose the **Modify** button from the **Select** panel to exit the **Air Terminal** tool.

10. Select the added air terminal and place the cursor at the center grip of the air terminal; the center grip turns red and the **Drag** tooltip appears. Press the left mouse button and drag the air terminal to fit it inside the ceiling tile, as shown in Figure 5-23.

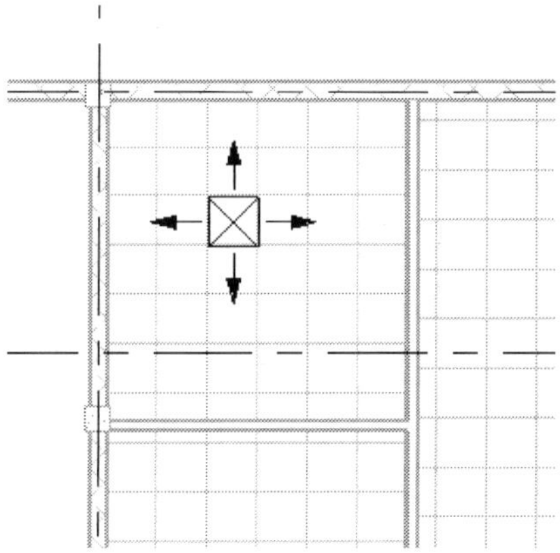

Figure 5-23 *Placing the supply air terminal inside the ceiling tile*

11. After placing the air terminal, release the left mouse button.

12. Next, ensure that the air terminal is selected and then choose the **Copy** tool from the **Modify** panel; a dashed box is displayed around the selected air terminal.

13. In the **Options Bar**, select the **Multiple** check box.

14. In the drawing area, place the cursor at the center of the air terminal and click when the **Midpoint** snap is displayed, as shown in Figure 5-24.

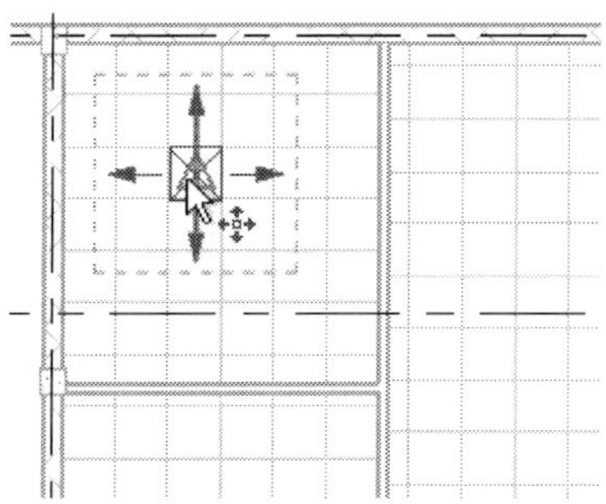

Figure 5-24 *Placing the cursor at the center of the air terminal*

15. Move the cursor vertically down and click when the vertical dimension appears as **12' (3658mm)** and the angle shows **90.000°**, as shown in Figure 5-25. Similarly, copy and place the other air terminals at required locations, as shown in Figure 5-26.

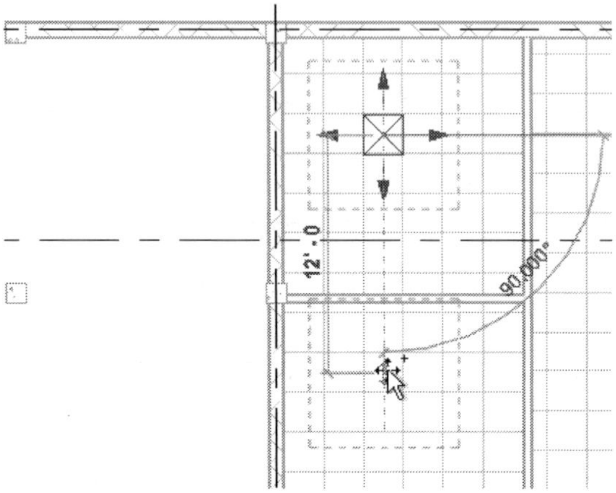

Figure 5-25 *Copying the air terminal to the other area*

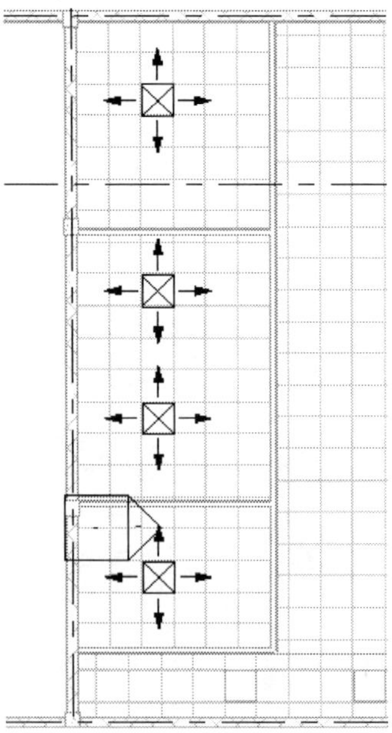

Figure 5-26 *Placing the air terminals in other areas*

Note

The placement of the air terminals may not be exact.

16. Choose the **Modify** tool from the **Select** panel to exit the tool.

Adding the Return and Exhaust Air Terminal

In this section, you will add the return and exhaust air terminal to the architectural layout.

1. Choose the **Air Terminal** tool from the **HVAC** panel of the **Systems** tab; the **Modify | Place Air Terminal** contextual tab is displayed.

2. Choose the **Load Family** tool from the **Mode** panel of the contextual tab; the **Load Family** dialog box is displayed.

3. In the dialog box, browse to **English-Imperial > Mechanical > MEP > Air Side Components > Air Terminals** folder (for Metric, browse to **English > Mechanical > MEP > Air Side Components > Air Terminals** folder) and then using the CTRL key, select the **Exhaust Diffuser - Hosted (M_Exhaust Diffuser-Hosted)** and **Return Diffuser - Hosted** families **(M_Return Diffuser-Hosted)** from the list displayed.

4. Choose **Open**; the selected families are loaded in the file and the **Load Family** dialog box is closed.

5. In the **Properties** palette, select the **Return Diffuser - Hosted: Workplane-based Return Diffuser** option (for Metric, select the **M_ Return Diffuser - Hosted: Workplane-based Return Diffuser** option) from the **Type Selector** drop-down list.

6. In the **Properties** palette, click in the value field corresponding to the **Flow** parameter and enter **300 CFM** (for Metric, **150 L/S**).

7. Choose the **Place on Face** button from the **Placement** panel in the **Modify|Place Air Terminal** contextual tab.

8. Next, in the drawing area, move the cursor in the Lounge area next to the CEO-Office room and then click inside the ceiling tile, as shown in Figure 5-27.

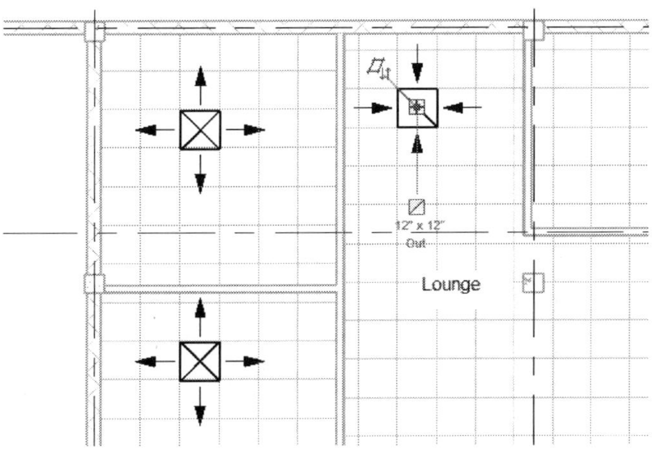

Figure 5-27 *Placing the return air terminal in the Lounge area*

9. Now, click inside the ceiling tile in the **Lounge** area to place the other return air terminals, as shown in Figure 5-28.

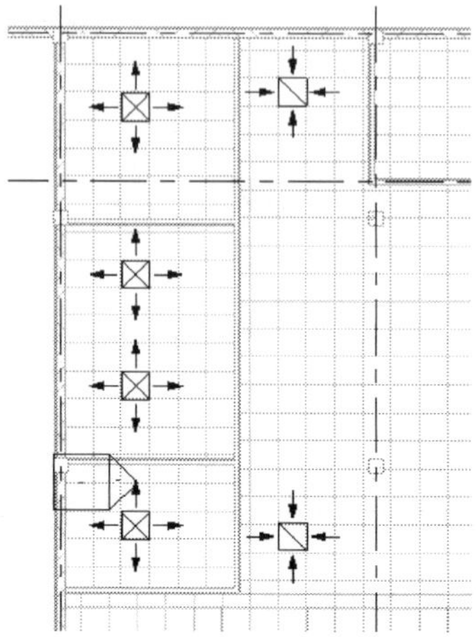

Figure 5-28 *Placing the other return air terminal in the Lounge area*

10. Next, in the **Properties** palette, select the **Exhaust Diffuser - Hosted: Workplane-based Exhaust Diffuser** type (for Metric, select the **M_Exhaust Diffuser - Hosted: Workplane-based Exhaust Diffuser**) from the **Type Selector** drop-down list.

11. In the **Properties** palette, click in the value field corresponding to the **Flow** parameter and enter **250 CFM** (for Metric, **120 L/S**).

12. In the drawing area, move the cursor to the Toilet 1 area, and click inside the ceiling tile at the location, as shown in Figure 5-29, to place the exhaust air terminal.

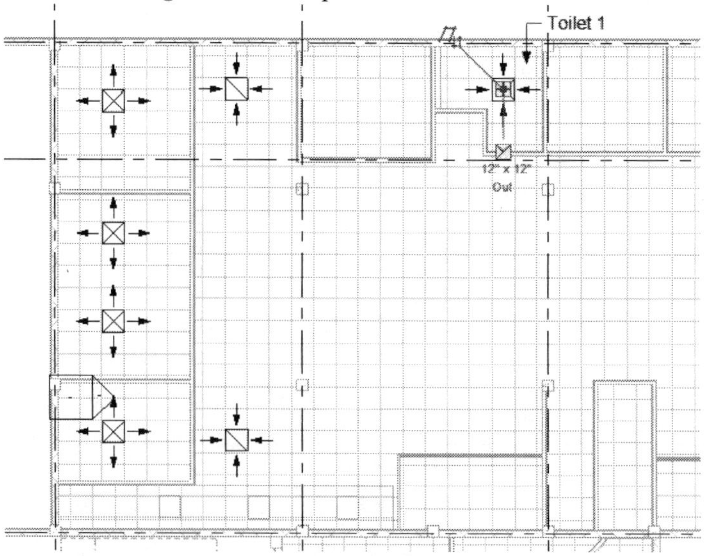

Figure 5-29 *Placing the exhaust air terminal in the Toilet 1 area*

13. Choose the **Modify** button from the **Select** panel to exit the tool.

Note

You can add more air terminals to complete the requirement of the HVAC system in the Office-Space area.

After you place the return and exhaust air terminals, remember to modify the airflow display arrows for air terminals that requires 2-way and 3-way blow patterns by setting the **UpArrow**, **RightArrow**, **LeftArrow**, and **DownArrow** parameters from the **Properties** palette

Saving and Closing the Project

In this section, you need to save the project and the settings using the **Save As** tool.

1. Choose **Save As > Project** from **File** menu. As you are saving the project for the first time, the **Save As** dialog box is displayed.

2. In this dialog box, browse to *C:|rmp_2025|c05_rmp_2025_tut* and in the **File name** edit box, enter **c05_Office-Space_tut1a** (**M_c05_Office-Space_tut1a**) and then choose the **Save** button to save the current project file with the specified name and to close the **Save As** dialog box.

3. Choose the **Close** option from the **File** menu.

EXERCISE

Exercise 1 HVAC System

Download the *c05_Conference-Center_exer1.rvt* (for Metric, *M_c05_Conference-Center_exer1.rvt)* file from *https://www.cadsofttech.com*. The path of the file is as follows: *Books > Civil/GIS BOOKS > Revit MEP > Revit MEP 2025 For Novices*Open the *c05_Conference-Center_exer1.rvt* file and create an HVAC system in the conference room for the *Conference-Center* project. While creating the HVAC system, you will add air terminals to the spaces, as shown in Figure 5-30. You will also create the duct system for the air terminals, as shown in Figure 5-31.

(Note : 1 CFM = 0.47 L/s)

1. Project view to be used :
   Floor Plans > Mechanical > 1ST FLOOR- HVAC DUCT

2. Family type to be used:
   Air Terminals
   For Imperial Rectangular Diffuser- Round Connection : 24*24 -10 Neck
   For Metric M_Supply Diffuser - Rectangular Face Round Neck : 600*600 - 250 Neck
   Duct System
   For Imperial Rectangular Duct : Galvanized
   For Metric Rectangular Duct : Galvanized

3. File name to be assigned: *c05_Conference-Center_exer1a.rvt*

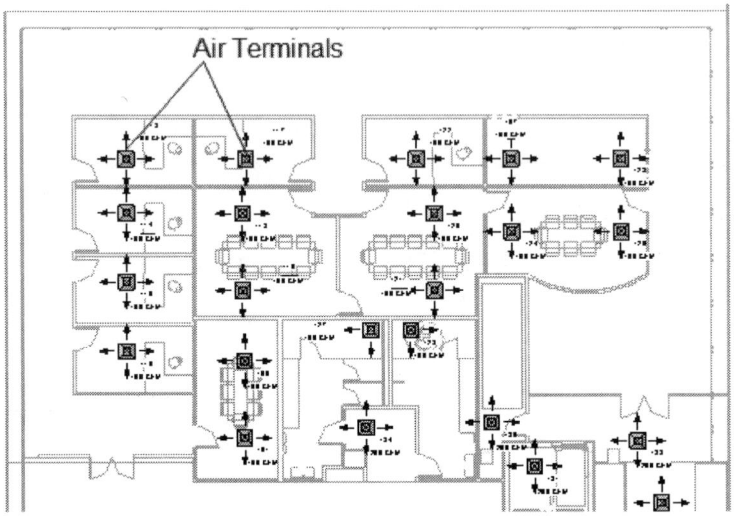

Figure 5-30 *Air Terminals added in the space*

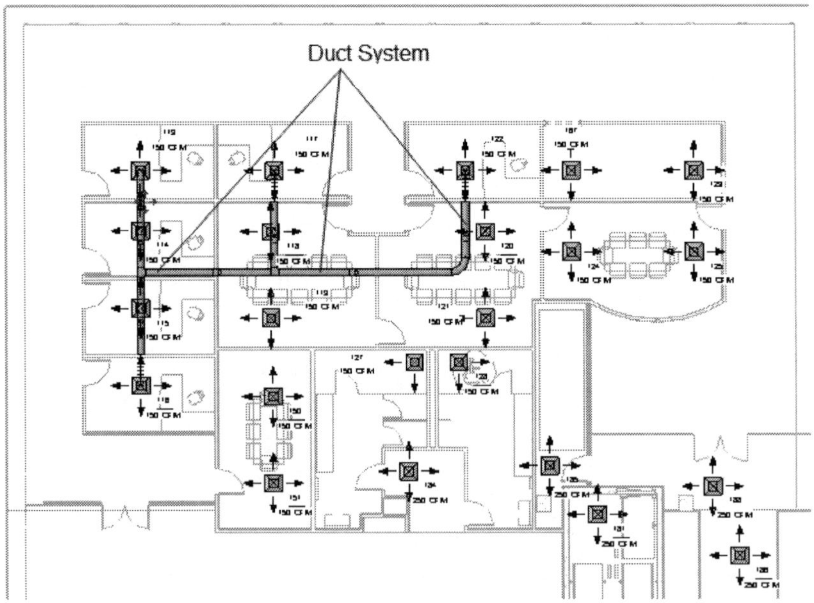

Figure 5-31 *Duct system created for the air terminals*

Chapter 6

Creating an Electrical System

Learning Objectives

After completing this chapter, you will be able to:

- *Add electrical equipments*
- *Add power and system devices*
- *Add lighting fixtures*
- *Specify electrical settings*
- *Create power distribution systems*
- *Perform lighting analysis*
- *Add electrical circuits and wires*

INTRODUCTION

In this chapter, you will learn about various procedures and tools available in Revit (MEP) to create an electrical system in a project. You will learn to add various power and system devices and lighting fixtures. You will also learn to modify various electrical settings, add conduits, cable trays, and specify the Power Distribution System for a project.

ADDING ELECTRICAL EQUIPMENT

In an MEP project, the electrical equipment consists of panels and transformers. This equipment plays an important role in the functioning of the building. The process of adding electrical equipment in a building is discussed next.

Adding Transformers

A transformer is an electrical equipment that is used to transform power from one circuit to another without changing the frequency of the input current. A transformer transfers electrical power between circuits without changing frequency, typically stepping voltage down for safe building distribution. In Revit MEP, transformers can be modeled in various sizes and voltage ratings, placed on floors or walls, and shown as symbols or physical models. Both wet and dry types are supported. **Figure 6-1** shows a dry-type transformer with specifications like 240x480 V primary voltage and NEMA 3R enclosure. To add a transformer, load the desired family using **Load Family** from the **Insert** tab, navigating to the appropriate folder (Imperial or Metric). Select from options like **Dry Type Transformer - NEMA Type 2 or 3R**, or **Wet Type Transformer - 12000-480Y277**. After loading, open a power floor plan, choose **Electrical Equipment** from the **Systems** tab, and select the transformer type in the **Properties** palette. You can then adjust its instance and type properties.

Figure 6-1 *A dry type transformer of 240 x 480 V, NEMA 3R*

Placing Switchboard Components

In an MEP project, a switchboard is an assembly of the following components: Circuit Breaker, Utility, Metering, and Transformer switchboards. Figure 6-2 shows an electrical switchboard with circuit breakers and safety switches.

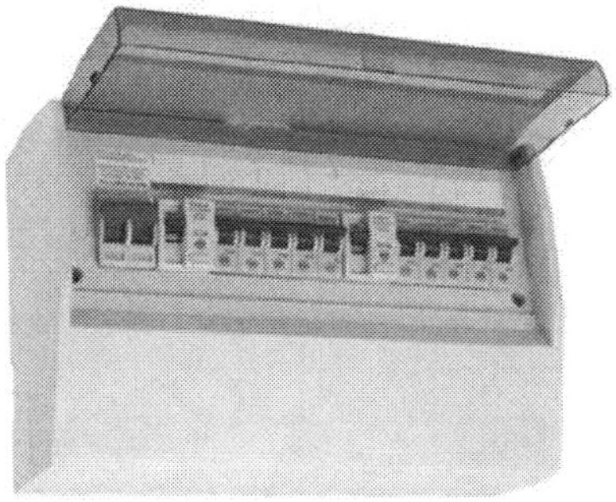

***Figure 6-2** An electrical switchboard with circuit breakers and safety switches*

Before placing switchboard components in a Revit project, load their families using **Load Family** from the **Insert** tab. Navigate to **Electrical > MEP > Electric Power > Distribution** (Imperial or Metric), and select required families like **Circuit Breaker**, **Metering**, **Utility**, or **Transformer Switchboard**. After loading, use the **Electrical Equipment** tool from the **Systems** tab to place them. Ensure **Tag on Placement** is selected to auto-tag components, and use **Rotate after placement** if needed. Select a switchboard type from the **Type Selector**, then adjust instance properties in the **Properties** palette and edit type properties via the **Edit** button, as shown in **Figure 6-3**. Place the switchboards in the drawing area using the cursor preview and rotate with the **SPACEBAR**. Add additional components to complete the switchboard assembly, as seen in **Figure 6-4**. After placement, assign a **Distribution System** in the **Properties** palette.

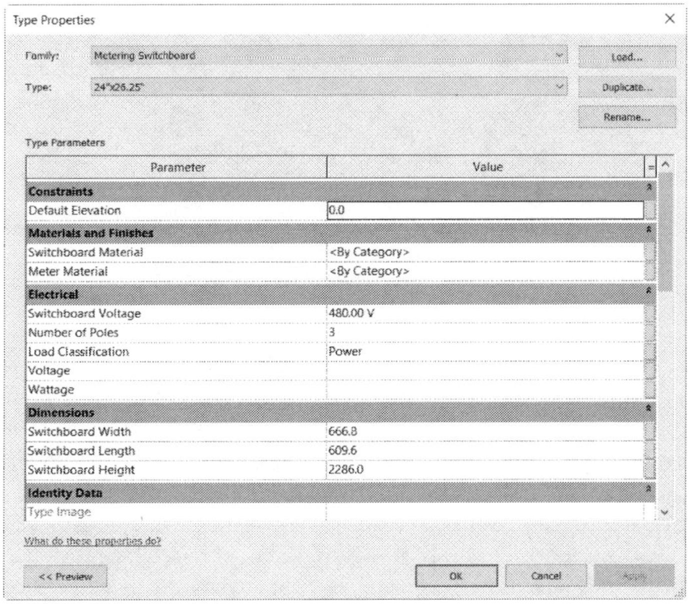

***Figure 6-3** The **Type Properties** dialog box of a Metering Switchboard*

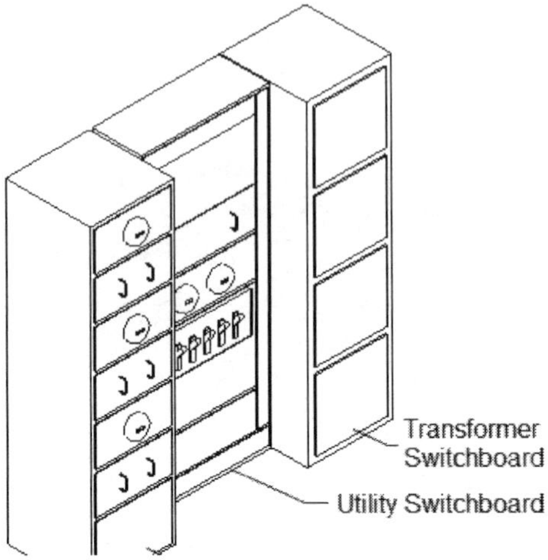

Figure 6-4 *The complete assembly of a switchboard*

Placing the Panel Board or Distribution Board

The panel board, also known as the distribution board, is a key component of an electrical system. It protects the building's electrical connections and distribution system from issues like overloading and short-circuiting. **Figure 6-5** illustrates an electrical panel board used in an office building.

Figure 6-5 *The electrical panel board for an office building*

A panel board, also known by various names such as breaker panel, fusebox, DB board (South Africa), or consumer unit (CU), is commonly referred to as a **panel** in this textbook. It consists of components like fuse links, bus bars, switches, and protective devices. In Revit MEP, to add a panel, use the **Electrical Equipment** tool from the **Systems** tab (Figure 6-1). First, load a panel family via the **Load Family** option in the **Modify | Place Equipment** tab (Figure 6-2). For instance, select **Lighting and Appliance Panelboard - 480V MCB** from the Distribution folder (Figure 6-3). After loading, choose the panel type in the **Properties** palette and adjust its type properties using the **Edit Type** button (Figure 6-4). Set parameters like Default Elevation and Load Classification, accessed through the Load Classifications dialog box (Figure 6-5). Other parameters like Voltage and Wattage can also be configured. Then, set instance parameters such as Offset, Mounting, and Circuit Naming in the Properties palette (Figure 6-6). Panels are typically placed on a vertical face using the default **Place on Vertical Face** tool, with a preview shown before placement (Figure 6-7). Use the SPACEBAR to flip placement, click to place the panel, and select **Modify** to finish.

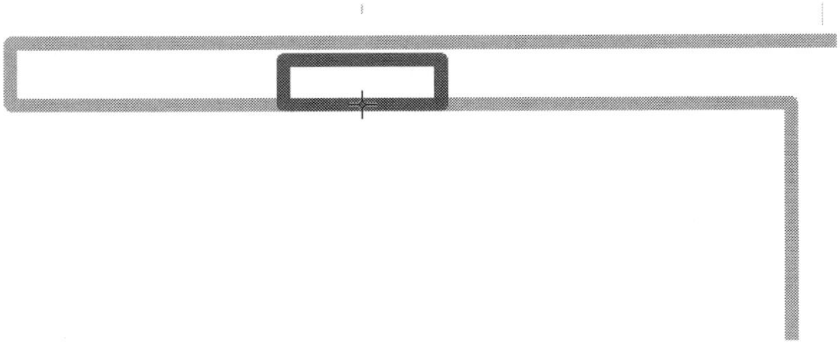

Figure 6-6 *The panel placed at the face of the wall*

ADDING POWER AND SYSTEM DEVICES

Devices consist of receptacles, switches, junction boxes, telephones, communications, and data terminal devices, nurse call devices, wall speakers, starters, smoke detectors, and fire alarm manual pull stations. Electrical devices are often hosted components (receptacles that must be placed on a wall or work plane). Figure 6-7 shows various electrical devices that can be used in an electrical system of a project. The procedure of adding various electrical fixtures to a project is discussed next.

Figure 6-7 *Various electrical devices*

Adding Electrical Fixtures

In an MEP project, to add electrical fixtures for power distribution, open the desired power floor plan (Figure 6-1) and activate the **Electrical Fixture** tool via **Systems > Electrical > Device**. In the **Modify | Place Devices** tab, ensure **Place on Vertical Face** is selected to position fixtures on a host element like a wall. Use the **Tag on Placement** option to attach a tag; when prompted, load the tag from **Annotations > Electrical** (Figure 6-2). In the **Options Bar**, set the alignment (Horizontal or Vertical) and select the appropriate tag type for fixtures using the **Loaded Tags** dialog box (Figure 6-3).

Next, from the **Properties** palette, select a fixture type, such as **Duplex Receptacle: Standard**. To load additional types (e.g., Simplex, Special Purpose, or Quadruplex), use the **Load Family** tool and browse to **Electrical > MEP > Electric Power > Terminals** (Figure 6-4). Loaded types appear in the **Type Selector** drop-down. Instance parameters like **Elevation**, **Offset**, and **Schedule Level** can be edited in the **Properties** palette (Figure 6-5). After adjusting values, place the fixture on the wall face using the SPACEBAR to flip orientation if needed (Figure 6-6). Click to place, then exit the tool using **Modify**. To edit type properties, use the **Edit Type** button in the **Properties** palette and adjust parameters like **Default Elevation**, **Switch Voltage**, and **Load** in the **Type Properties** dialog box (Figure 6-8).

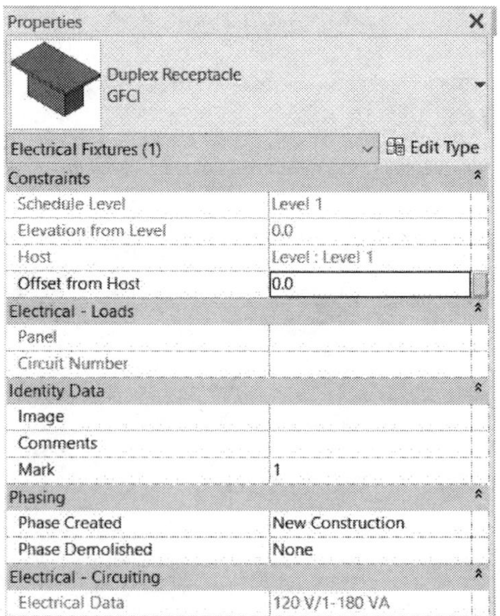

Figure 6-8 *The various instance parameters for the electrical fixture*

Adding Lighting Devices

In an MEP project, to add lighting devices like switches and sensors, activate the desired power floor plan and use the **Lighting tool** from **Systems > Electrical > Device** (Figure 6-1). By default, the **Place on Vertical Face** option is selected, allowing placement on wall faces. Alternatively, choose **Place on Face** or **Place on Work Plane** (Figure 6-2). Use **Tag on Placement** to add tags; if tags aren't loaded, select **Yes** in the **No Tag Loaded** window to open the **Load Family** dialog and load the appropriate tag family from **English-Imperial > Annotations > Electrical** (Figure 6-3). You can specify tag alignment and type using the **Options Bar** and **Loaded Tags** dialog (Figure 6-4). From the **Type Selector** in the **Properties palette**, choose a fixture type like **Lighting Switches: Three Way**. Load additional fixture types via the **Load Family** tool and select them from the **Electric Power > Terminals** folder (Figure 6-5). Instance parameters like **Elevation**, **Switch Voltage**, and **Switch ID** can be edited in the **Properties palette** (Figure 6-6). After selecting and verifying properties, place the fixture by clicking on the desired wall face. Use the **SPACEBAR** to flip its orientation, and **Modify** to exit the tool.

To adjust type properties, select the fixture and choose **Edit** in the **Properties palette** to open the **Type Properties** dialog (Figure 6-7). Modify parameters like **Default Elevation**, **Load Classification**, and **Apparent Load** based on the fixture type. After editing, select **OK** to apply changes.

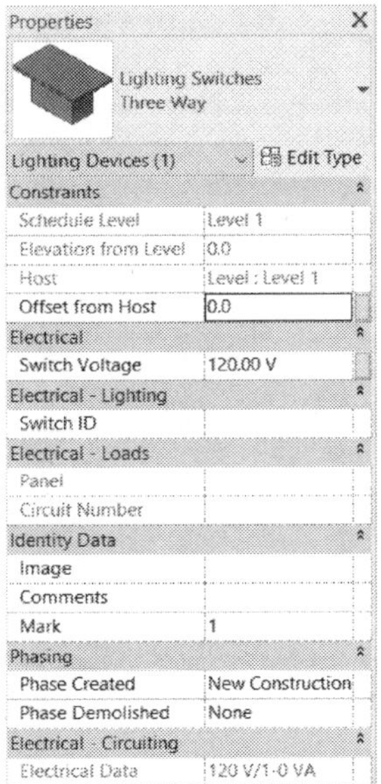

Figure 6-9 *The instance parameters for the lighting device*

Adding Communication Devices

In an MEP project, to add communication devices like Intercom, Substation, and Speakers, you can use the **Communication tool**. First, activate the desired power floor plan from the Project Browser, then go to **Systems > Electrical > Device** and select the Communication tool. The **Modify | Place Communication Device** tab opens, with **Place on Vertical Face** selected by default, allowing placement on wall faces. Alternatively, you can use **Place on Face** or **Place on Work Plane**

To tag the device, click **Tag on Placement**; if no tag is loaded, choose **Yes** to open the **Load Family** dialog, then browse to **English-Imperial > Annotations > Electrical**, select a tag, and load it.

In the **Options Bar**, set tag alignment as **Horizontal** or **Vertical**, then click **Tags** to open the **Loaded Tags** dialog and assign a tag type. In the **Properties palette**, use the **Type Selector** to choose the communication device type. To add more device types, use **Load Family** again and browse to **Electrical > MEP > Information and Communication > Communication**, then load families like **Speaker - Round**, **TV - Square**, etc. Once selected, you can modify instance parameters in the **Properties palette**. Place the fixture by positioning it on the wall, adjusting it with the **SPACEBAR** if needed, and clicking to place. Finish by selecting the **Modify** tool to exit the command.

ADDING LIGHTING FIXTURES

In a project, a **light fixture** (also known as light fitting or luminaire) is an electrical device used to create artificial light using electric lamps. It typically consists of a fixture body, a light socket for holding and replacing the lamp, a switch for control, and a connection to a power source. Fixtures can be fixed or moveable, and may include additional components such as reflectors, lenses, housings, or ballasts. The classification of light fixtures depends on installation, function, and lamp type. **Figure 6-10** illustrates various ceiling lighting fixtures used in projects. In Revit (MEP), lighting fixtures are generally hosted components placed on ceilings or walls. To insert one, select the appropriate view (floor or ceiling plan), load the desired family from the relevant folder, and choose the fixture type. Figure 6-11 shows an example list of internal lighting fixture families available for selection.

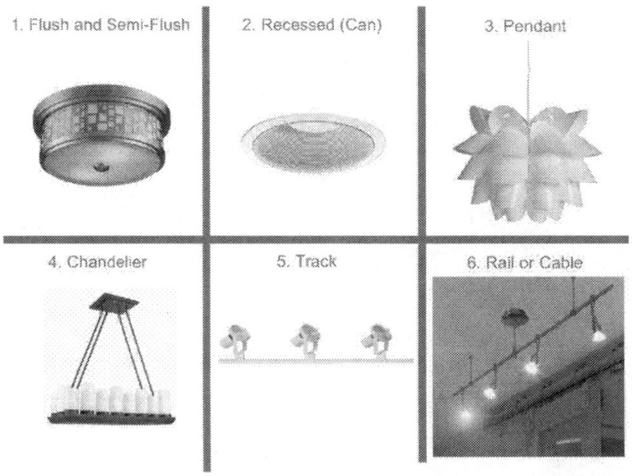

Figure 6-10 *The various types of ceiling lighting fixtures*

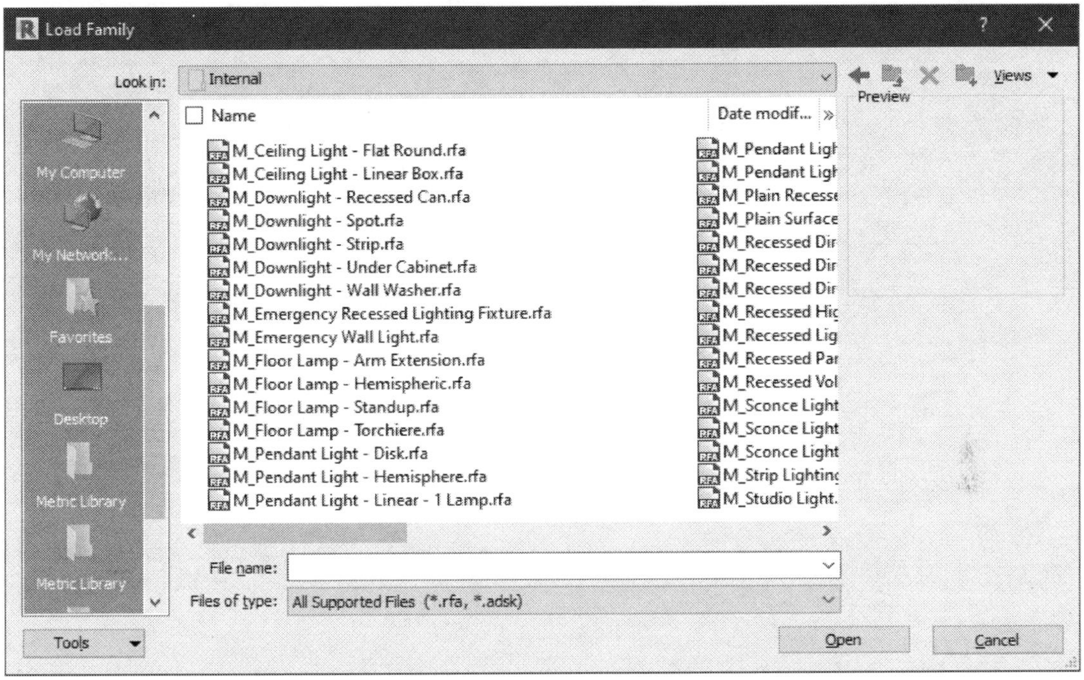

*Figure 6-11 The **Load Family** dialog box displaying the families for internal lighting fixtures*

SPECIFYING THE ELECTRICAL SETTINGS

After placing fixtures, devices, and equipment in the electrical system, you must configure various settings to properly connect these elements and control how wiring information is displayed. In the **Electrical Settings** dialog box (Figure 6-12), accessible from the **Electrical** panel of the **Systems** tab, you can define voltage types, distribution system characteristics, wiring visibility (including tick marks for wire counts), and other project-specific settings. These configurations help ensure correct connections and can be saved as templates for future use.

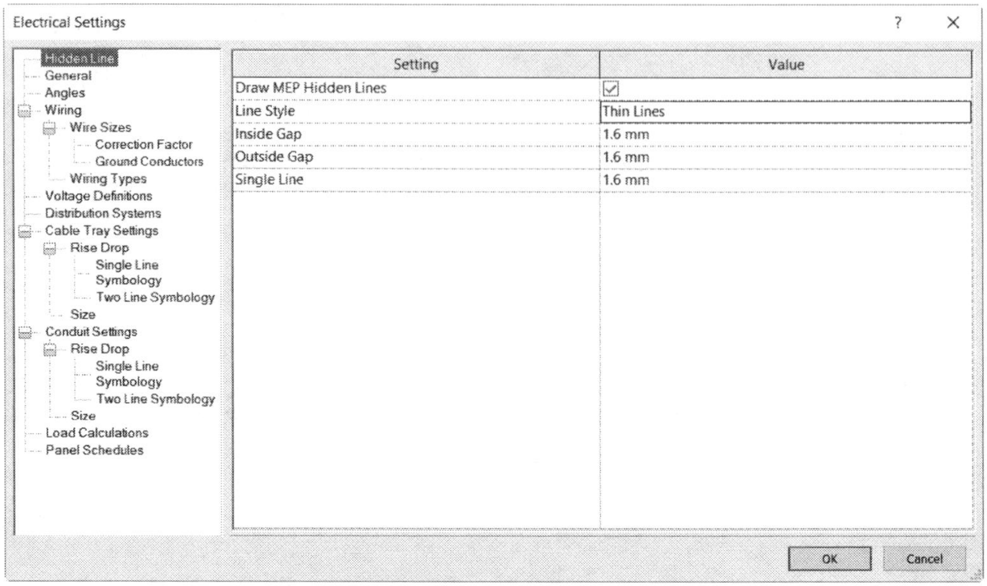

*Figure 6-12 The **Electrical Settings** dialog box*

Setting the Wires

In the **Electrical Settings** dialog box, you can specify various settings of the wire such as its display in the plan view, size in the circuit, and various settings for its correction factor and specifications for its ground conductors. To specify the settings of the wires in the **Electrical Settings** dialog box, click on the **Wiring** option displayed in the left pane of the dialog box; various options for setting the specification of the wiring are displayed in a table in the right pane of the dialog box, as shown in Figure 6-13.

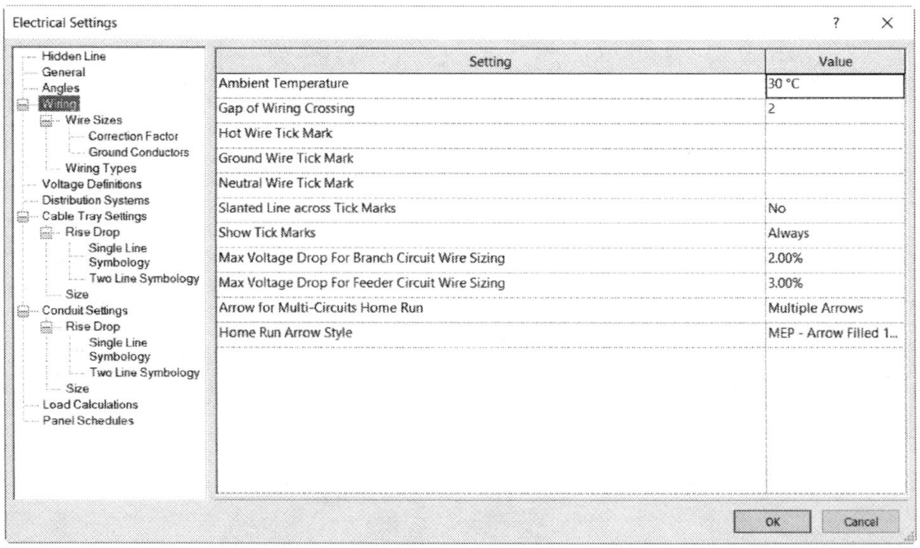

*Figure 6-13 The various options for the **Wiring** option in the **Electrical Settings** dialog box*

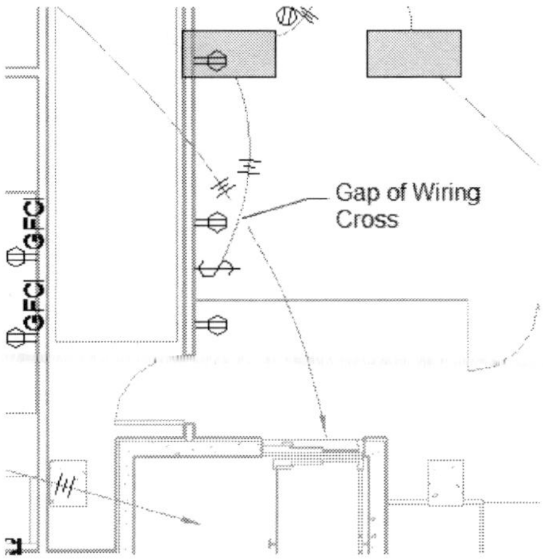

Figure 6-14 *The gap created in wiring cross*

Next, after setting the options for the wirings, you can set the correction factor for the wiring sizes. The correction factor is a factor based on temperature to be used in load calculations. You can set correction factors for the three wire temperature ratings for both aluminum and copper wires. To specify the correction factors for the wires, click **Wiring > Wiring Sizes > Correction Factor** node from the left pane of the **Electrical Settings** dialog box; various options to set the correction factors for the wiring of different sizes are displayed in the right pane, as shown in Figure 6-17.

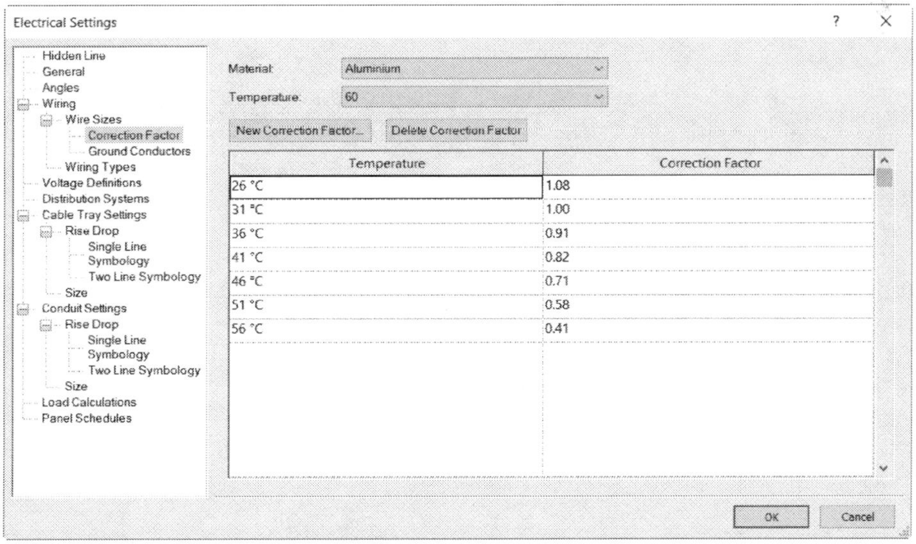

Figure 6-15 *The* ***Electrical Settings*** *dialog box displaying various options for correction factor*

In the right pane of the Electrical Settings dialog box, you can specify the material of the wiring by selecting either **Aluminum** (default) or **Copper** from the **Material** drop-down list to set the appropriate correction factors; once the material is selected, choose a temperature from the **Temperature** drop-down list to view the corresponding correction factor table. To add a new entry, click the **New Correction Factor** button, enter the temperature and correction factor values in the dialog box, and choose **OK** to add them to the table. To delete an entry, select it from the table, click the **Delete Correction Factor** button, and confirm the deletion in the dialog box. After setting the correction factors, navigate to **Wiring > Wire Sizes > Ground Conductors** in the left pane to configure the ground conductor settings, as illustrated in **Figure 6-18**.

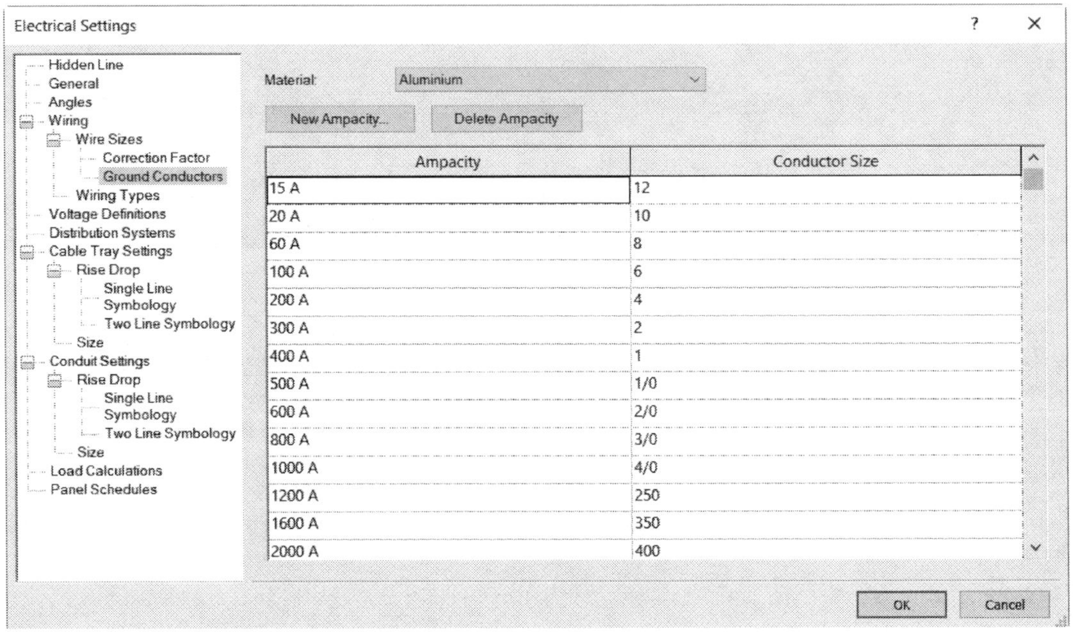

Figure 6-17 *The* ***Electrical Settings*** *dialog box displaying various options for ground conductors*

In the right pane, select either **Aluminum** or **Copper** from the **Material** drop-down list to choose the material for the ground conductor; this displays a table of different ground conductor sizes and their ampacities. To add a new row, click **New Ampacity**, enter the desired **Ampacity** (in amperes) and **Wire Size** (in 1/8" multiples or mm for Metric) in the dialog box (e.g., entering 12 sets the wire size to 12/8" or #12), and then click **OK** to add the values to the table. To delete a row, select it and click **Delete Ampacity**, then confirm in the dialog box. After configuring the ground conductor, navigate to **Wiring > Wiring Types** in the left pane of the **Electrical Settings** dialog box; the right pane will show a table of wire types and their specifications, such as material, temperature rating, insulation, and neutral requirements, as illustrated in **Figure 6-19**.

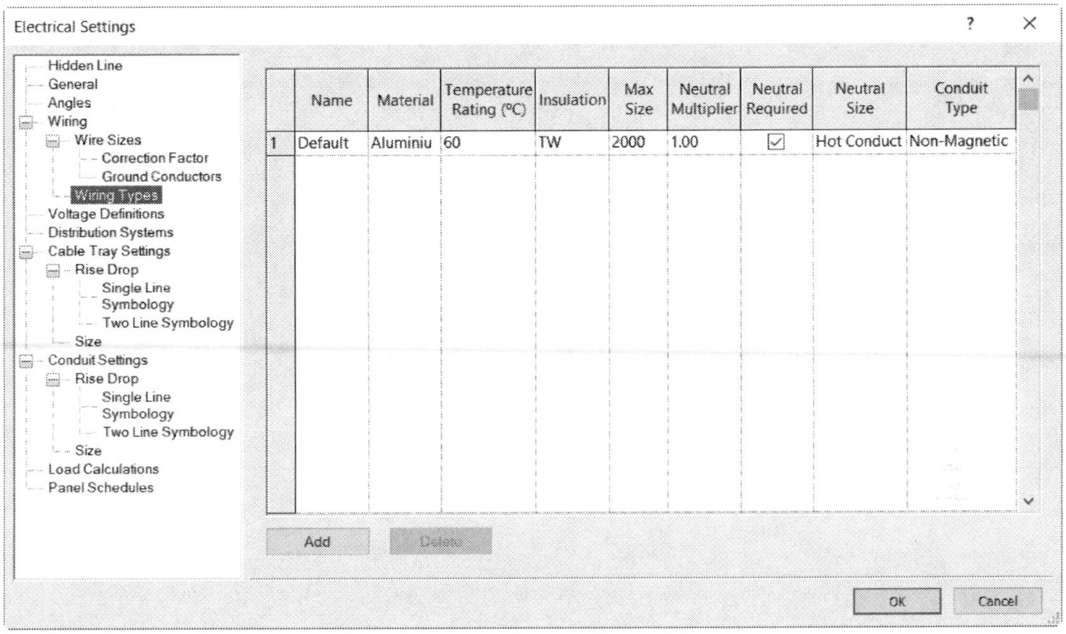

*Figure 6-18 The **Electrical Settings** dialog box displaying various options for wiring types*

To modify the specification of any wire type, click on the corresponding cell in the table (e.g., the **Temperature Rating** column) and select the desired option from the drop-down list to update the setting Other columns can be edited similarly to configure the required settings. To add a new wire type, click the **Add** button at the bottom of the right pane, which inserts a new row with default values that can be customized. To remove a wire type, select its row and click the **Delete** button. After finalizing the wiring specifications, including correction factors, ground conductor settings, and wire types, proceed by selecting the **Voltage Definition** node in the left pane of the **Electrical Settings** dialog box to configure the voltage definition of the electrical system.

Setting the Voltage Definition

The voltage definition in an electrical system refers to the specification of minimum and maximum voltage values used in a project, allowing for the assignment of voltage ratings to various devices or equipment. These definitions help establish different distribution system configurations. By selecting the Voltage Definitions node in the Electrical Settings dialog box, a table appears in the right pane (Figure 6-19), displaying voltage definitions with columns such as Name, Value, Minimum, and Maximum. To add a new voltage definition, click the Add button, which inserts a new row into the table.

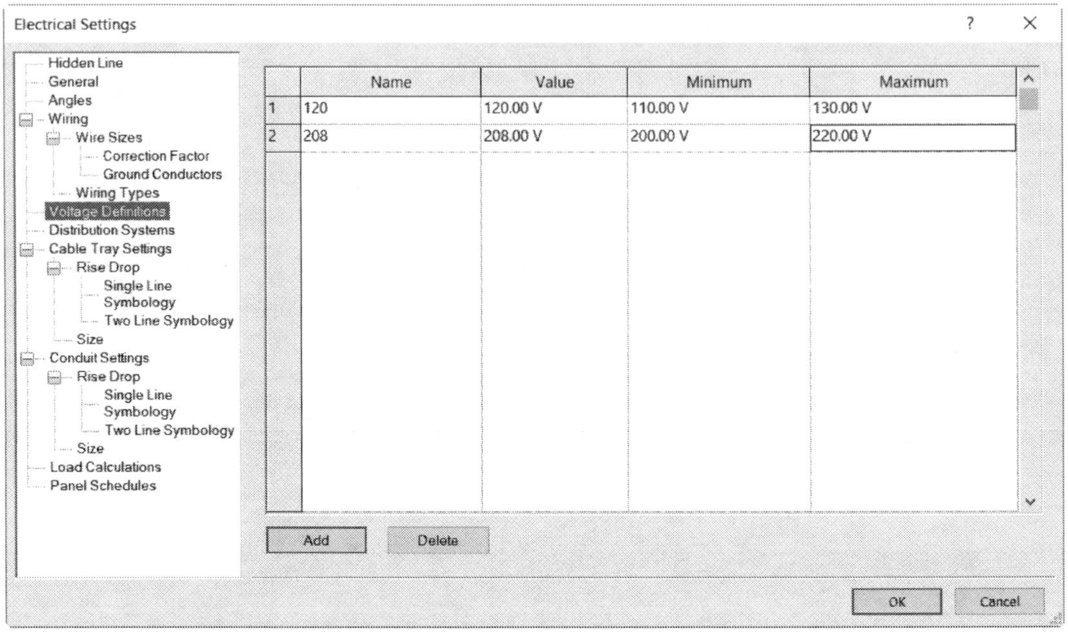

Figure 6-19 *The* ***Electrical Settings*** *dialog box displaying various options for voltage definitions*

You can click any cell in the voltage specification table, such as those in the "Maximum" column, to enter the desired value for a voltage definition. To delete a voltage definition, select the corresponding row and click the "Delete" button to remove it from the table. Once you've defined the voltages for the electrical system, you can proceed to set the specifications for the project's distribution system, which will be discussed next.

Setting the Distribution System

To define distribution systems in your project, open the Electrical Settings dialog box and click on the Distribution Systems node; a table will appear in the right pane showing columns such as Name, Phase, Configuration, Wires, L-L Voltage (Line to Line), and L-G Voltage (Line to Ground). You can edit any cell by selecting or entering values, for instance, choosing an option from the drop-down list in the L-L Voltage column. To add a new distribution system, click the Add button, then fill in the new row by specifying the Name and other parameters. Once defined, you can proceed to configure cable tray, conduit, load calculations, and panel schedule settings.

Setting the Load Calculations

To specify load classification and demand factor settings for circuits, click on the Load Calculations node in the Electrical Settings dialog box. In the right pane, select the Load Classifications button, which will open the Load Classifications dialog box, as shown in Figure 6-20.

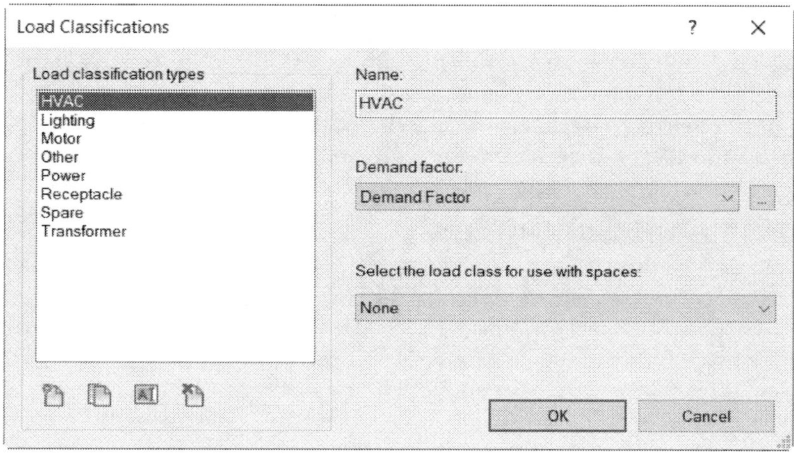

Figure 6-20 *The* ***Load Classifications*** *dialog box*

In the Load Classifications dialog box, select a type from the Load classification types area, and its specifications will appear on the right. You can create a new classification by choosing the New button, entering a name, and selecting **OK**. Afterward, select the new type, choose a demand factor from the drop-down list, or browse for additional options in the Demand Factors dialog box (Figure 6-21). In the Demand Factors dialog, you can select a type, set the calculation method, and adjust related options. After configuring the demand factor, click **OK**, and return to the Load Classifications dialog box to choose a load class for the system.

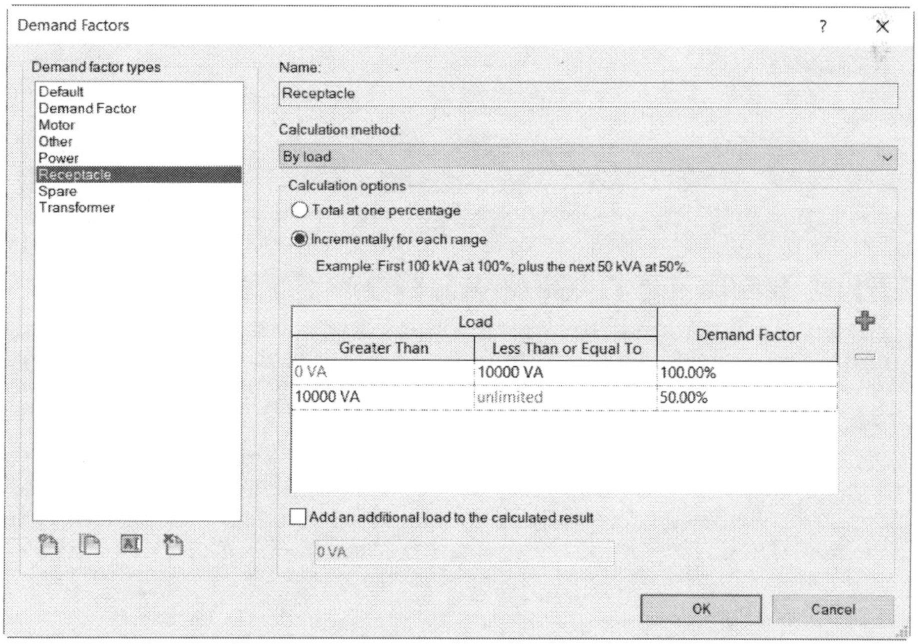

Figure 6-21 *The* ***Demand Factors*** *dialog box*

In the Load Classifications dialog box, you can create, delete, or rename load classification types using the **New**, **Delete**, or **Rename** buttons. In the **Electrical Settings** dialog box, clicking the **Demand Factors** button opens the **Demand Factors** dialog, where you can adjust options and click **OK** to confirm. To enable load calculations for loads in spaces, select the "Run calculations for loads in space" checkbox. After adjusting the load calculation options, click on the Panel Schedules node to modify panel schedule settings in the project.

Setting the Panel Schedules

To set panel schedule settings, click on the **Panel Schedules** node in the **Electrical Settings** dialog box. Here, you can specify the default label for spares using the Spare Label parameter, choose to include spares in the panel totals with the Include Spares in Panel Totals checkbox, and merge multi-pole circuits into a single cell using the Merge multi-poled circuits into a single cell checkbox.

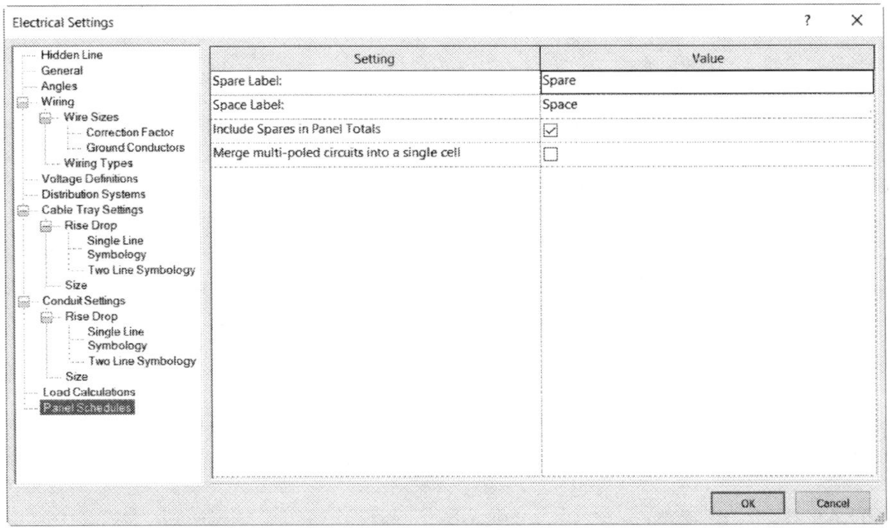

*Figure 6-22 The options for **Panel Schedules** option*

CREATING POWER DISTRIBUTION SYSTEM

To create a power distribution system between a transformer and two panels (Figure 6-24), first place the transformer using the **Electrical Equipment** tool, selecting the appropriate transformer type and distribution system settings. Next, add two panels by selecting their types from the Properties palette and placing them in the floor plan. Once the panels are added, select the L1 panel and create a power circuit by choosing the Power option in the **Create Systems** panel (Figure 6-25), then link it to the transformer by selecting the T1 panel from the dropdown. Finally, right-click the transformer connector symbol, select the Create Power Circuit option, and link it to the H1 panel to complete the distribution system setup.

Electrical Analysis

In Revit 2025, electrical design requirements can be specified directly within the Revit file, eliminating the need for separate spreadsheets or PDFs. The new workflow allows the use of linked DWGs, PDFs, or Revit models for load takeoffs, which can be used to determine preliminary building loads and equipment component loads (Figure 6-23). Tools in the Electrical Analysis panel of the **System** tab, such as the **Area Based Load Boundary** and **Area Based Load** tools, allow you to define and specify area loads (Figure 6-25). Additionally, the **Equipment Load** tool can be used to assign equipment loads to the system (Figure 6-26).

PERFORMING LIGHTING ANALYSIS

Lighting analysis helps plan the type of light fixtures for the project by creating a schedule that displays both the fixtures and lighting criteria. This schedule allows you to compare the planned and actual illumination levels as fixtures are placed. Figure 6-23 shows a sample schedule, where the last column calculates the difference between required and actual lighting levels. If the difference exceeds 6 footcandles, the cell turns brown. Since no lights are in the model yet, all cells appear brown initially, with the goal being to achieve a schedule with no brown cells in the final stage. The lighting analysis procedure is detailed in Tutorial 1 of this chapter.

<Lighting Analysis>

A	B	C	D	E	F	G	H	I
Required Illuminati	Average Estimated Illumination	Ceiling Reflectanc	Floor Reflectance	Lighting Calculatio	Name	Number	Wall Reflectance	Delta
30 fc	15 fc	0.75	20.00%	2' - 6"	CEO-Office	101	0.50	[illegible]
30 fc	0 fc	0.75	20.00%	2' - 6"	V.P. Marketing	102	0.50	[illegible]
30 fc	0 fc	0.75	20.00%	2' - 6"	Toilet-1	103	0.50	[illegible]
30 fc	0 fc	0.75	20.00%	2' - 6"	Rest Room	104	0.50	[illegible]
30 fc	0 fc	0.75	20.00%	2' - 6"	Cafeteria	105	0.50	[illegible]
20 fc	0 fc	0.75	20.00%	2' - 6"	Store	106	0.50	[illegible]
30 fc	0 fc	0.75	20.00%	2' - 6"	Purchase	107	0.50	[illegible]
30 fc	0 fc	0.75	20.00%	2' - 6"	International	108	0.50	[illegible]
30 fc	0 fc	0.75	20.00%	2' - 6"	H.R.	109	0.50	[illegible]
45 fc	0 fc	0.75	20.00%	2' - 6"	Central Area	110	0.50	[illegible]
30 fc	0 fc	0.75	20.00%	2' - 6"	Quality	111	0.50	[illegible]
20 fc	0 fc	0.75	20.00%	2' - 6"	Open	112	0.50	[illegible]
35 fc	0 fc	0.75	20.00%	2' - 6"	Conference-2	113	0.50	[illegible]
35 fc	0 fc	0.75	20.00%	2' - 6"	Conference	114	0.50	[illegible]
25 fc	0 fc	0.75	20.00%	2' - 6"	Lounge	115	0.50	[illegible]
35 fc	0 fc	0.75	20.00%	2' - 6"	Meeting Room	116	0.50	[illegible]
30 fc	0 fc	0.75	20.00%	2' - 6"	Service-1	117	0.50	[illegible]
20 fc	0 fc	0.75	20.00%	2' - 6"	Shaft	118	0.50	[illegible]
20 fc	0 fc	0.75	20.00%	2' - 6"	Service-2	119	0.50	[illegible]
30 fc	0 fc	0.75	20.00%	2' - 6"	Accounts	120	0.50	[illegible]

Figure 6-23 *A typical lighting schedule*

CREATING CIRCUITS

In a project, circuits connect devices or fixtures in the electrical system, while wires symbolically represent these connections. To create a circuit, select a device, such as a receptacle, and use the Create Systems tool to generate the appropriate circuit type (e.g., Power-Balance, Communication, Fire Alarm). In the Modify | Electrical Circuits tab, use the Edit Circuit tool to add devices to the circuit and finish editing. The circuit will appear as a dotted line, and the circuit name will initially display as <unnamed> in the System Selector. To complete the circuit, select a panel from the Panel drop-down list, ensuring the panel's distribution system matches the device's connector properties. Once selected, the circuit will be finalized, and the <unnamed> option will be replaced by the new circuit name.

ADDING WIRES TO THE CIRCUIT

To manually draw wires in a project, use the Arc Wire, Spline Wire, or Chamfered Wire tools from the **Systems > Electrical > Wire drop-down menu**, as shown in Figure 6-24. To create an arc wire, select the Arc Wire tool, then choose Tag on Placement in the Tag panel to display tags with the wires. In the Properties palette, select the wire type from the Type **Selector**, click to define the start, mid, and end points in the drawing view, and the wire will be created. You can adjust its instance properties in the Properties palette. Similarly, use the Spline Wire or **Chamfered Wire** tools to draw other wire types.

TUTORIALS

General instructions for downloading tutorial files:

1. Download the *c06_rmp_2025_tut.zip* file for this tutorial from *https://www.cadsofttech.com*. The path of the file is as follows: *Books > Civil/GIS BOOKS > Revit MEP > Revit MEP 2025 For Novices*

2. Now, save and extract the downloaded folder at the following location: *C:\rmp_2025\c06_rmp_2025_tut*

Tutorial 1 Planning the Electrical Systems

In this tutorial, you will specify the settings for the electrical system and also load the family components required for the project. **(Expected time: 45 min)**

1. File to be used:

   For Imperial: *c06_archi_elec_tut1.rvt*
   For Metric: *M_c06_archi-elec_tut1.rvt*

2. File name to be assigned:

   For Imperial *c06_Office-Space_tut1a.rvt*
   For Metric *M_c06_Office-Space_tut1a.rvt*

The following steps are required to complete this tutorial:

a. Open the *c06_archi_elec_tut1.rvt* (For Metric *M_c06_archi_elec_tut1.rvt*) project file.
b. Specify the electrical settings.
c. Load family components.
d. Save the project.
e. Close the project.

Opening the Project File

1. Choose **Open > Project** from the **File** menu; the **Open** dialog box is displayed.

2. In the dialog box, browse to *c:\rmp_2025\c06_rmp_2025_tut* folder and then select the *c06_archi_elec_tut1.rvt* (*M_ c06_archi_elec_tut1.rvt*) file. Next, choose the **Open** button; the project file is opened.

Note

The architectural model named c04_archi_spaces_rmp_2025.rvt (for Metric, M_c04_archi_spaces_rmp_2025.rvt) linked in this tutorial file is located in the c04_rmp_2025_tut folder. It is recommended to maintain the relative path to the architectural model. However, if the link is lost, you need to reload the file using the ***Manage Links*** *dialog box which has already been discussed in the previous chapters.*

Specifying Electrical Settings

1. Choose the **Electrical Settings** button from the **Electrical** panel of the **Systems** tab; the **Electrical Settings** dialog box is displayed.

2. In the left pane of this dialog box, click on **Wiring > Wiring Types** node; a table is displayed.

3. Choose the **Add** button in the dialog box; a new row is displayed in the table.

4. In this row, specify the settings as follows:

   Name: **AL-THHN** Material: **Aluminium** Temperature Rating: **75**
   Insulation: **THHN** Max Size: **500** Neutral Multiplier: **1.00**
   Neutral Required: Select the check box Conduit Type: **Steel**
   Neutral Size: **Hot Conductor Size**

5. Next, in the left pane of the dialog box, select the **Voltage Definitions** node; a table is displayed.

6. In the table, retain the default settings for all the parameters, refer to Figure 6-24.

	Name	Value	Minimum	Maximum
1	120	120.00 V	110.00 V	130.00 V
2	208	208.00 V	200.00 V	220.00 V
3	240	240.00 V	220.00 V	250.00 V
4	277	277.00 V	260.00 V	280.00 V
5	480	480.00 V	460.00 V	490.00 V

Figure 6-24 *The table displaying the specified parameter for the* ***Voltage Definitions*** *node*

7. In the left pane of the dialog box, select the **Distribution Systems** node; a table is displayed in the right pane.

8. Retain the default settings in this table, as shown in Figure 6-25.

	Name	Phase	Configuration	Wires	L-L Voltage	L-G Voltage
1	120/208 Wye	Three	Wye	4	208	120
2	120/240 Single	Single	None	3	240	120
3	480/277 Wye	Three	Wye	4	480	277

Figure 6-25 *The table displaying the specified parameter for the* ***Distribution Systems*** *node*

9. In the left pane of the dialog box, click on the **Load Calculations** node; various options related to the calculations of the electrical load are displayed in the right pane.

10. In the right pane, choose the **Load Classifications** button; the **Load Classifications** dialog box is displayed, as shown in Figure 6-26.

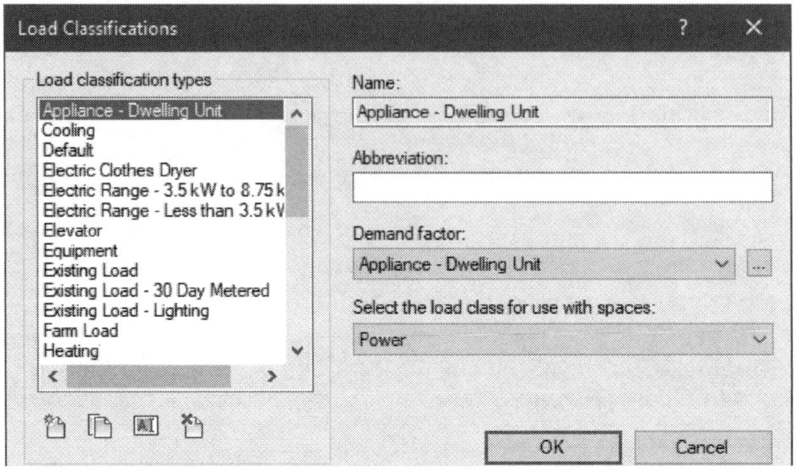

Figure 6-26 *The* ***Load Classifications*** *dialog box*

11. In the **Load classification types** area of this dialog box, select the **Other** option and then in the right side of the dialog box, ensure that the **Other** option is selected in the **Demand Factor** drop-down list.

12. Choose the browse button displayed next to the **Demand factor** drop-down list; the **Demand Factors** dialog box is displayed, as shown in Figure 6-27.

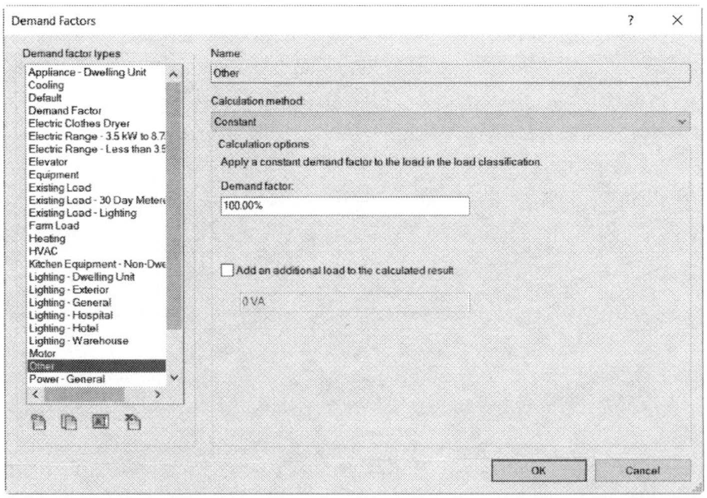

Figure 6-27 *Partial view of the* ***Demand Factors*** *dialog box*

13. In this dialog box, ensure that the **Other** option is selected in the **Demand factor types** area and then select the **By load** option from the **Calculation method** drop-down list; a table is displayed below it.

14. In the **Calculation options** area, select the **Incrementally for each range** radio button and then choose the **Split the selected row** button (icon displayed in '+' symbol) twice. On doing so, two rows are added to the table, refer to Figure 6-28.

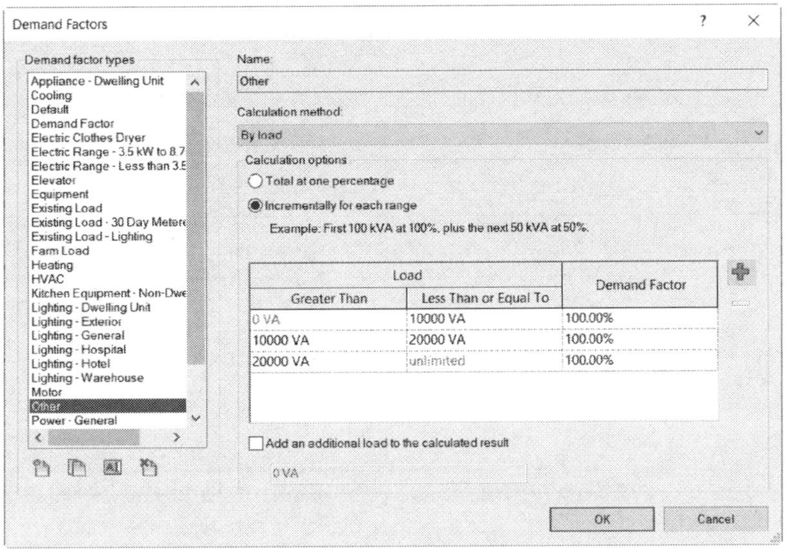

Figure 6-28 The rows added in the table

15. Now, assign the settings in this table, as given in Figure 6-29.

Load		Demand Factor
Greater Than	Less Than or Equal To	
0 VA	2000 VA	100.00%
2000 VA	70000 VA	50.00%
70000 VA	unlimited	30.00%

Figure 6-29 The settings specified in the rows added

16. Choose **OK**; the **Demand Factors** dialog box is closed. Next, choose **OK** twice in the **Load Classifications** and **Electrical Settings** dialog boxes to close them.

Loading the Electrical Components

1. Choose the **Load Family** tool from the **Load from Library** panel of the **Insert** tab; the **Load Family** dialog box is displayed.

2. In the dialog box, browse to **English-Imperial > Electrical > MEP > Electric Power > Distribution** folder (for Metric, browse to **English > Electrical > MEP > Electric Power > Distribution)** and then select the *Lighting and Appliance Panelboard - 208V MLO - Surface.rfa* (for Metric, *M_ Lighting and Appliance Panelboard - 208V MLO - Surface.rfa*) file from it.

3. Press and hold the CTRL key, and then select the *Lighting and Appliance Panelboard - 480V MCB - Surface.rfa* (for Metric, *M_Lighting and Appliance Panelboard - 480V MCB - Surface.rfa*) file.

4. Choose the **Open** button; the **Load Family** dialog box is closed and the selected families are loaded in the project.

5. Repeat Steps 1 and 2 and select the other family files required for this tutorial from their respective folders. Refer to the table given next for the name of the file and their folder locations.

Name of the Family(ies)	Folder Location
Duplex Receptacle, Lighting Switches (for Metric M_Duplex Receptacle, M_Lighting Switches)	English-Imperial > Electrical > MEP > Electric Power > Terminals (for English > US > Electrical > MEP > Electric Power > Terminals)
Dry Type Transformer - 480-208Y120 - NEMA Type 2 (for Metric M_Dry Type Transformer - 480-208Y120 - NEMA Type 2)	English-Imperial > Electrical > MEP > Electric Power > Generation and Transformation (English > US > Electrical > MEP > Electric Power > Generation and Transformation)
Troffer Corner Insert (for Metric M_ Troffer Corner Insert)	English-Imperial > Lighting > MEP > Internal (English > US > Lighting > MEP > Internal)

Saving and Closing the Project

In this section, you need to save the project and the settings.

1. Choose **Save As > Project** from the **File** menu; the **Save As** dialog box is displayed.

2. In this dialog box, browse to the *C:\rmp_2025\c06_rmp_2025_tut* folder and then in the **File name** edit box, enter the text **c06_Office-Space_tut1a** (for Metric, **M_c06_Office-Space_ tut1a**) and then choose the **Save** button to save the current project file with the specified name and to close the **Save As** dialog box.

3. Choose the **Close** option from **File** menu; the file is closed.

EXERCISE

Exercise 1 Placing Lighting Fixtures and Switches

Download the *c06_Conference-Center_exer1.rvt* file (for Metric *M_c06_Conference-Center_exer1.rvt*) from *https://www.cadsofttech.com*. The path of the file is as follows: *Books > Civil/GIS BOOKS > Revit MEP > Revit MEP 2025 For Novices.*

In this exercise, you will add the lighting fixtures, switches, and junction boxes for spaces in the 1ST Floor-ELEC LIGHT'G view. Refer to Figure 6-30 for the placement of various components.

1. Project view to be used:
   Floor Plans > Electrical > 1ST Floor-ELEC LIGHT'G
2. Family type to be used:
   Lighting Fixtures
   For Imperial Duplex Receptacle: Standard and M_Duplex Receptacle: GFCI
   For Metric M_Duplex Receptacle: Standard and M_Duplex Receptacle: GFCI
   Switches
   For Imperial Lighting Switches: Single Pole
   For Metric M_ Lighting Switches: Single Pole
   Junction Boxes
   For Imperial Junction Boxes-Load: 4" Square 120-1
   For Metric M_Junction Boxes-Load: 4" Square 120-1
3. Elevation
   Switches: 3' 9" (1145mm)
   Receptacles: 1'3" (381mm)
   Junction Boxes: 8' (2438mm)
4. File name to be assigned: *c06_Conference-Center_exer1a.rvt* (for Metric *M_ c06_ Conference-Center_exer1a.rvt)*

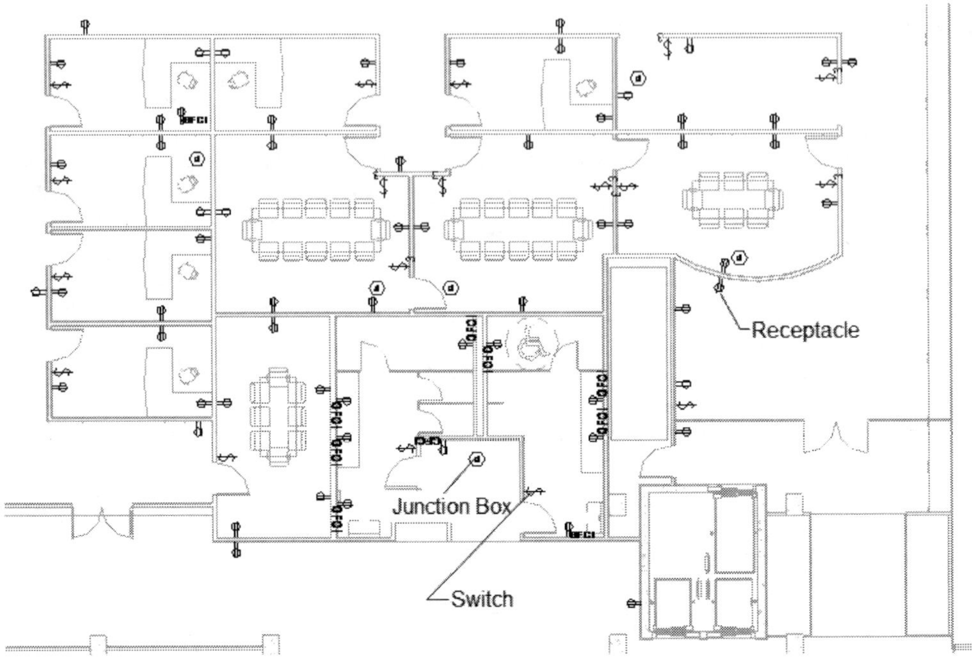

Figure 6-30 *The placement of various electrical components in the project*

Chapter 7

Creating Plumbing Systems

Learning Objectives

After completing this chapter, you will be able to:

- *Add plumbing fixtures*
- *Create pipe settings*
- *Auto route the pipe*
- *Manually route the pipe*

INTRODUCTION

A plumbing system is a network of pipes, drain fittings, valves, valve assemblies, and devices that are installed in a building to distribute water for drinking, washing, and heating, and also to drain the waste out of the building. In this chapter, you will learn various steps involved in creating a plumbing system in a building.

CREATING A PLUMBING SYSTEM

To create a plumbing system in Revit, you will need to add plumbing fixtures, define various pipe settings, and then route the pipe to connect the fixtures to supply water or drain the waste out of it. Figure 7-1 shows a typical plumbing system in a building. In Revit, you can create the plumbing system under the **Plumbing** discipline. To start working in a plumbing system, expand **Plumbing > Floor Plans** node in the **Project Browser** and then click on the desired level. In the **Project Browser**, the following floor plans are available under the **Floor Plans** node: **1-Plumbing** and **2-Plumbing**. You can select any of these floor plans and start creating the plumbing system. The steps involved in creating a plumbing system are discussed next.

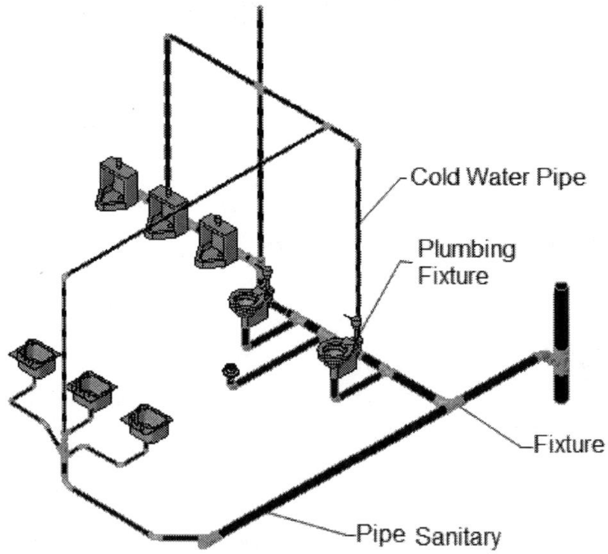

***Figure 7-1** A typical plumbing system in a building project*

Adding Plumbing Fixtures

A plumbing fixture is a device connected to the plumbing system to supply and drain water, commonly including bathtubs, sinks, toilets, urinals, and showers. Each fixture has water outlets, a drain, a flood rim to prevent overflow, and traps that create a water seal to block foul gases from entering the building. To add a plumbing fixture in a project, select the **Plumbing Fixture** tool from the **Plumbing & Piping** panel on the **Systems** tab. Then, use the **Load Family** tool to open the **Load Family** dialog box. Browse to the appropriate folder (e.g., **English-Imperial > Plumbing > MEP > Fixtures** or **English > Plumbing > MEP > Fixtures** for Metric), select and load the desired fixture family (see **Figure 7-2**). Use the **Type Selector** in the **Properties** palette to choose the fixture type, which may be wall-hosted, like the **Urinal - Wall Hung: 3/4" Flush Valve**, or face-hosted, such as the **Single - Island - Single 18" x 18" - Private** type.

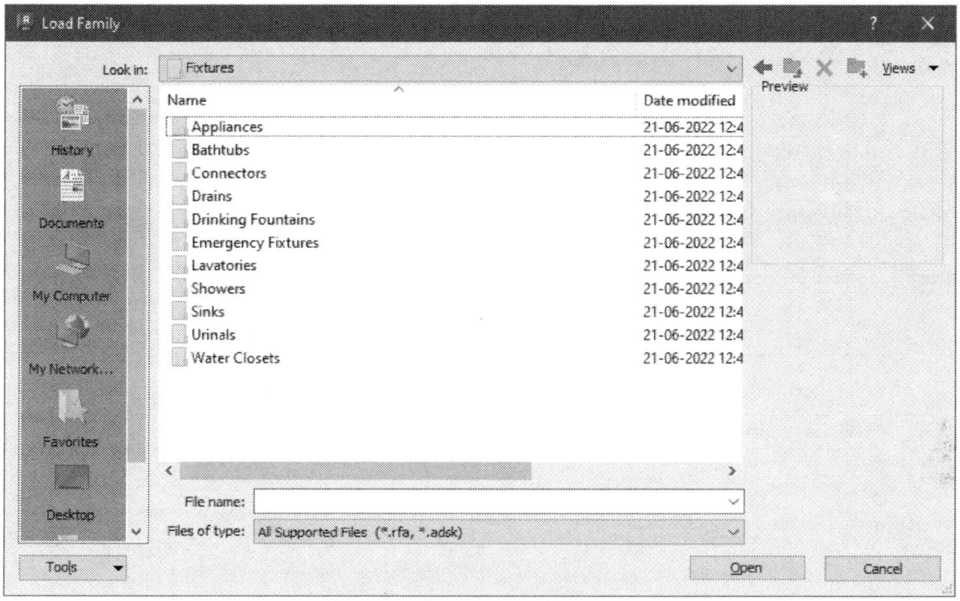

***Figure 7-2** The **Load Family** dialog box displaying different folders for the fixtures*

For example, to add a plumbing fixture, you can select the **Bath Tub : 42" x 30" - Private** type from the **Type Selector** drop-down list. For Metric system, select the **M_Bath Tub : 1065 mm x 760 mm - Private** type. Next, in the **Properties** palette, you can modify various instance properties of the fixture. To modify the type properties of the selected plumbing fixture, choose the **Edit Type** button in the **Properties** palette; the **Type Properties** dialog box will be displayed, as shown in Figure 7-3. You can use this dialog box to edit type properties of the selected fixture type. In this case, it is **Bath Tub : 42" x 30" - Private** type (or for Metric **M_Bath Tub : 1065 mm x 760 mm - Private** type).

Tip

You can download various family types related to plumbing design and parametric BIM objects uploaded by manufacturers and developers across the world from the link www.bimobject.com.

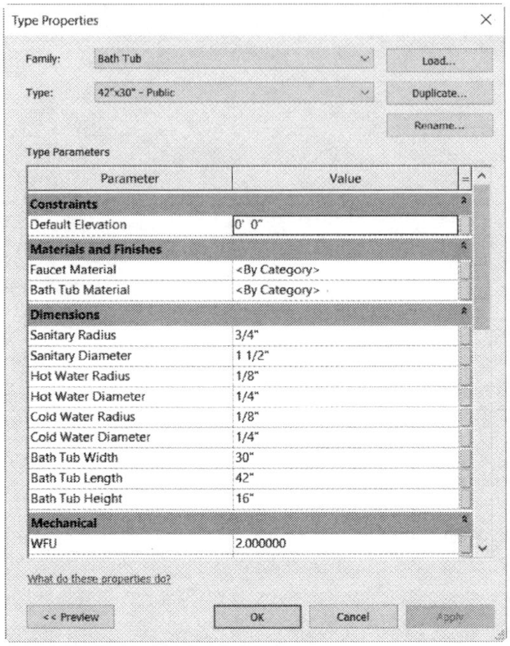

Figure 7-3 *The* ***Type Properties*** *dialog box for the* ***Bath Tub : 42" x 30" - Private*** *type*

After setting the type properties, click **OK** to close the dialog box. You can then modify the instance properties of the plumbing fixture in the **Properties** palette. Once configured, move the cursor in the drawing area to preview the fixture—such as a bathtub—and click to place it at the desired location. A 3D view of the bathtub with connectors is shown in **Figure 7-4**.

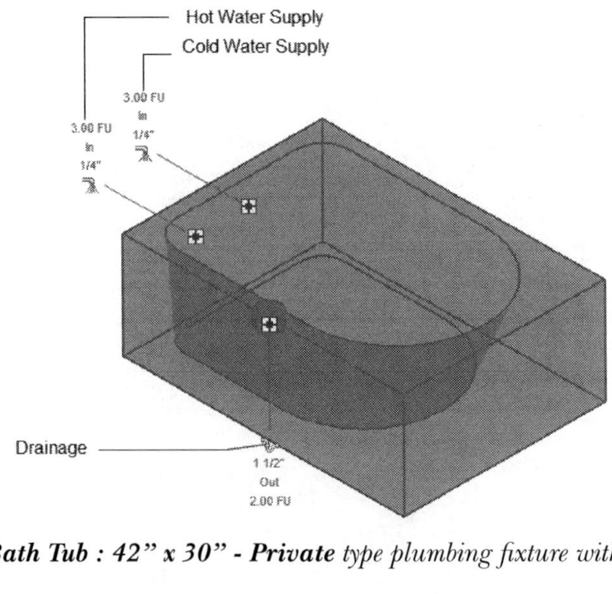

Figure 7-4 *The* ***Bath Tub : 42" x 30" - Private*** *type plumbing fixture with various connectors*

Tip

In a plumbing system, fixture unit is an important aspect of plumbing design used to determine the size of the pipes connected to the fixtures in the system for both water supply and waste water.

Specifying the Pipe Settings

After placing the fixtures at their required location in the drawing, you now need to create a pipe system that will connect the fixtures with the pipes. The pipe system consists of a network of pipes and fixtures to supply water to the building and also to drain the waste out of the building. Before you create the network for these utilities, you need to set the specification of the pipe and pipe fixtures to be used in the pipe system. In Revit, you can specify the type of pipe, different sizes of pipe, and the material of the pipe before creating network of the pipe system. The various procedures to set the pipe system are discussed next.

Routing Pipes in the Pipe System

While creating a pipe system, routing of the pipe is the most important task. Routing of a pipe involves connecting the fixtures with it such that the waste from the building system is easily drained to the sewer and the water is supplied to the building as per the requirement. In the Plumbing discipline of an MEP project, the pipes are routed above ceilings, in walls, in chases, and under floor slabs. The routing of the pipes should be correctly designed in coordination with the structural and architectural model of the building.

In Revit, you can route a pipe using three methods: Auto Routing, Manual Routing, and Slope. The process of routing pipes using all the three methods is discussed next.

Modifying a Pipe Segment

To modify a pipe segment, select it from the drawing; the **Modify | Pipes** contextual tab and the **Properties** palette will be displayed with the values of the instance parameters of the pipe segment selected. Also, you will notice that the selected pipe is highlighted in blue color and the connectors are displayed at its two endpoints along with the value of offset of the endpoints. In the **Edit** panel of the **Modify|Pipes** contextual tab, choose options to modify the justification, slope constraints, and capping at the open end of the pipe segment. Using the tool from the **Pipe Insulation** panel, you can add insulation cover to a pipe segment. To do so, choose the **Add Insulation** tool; the **Add Pipe Insulation** dialog box will be displayed. In this dialog box, select the type of insulation from the **Insulation Type** drop-down list and then specify a thickness value in the **Thickness** edit box. Next, choose the **Edit Type** button next to the **Insulation Type** drop-down list; the **Type Properties** dialog box will be displayed. You can use this dialog box to edit the type of insulation applied to the pipe. In the **Type Properties** dialog box, click in the **Value** field corresponding to the **Material** parameter and then choose the browse button; the **Material Browser** dialog box will be displayed. In this dialog box, specify the material to be used for insulation and then choose the **OK** button; the **Material Browser** dialog box will be closed.

In the **Type Properties** dialog box, you can also edit other type properties of the insulation such as its keynote, model, manufacturer, cost, and so on. After modifying the parameters, choose **OK** to close the **Type Properties** dialog box. Similarly, choose the **OK** button to close the **Add**

Pipe Insulation dialog box; the insulation cover will be added to the pipe segment. Next, you can choose the **Edit Insulation** tool from the **Pipe Insulation** panel to modify the insulation added. Also, you can choose the **Remove Insulation** tool from the **Pipe Insulation** panel to remove the pipe insulation from the pipe segment. Next, you can modify the instance properties of the selected pipe segment from the **Properties** palette. For example, you can modify various parameters such as **Horizontal Justification**, **Vertical Justification**, **Reference Level**, **Offset**, **System Type**, **Diameter**, **Pipe Segment**, and so on in this palette. After specifying various instance parameters, choose **Modify** from the **Select** panel to exit the modification process.

Placing Fittings

After placing the pipe, you can add various fittings to it. The fittings that you can add to a pipe include elbows, tees, wyes, crosses, unions and other types of fittings. Some of the fittings can be placed at endpoints of the pipe segment or directly at a point along the length of the pipe segment. To add a pipe fitting to the pipe segment, choose the **Pipe Fittings** tool from the **Plumbing & Piping** panel of the **Systems** tab; the **Modify | Place Pipe Fittings** contextual tab will be displayed. In the **Level** drop-down list of the **Options Bar**, select the level to which the pipe fitting will be associated. Also, you can select the **Rotate after placement** check box if you want the pipe fitting be rotated at the time of its placement.

Next, in the **Properties** palette, select a type for the fittings from the **Type Selector** drop-down list. Also, in the **Properties** palette, you can specify various instance properties of the pipe fitting such as **Offset**, **Insulation Thickness**, **Nominal Radius**, and more. After specifying the instance properties of the fitting, place the cursor at the end of the pipe fitting or at the desired location along the pipe. Notice that the preview of the fitting will be displayed with the cursor. For example, if you select the **M_Pipe Elbow : Standard** type from the **Type Selector** drop-down list, the preview of the selected type will be displayed with the cursor in the drawing area.

As you move the cursor and place it at the endpoint of the pipe, the elbow will snap to the end of the pipe and will resize according to the diameter of the pipe. Click when the elbow snaps to the endpoint; the elbow will be placed and connected to the pipe segment, as shown in Figure 7-5.

Now, choose **Modify** from the **Select** panel of the **Modify | Place Pipe Fittings** contextual tab to exit the **Pipe Fittings** tool. You can now modify the properties and connection of the pipe fitting. To do so, select the added fitting in the drawing; the **Modify | Pipe Fittings** tab will be displayed and also instance properties related to the fittings will be displayed in the **Properties** palette. In the **Modify | Pipe Fittings** tab, you can choose the tools from the **Layout** panel to connect the pipe fittings with other pipe segments. In the **Edit** panel, you can use various tools to modify the slope, justification, and end cap of the pipe fitting. And, in the **Pipe Insulation** panel, you can use various tools to add an insulation to the pipe fitting.

***Figure 7-5** The elbow placed at the end of the pipe segment*

Next, in the **Option Bar**, you can select an option from the **Diameter** drop-down list to modify the diameter of the pipe fixture. Also, you can select or specify an option in the **Offset** edit box to modify the existing offset value of the pipe fixture. In the **Properties** palette, you can modify various instance properties of the selected pipe fitting. Next, choose **Modify** from the **Select** panel to exit the modification.

Placing Pipe Accessories

Similar to placing pipe fittings, you can add pipe accessories to the pipe segment. The pipe accessories include valves, connectors, venturi flow meter, pressure gauge, temperature gauge, air separator, and so on. The families of pipe accessories can be loaded from the **Accessories** and the **Valves** folders which are located at **English-Imperial > Pipe** folder location (for Metric **English > Pipe** folder location) by using the **Load Family** tool from the **Load from Library** panel in the **Insert** tab. After loading the required families, choose the **Pipe Accessories** tool from the **Plumbing & Piping** panel of the **Systems** tab. On doing so, the **Modify | Place Pipe Accessory** tab will be displayed. In the **Options Bar**, select the desired level from the **Level** drop-down list and then in the **Properties** palette, select a type of accessory from the **Type Selector** drop-down list. For example, you can select the **Double Check Valve - 2.5-10 Inch** type (for Metric **Double Check Valve - 65-250mm)** from the **Type Selector** drop-down list and then in the **Properties** palette, specify the instance properties of the accessory such as **Pressure Drop**, **Flow, Loss Method**, and more. After specifying the instance properties, move the cursor in the drawing area and place it over the pipe at a location where you want to add the accessory. As you move the cursor, you will notice that the preview of the selected type of accessory will be displayed with it. Place the cursor and click when it snaps to the pipe at the desired location; the pipe accessory will be placed, as shown in Figure 7-6. Choose **Modify** from the **Select** panel of the **Modify | Place Pipe Accessory** tab to exit the tool.

After placing the accessory, you can select it and modify its properties and connections to the pipe. On selecting the pipe accessory, its various controls will be displayed, as shown in Figure 7-7. You can use these controls to rotate and flip the accessory to the desired direction. Also, you can use the controls to add pipe segments to the inlet and outlet connectors of the accessory, if required.

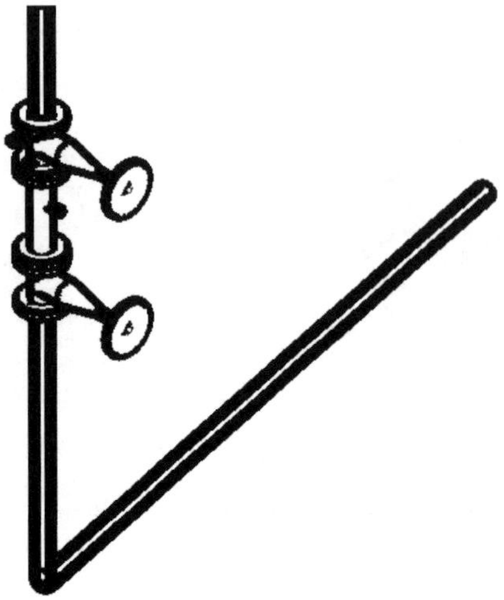

Figure 7-6 The ***Double Check Valve - 2.5-10 Inch*** *type pipe accessory placed at the pipe*

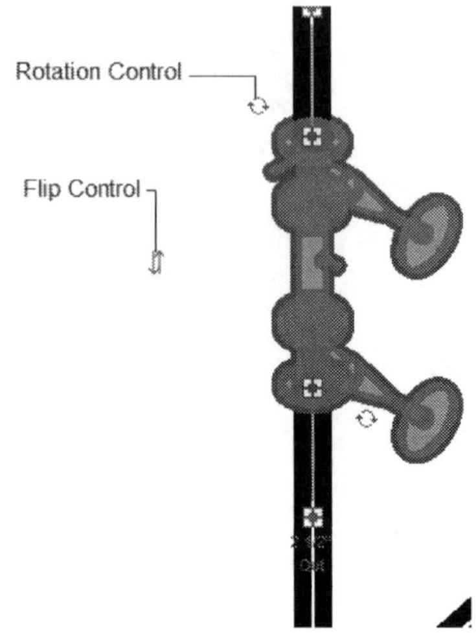

Figure 7-7 The various controls of the selected pipe accessory

Adding Plumbing Equipments

Similar to adding plumbing fixtures, you can add plumbing equipments in the project. The plumbing equipments include hot water heaters, sewage pumps, waste water pumps, drainage pumps, and more. The families of the plumbing equipments can be loaded from the **Equipment** folder which is located at **English-Imperial > Plumbing > MEP** folder location (for Metric, **US > English >Plumbing > MEP** folder location) by using the **Load Family** tool from the **Load from Library** panel in the **Insert** tab. After loading the required families, choose the **Plumbing Equipment** tool from the **Plumbing & Piping** panel of the **Systems** tab. On doing so, the **Modify | Place Plumbing Equipment** tab will be displayed. In the **Options Bar**, select the desired option and then in the **Properties** palette, select a type of equipment from the **Type Selector** drop-down list. For example, you can select the **Water Heater 40 Gallon** type (for Metric, **M_Water Heater 152 L** from the **Type Selector** drop-down list and then in the **Properties** palette, specify the instance properties of the equipment such as **Level**, **Offset from Host**, and others. After specifying the instance properties, move the cursor in the drawing area and place equipment at a desired location.

TUTORIAL

General instructions for downloading tutorial files:

1. Download the *c07_rmp_2025_tut.zip* file for this tutorial from *https://www.cadsofttech.com*. The path of the file is as follows: *Books > Civil/GIS BOOKS > Revit MEP > Revit MEP 2025 For Novices*

2. Now, save and extract the downloaded folder at the following location:
   C:\rmp_2025\`

Tutorial 1 Office-Space

In this tutorial, you will create a plumbing system for the Office-Space project with the following parameters and project specifications: **(Expected time: 45 min)**

1. File name to be used:
   For Imperial *c07_Office-Space_tut1.rvt*
   For Metric *M_c07_Office-Space_tut1.rvt*
2. File name to be assigned:
   For Imperial *c07_Office-Space_tut1a.rvt*
   For Metric *M_c07_Office-Space_tut1a.rvt*

The following steps are required to complete this tutorial:

a. Open the project file.
b. Load the plumbing component families.
c. Configure the piping systems.
d. Adding the plumbing fixtures.
e. Set the view range.
f. Place the floor drain.
g. Create the sanitary system.
h. View the plumbing system in 3D.
i. Save the project by using the **Save As** tool.
   For Imperial *c07_Office-Space_tut1a.rvt*
   For Metric *M_c07_Office-Space_tut1a.rvt*
j. Close the project by using the **Close** tool.

Opening the Project

1. Choose **Open > Project** from the **File** menu; the **Open** dialog box is displayed.
2. In the **Open** dialog box, browse to *C:\rmp_2025\c07_rmp_2025_tut* and select the desired file.
   For Imperial *c07_Office-Space_tut1*
   For Metric *M_c07_Office-Space_tut1*
3. Now, choose **Open**; the selected file opens in the Revit interface.

Loading the Plumbing Component Families

1. Choose the **Load Family** tool from the **Load from Library** panel of the **Insert** tab; the **Load Family** dialog box is displayed.
2. In this dialog box, browse to **English-Imperial > Plumbing > MEP > Fixtures > Urinals** folder and then select the **Urinal - Wall Hung.rfa** file and choose **Open**; the selected family is loaded in the current project. For the Metric system, browse to **English > US > Plumbing > MEP > Fixtures > Urinals** folder and then select the **M_Urinal - Wall Hung.rfa** file.
3. Repeat the procedure followed in steps 1 and 2 and select the other family files required for this tutorial from their respective folders. Refer to the table given next for the name of the files and their folder location.

Name of the Family Catagory(ies)	Folder Location (Imperial)	Folder Location (Metric)
Water Closet - Flush Valve - Floor Mounted (Imperial) M_Water Closet - Flush Valve - Floor Mounted (Metric)	English-Imperial > Plumbing > MEP > Fixtures > Water Closets	English > Plumbing > MEP > Fixtures > Water Closets
Sink - Island - Single.rfa (Imperial) M_Sink - Island - Single. rfa (Metric)`	English-Imperial > Plumbing > MEP > Fixtures > Sinks	English > Plumbing > MEP > Fixtures > Sinks
Floor Drain - Round(Imperial) M_Floor Drain - Round (Metric)	English-Imperial > Plumbing > MEP > Fixtures > Drains	English > Plumbing > MEP > Fixtures > Drains
Tee - PVC - Sch 40 (Imperial) M_Tee - PVC - Sch 40 (Metric)	English-Imperial > Pipe > Fittings > PVC > Sch 40 > Socket-Type	English > Pipe > Fittings > PVC > Sch 40 > Socket-Type
Bend - PVC - Sch 40 - DWV (Imperial) M_Bend - PVC - Sch 40 - DWV (Metric)	English-Imperial > Pipe > Fittings > PVC > Sch 40 > Socket-Type > DWV	English > Pipe > Fittings > PVC > Sch 40 > Socket-Type > DWV

Note

*In case, the families are already present in the project, the **Family Already Exists** dialog box is displayed. Choose the **Overwrite the existing version and its parameter values** option; the family is loaded.*

Configuring the Piping Systems

1. In the **Project Browser**, double-click on the **PVC-DWV** sub-node from **Families > Pipes > Pipe Types** node; the **Type Properties** dialog box is displayed.
2. In this dialog box, choose the **Duplicate** button; the **Name** dialog box is displayed.
3. In this dialog box, type the **Sanitary-PVC** text in the **Name** edit box.
4. Choose **OK**; the **Name** dialog box is closed and the **Type Properties** dialog box appears.
5. In the **Type Properties** dialog box, ensure that the **Sanitary-PVC** option is selected in the **Type** drop-down list and then choose the **Edit** button displayed in the **Value** field corresponding to the **Routing Preferences** parameter; the **Routing Preferences** dialog box is displayed.

6. In this dialog box, click in the desired field and select the required options from the drop-down list displayed. The options to be selected are given in Figure 7-8.

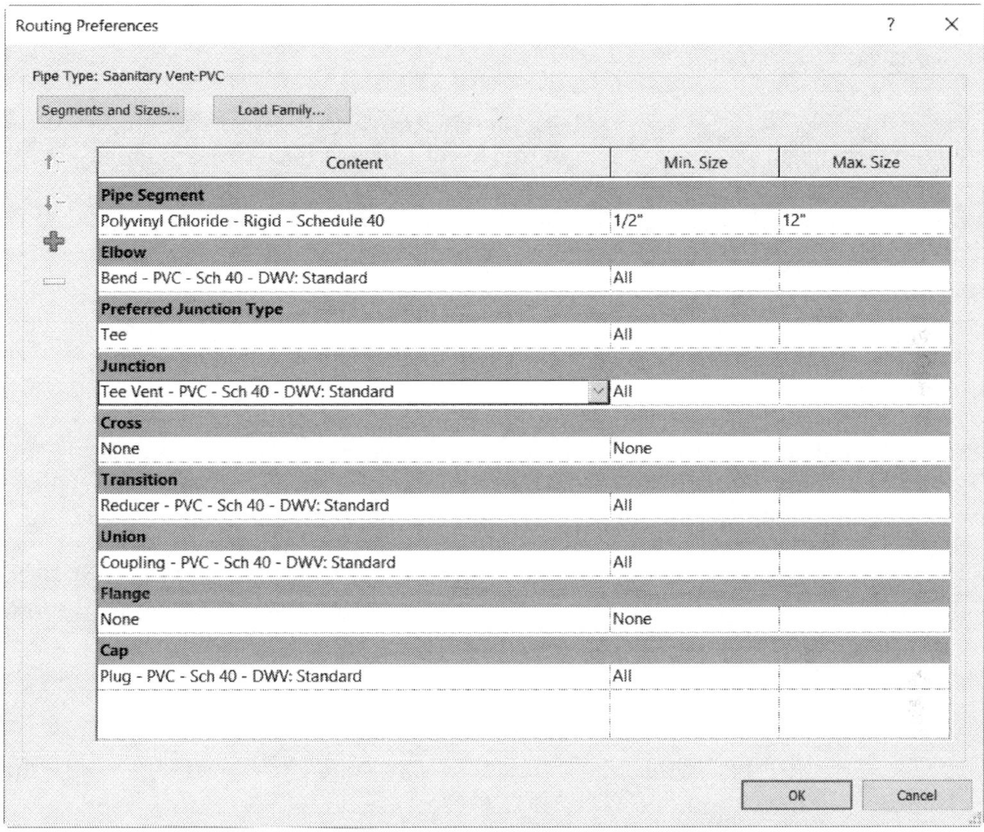

***Figure 7-8** The **Routing Preferences** dialog box*

7. After assigning the specified values in the **Routing Preferences** dialog box, choose **OK** to close it.

8. In the **Type Properties** dialog box, again choose the **Duplicate** button; the **Name** dialog box is displayed. In this dialog box, type **Sanitary Vent-PVC** in the **Name** edit box and choose **OK**; the **Name** dialog box is closed.

9. In the **Type Properties** dialog box, ensure that the **Sanitary Vent-PVC** option is selected by default and then choose the **Edit** button corresponding to the **Routing Preference** parameter; the **Routing Preferences** dialog box is displayed.

10. In this dialog box, choose the **Load Family** button; the **Load Family** dialog box is displayed.

11. In the **Load Family** dialog box, browse to **English-Imperial > Pipe > Fittings > PVC > Sch 40 > Socket-Type > DWV** folder and then select the **Tee Vent - PVC - Sch 40 - DWV** family from the list. For Metric system, browse to **English > US > Pipe > Fittings > PVC**

> Sch 40 > Socket-Type > DWV folder and then select the **M_Tee Vent - PVC - Sch 40 - DWV** family from the list. Choose **Open**; the **Load Family** dialog box is closed. In the **Content** column of the **Routing Preferences** dialog box, click on the field corresponding to the **Junction** option and select the **Tee-Vent PVC - Sch 40 - DWV: Standard** option for Imperial or **M_Tee Vent - PVC - Sch 40 - DWV : Standard** for Metric from the drop-down list displayed.

12. Keep rest of the settings unchanged in the **Routing Preferences** dialog box, refer to Figure 7-29 and choose the **OK** button; the **Routing Preferences** dialog box is closed.

13. In the **Type Properties** dialog box, choose **OK** to close it.

Adding the Plumbing Fixtures

In this section, you need to add 1 toilet, 2 urinals, 2 sinks, and a floor drain to the floor plan of the Office-Space project.

1. Choose the **Plumbing Fixture** tool from the **Plumbing & Piping** panel of the **Systems** tab; the **Modify | Place Plumbing Fixture** contextual tab is displayed. In this tab, choose the **Tag on Placement** button from the **Tag** panel, if it is not chosen by default. This enables the placing of the tag along with the fixture.

2. In the **Properties** palette, select the **Water Closet-Flush Valve -Floor Mounted Private-1.6 gpf** option from the **Type Selector** drop-down list. For Metric system, select **M_Water Closet-Flush Valve -Floor Mounted Private-6.1 Lpf**.

3. In the **Options Bar**, select the **Rotate after placement** check box.

4. Next, zoom in the toilet area in the south. Now, move the cursor inside the toilet area and click at the location shown in Figure 7-9.

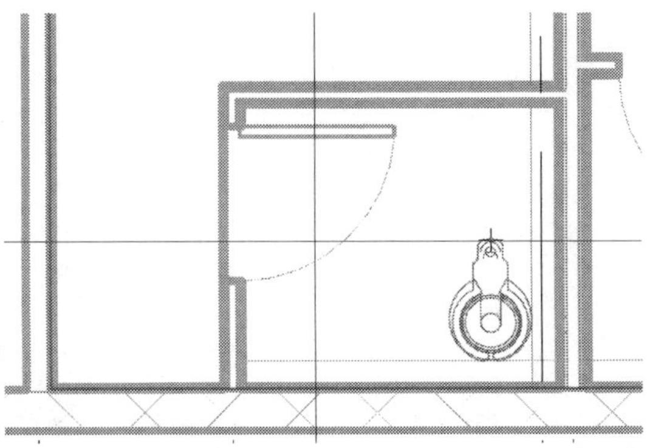

Figure 7-9 *The fixture placed at the specified location*

5. Next, in the **Options Bar**, enter **-90** in the **Angle** edit box and press ENTER; the fixture is placed at the specified location, as shown in Figure 7-10.

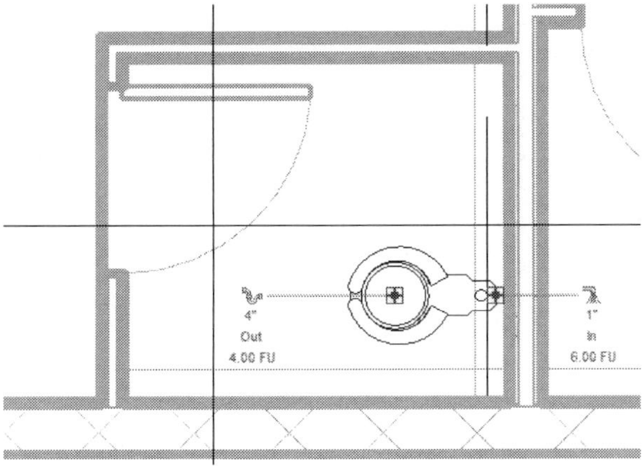

Figure 7-10 *The fixture placed after rotation*

6. Choose **Modify** from the **Select** panel to exit the **Plumbing Fixture** tool.

7. Next, choose the **Plumbing Fixture** tool from the **Plumbing & Piping** panel of the **Systems** tab; the **Modify | Place Plumbing Fixture** contextual tab is displayed.

8. In the **Properties** palette, select the **Urinal Wall Hung 3/4" Flush Valve** from the **Type Selector** drop-down list. Also, click in the value field of the **Elevation** parameter and type **1'6"**. For Metric system, select the **Urinal Wall Hung 20 mm Flush Valve** from the **Type Selector** drop-down list. Then, in the **Properties** palette, click in the value field of the **Elevation** parameter and type **450**.

9. In the **Properties** palette, ensure that the **Level 1** option is specified corresponding to the **Schedule Level** parameter. Also, make sure that the Place on Vertical Face option is chosen in the **Placement** panel of the **Modify | Place Plumbing Fixture** contextual tab

10. Now, in the **Placement** panel of the **Modify | Place Plumbing Fixture** contextual tab, ensure that the **Place on Vertical Face** button is chosen. Next, place the cursor at a location in the toilet area, as shown in Figure 7-11 Click to place the fixture at the specified location.

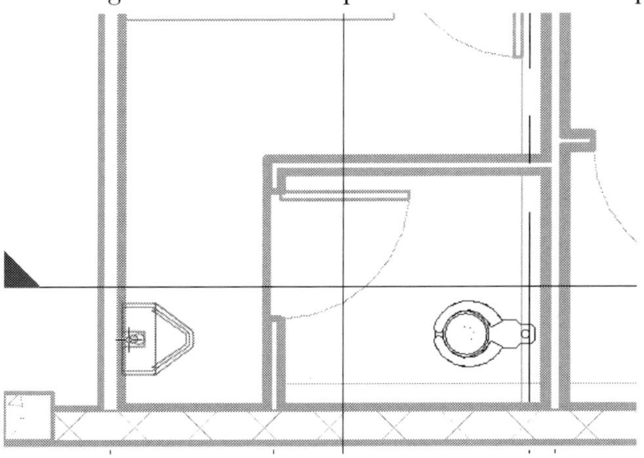

Figure 7-11 *The urinal fixture placed at the desired location*

11. Repeat the procedure given in step 10 to place another urinal of the same type, as shown in Figure 7-12.

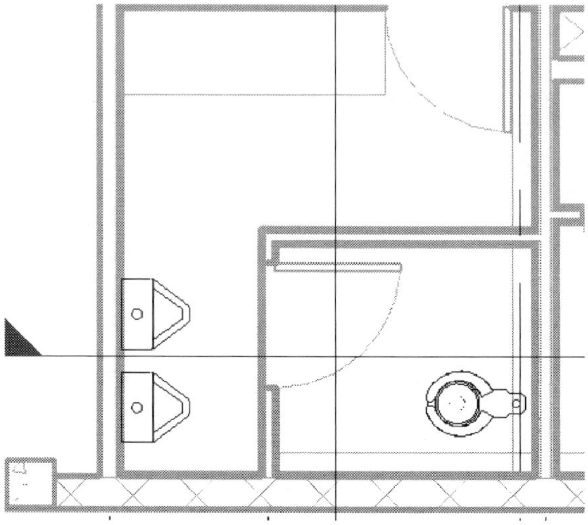

Figure 7-12 *The second urinal fixture placed at the specified location*

12. After placing the urinals, choose the **Modify** tool to exit the selection.

13. Choose the **Plumbing Fixture** tool from the **Plumbing & Piping** panel of the **Systems** tab; the **Modify | Place Plumbing Fixture** contextual tab is displayed.

14. In the **Properties** palette, select the **Sink-Island-Single 18"x18"-Private** type from the **Type Selector** drop-down list. For Metric system, select the **M_Sink-Island-Single 455 x455mm-Private** type.

15. Now, choose the **Place on Face** tool from the **Placement** panel of the **Modify | Place Plumbing Fixture** contextual tab.

16. Move the cursor at the face of the slab and click, refer to Figure 7-13; the sink is placed.

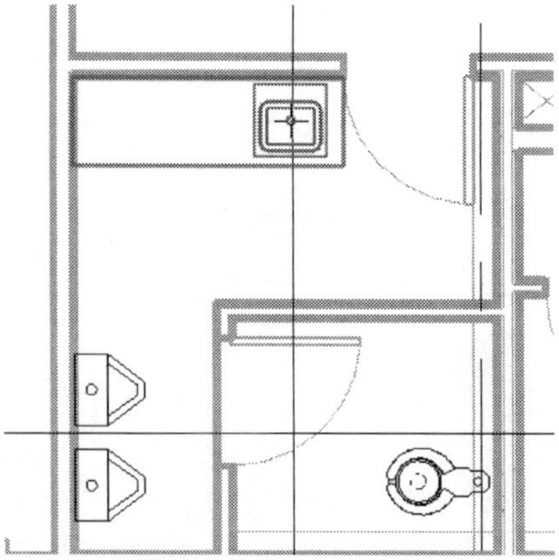

Figure 7-13 *The preview of the sink fixture displayed*

17. Add one more sink as done in previous steps. For placing the sink, refer to Figure 7-14.

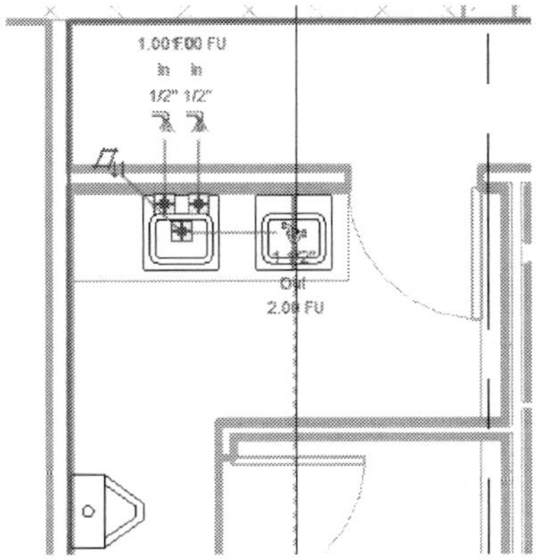

Figure 7-14 The second sink fixture placed

18. Choose **Modify** from the **Select** panel of the **Modify | Place Plumbing Fixture** contextual tab.

Setting the View Range

In this section, you will set the view range of the floor plan view.

1. In the **Properties** palette, choose the **Edit** button corresponding to the **View Range** parameter; the **View Range** dialog box is displayed.

2. In the **Primary Range** area of this dialog box, ensure that the **Associated Level (Level1)** option is selected in the **Bottom** drop-down list and enter **-5' (-1500 mm)** in the **Offset** edit box displayed next to it.

3. In the **View Depth** area of this dialog box, ensure that the **Associated Level (Level1)** option is selected in the **Level** drop-down list and then enter **-5' (-1500 mm)** in the **Offset** edit box displayed next to it.

4. Choose **Apply** and then **OK** to close the **View Range** dialog box.

Placing the Floor Drain

In this section, you will place the floor drain in the view.

1. Choose the **Plumbing Fixture** tool from the **Plumbing & Piping** panel of the **Systems** tab; the **Modify | Place Plumbing Fixture** contextual tab is displayed.

2. In the **Properties** palette, select the **Floor Drain - Round: 5" Strainer - 2" Drain** from the **Type Selector** drop-down list. For Metric system, select the **Floor Drain - Round: 125mm Strainer - 50mm Drain** option.

3. Choose the **Place on Face** tool from the **Placement** panel and then move the cursor to the location shown in Figure 7-15, and click; the floor drain is placed.

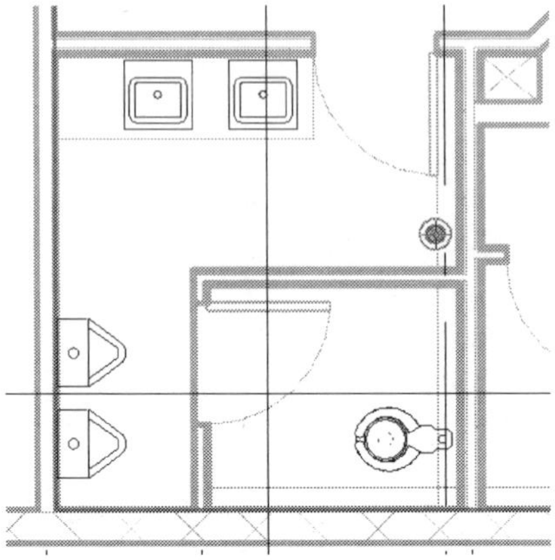

Figure 7-15 The location of the floor drain

4. Now, choose the **Modify** button from the **Select** panel to exit the **Plumbing Fixture** tool.

Creating the Sanitary System

1. Choose the **Analyze** tab and then choose the **Show Disconnects** button from the **Check Systems** panel; the **Show Disconnects Options** dialog box is displayed.

2. In this dialog box, select the **Pipe** check box and choose the **OK** button; the **Show Disconnects Options** dialog box is closed and all the open connectors in the fixtures are displayed as warning symbols.

3. Choose the **Visual Style** button in the **View Control Bar** and then choose the **Wireframe** option from the flyout displayed.

4. Choose the **Mechanical Settings** button from the **Plumbing & Piping** panel of the **Systems** tab; the **Mechanical Settings** dialog box is displayed.

5. In the left-pane of this dialog box, choose **Pipe Settings > Conversions**; the right-pane displays various properties related to the pipe settings, refer to Figure 7-16.

6. In the right pane, select the **Sanitary** option from the **System Classification** drop-down list.

7. In the **Main** area, click in the **Value** field corresponding to the **Pipe Type** parameter and select the **Pipe Types : PVC - DWV** option from the drop-down list displayed.

8. Specify the other settings in the **Main** and **Branch** areas, as shown in Figure 7-17. For Metric system, specify **-305** for the **Middle Elevation** parameters.

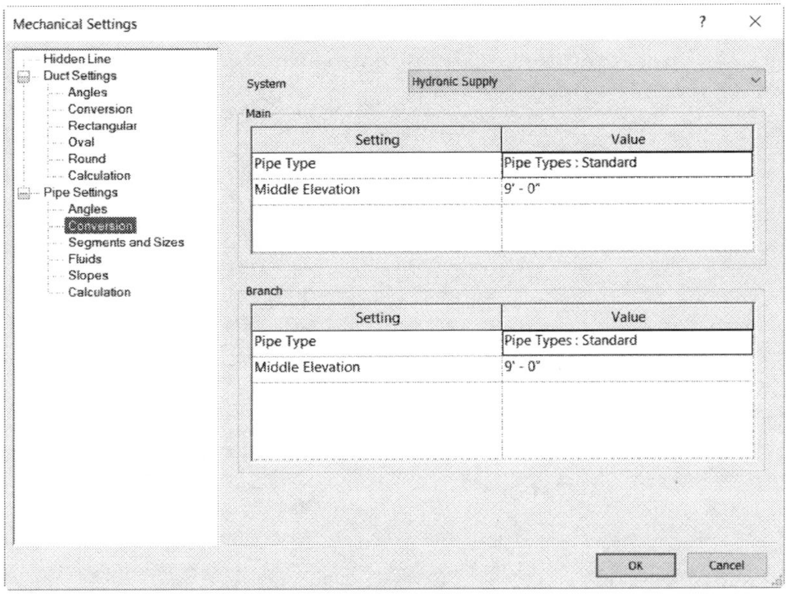

Figure 7-16 *Various properties displayed related to pipe settings*

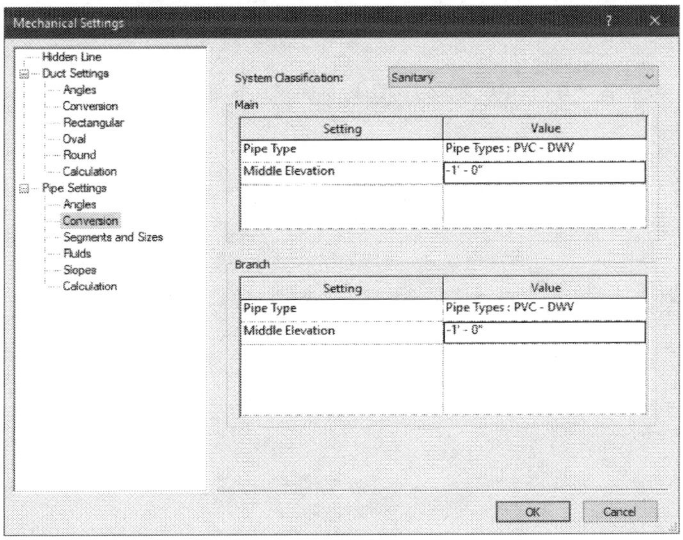

Figure 7-17 *The pipe settings for the* ***Sanitary*** *system*

9. Choose **OK**; the **Mechanical Settings** dialog box is closed.

10. Select the water closet and the floor drain fixtures, as shown in Figure 7-18; the **Modify | Plumbing Fixtures** tab is displayed.

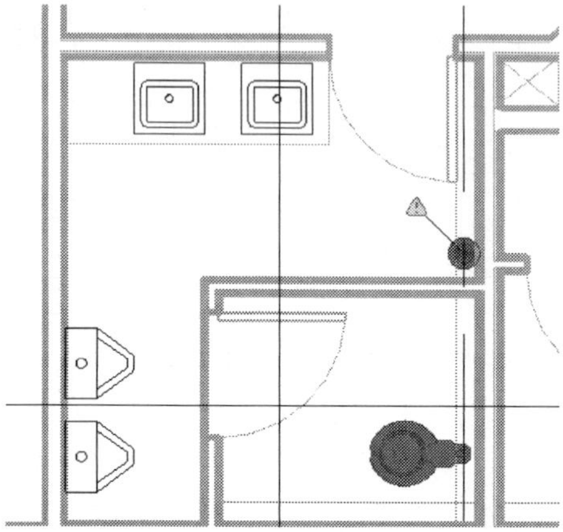

Figure 7-18 *The floor drain and the water closet fixtures selected*

11. In the **Create Systems** panel, choose the **Piping** tool; the **Create Piping System** dialog box is displayed.

12. In this dialog box, ensure that the **Sanitary** option is selected in the **System type** drop-down list.

13. In the **System name** edit box, type **Drainage System - Toilet 1** and choose **OK**; the fixtures in the system are bounded in a rectangular box, as shown in Figure 7-19.

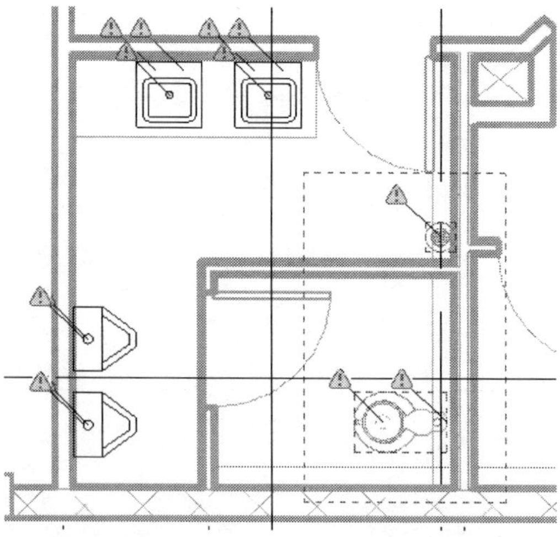

Figure 7-19 *The fixtures in the system bounded in a rectangular box*

14. Choose the **Generate Layout** tool from the **Layout** panel; the **Generate Layout** contextual tab is displayed.

15. In the **Options Bar**, ensure that the **Network** option is selected in the **Solution Type** drop-down list and then choose the **Next Solution** button, "**>**"; the text **2 of 5** is displayed before it and a solution is displayed in the drawing, as shown in Figure 7-20.

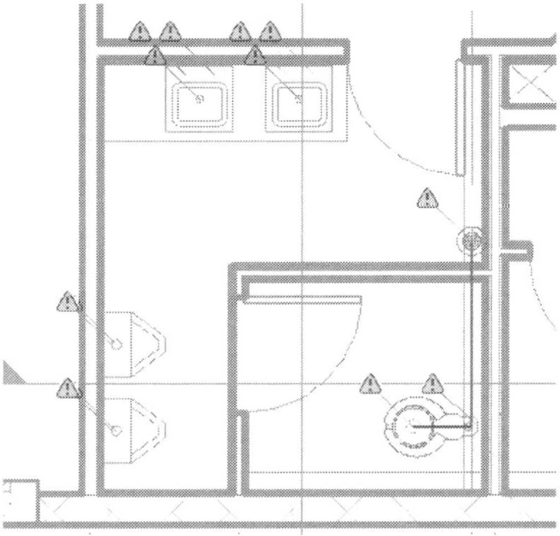

Figure 7-20 *The layout solution for the system*

16. In the **Options Bar**, choose the **Settings** button; the **Pipe Conversion Settings** dialog box is displayed.

17. In the left pane of this dialog box, ensure that the **Main** option is selected and then in the right pane, enter **-1'** (**-305 mm**) in the **Offset** parameter. Next, select the **Pipe Types: PVC - DWV** option in the **Pipe Type** parameter.

18. In the left pane of the **Pipe Conversion Settings** dialog box, select the **Branch** option. In the right pane, enter **-1'** (**-305 mm**) for the **Offset** parameter and select **Pipe Types: PVC - DWV** for the **Pipe Type** parameter.

19. Choose **OK**; the **Pipe Conversion Settings** dialog box is closed.

20. Choose the **Finish Layout** tool from the **Generate Layout** panel of the **Generate Layout** tab; the pipe system is created between the fixtures. Now, choose the **Detail Level** option in the **View Control Bar**; a flyout is displayed. Select the **Fine** option from the displayed flyout.

21. Next, select the floor drain fixture and then choose the **Piping Systems** tab; the options in this tab are displayed.

22. In the **Piping Systems** tab, ensure that the **Drainage System-Toilet 1** option is selected in the **System Selector** drop-down list in the **System Tools** panel and then choose the **Edit System** tool; the **Edit Piping System** contextual tab is displayed.

23. In this tab, ensure that the **Add to System** button is chosen in the **Edit Piping System** panel and then select the two urinals and sinks, refer to Figure 7-21.

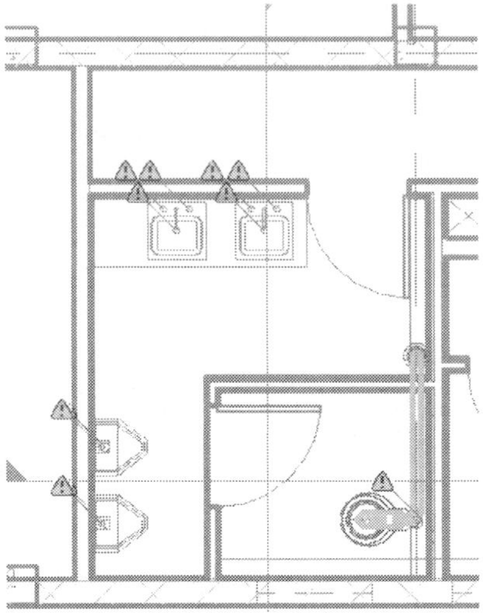

Figure 7-21 *The two urnials and sinks selected*

24. Choose the **Finish Editing System** tool from the **Mode** panel; the urinals and the sinks are added to the system.

25. Next, select any of the two urinals added to the system and then choose the **Generate Layout** tool from the **Layout** panel of the **Modify | Plumbing Fixture** tab; the default layout is displayed in the drawing area.

26. In the **Options Bar**, ensure that the **Network** option is selected in the **Solution Type** drop-down list and the choose the **Next Solution** button three times; the text **4 of 4** is displayed before the button and the solution layout is displayed in the drawing area, as shown in Figure 7-22.

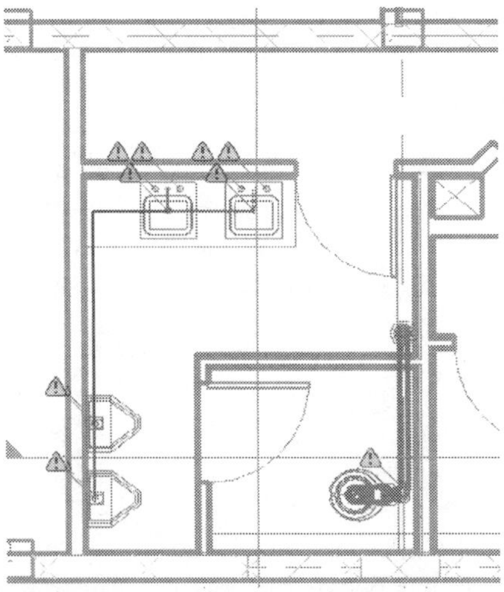

Figure 7-22 *The solution layout of the added fixture*

27. Choose the **Finish Layout** tool from the **Generate Layout** panel of the **Generate Layout** tab; the layout is generated.

Viewing the Plumbing System in 3D

In this section, you will display the 3D view of the plumbing system.

1. Choose the **Manage Links** tool from the **Manage Project** panel of the **Manage** tab; the **Manage Links** dialog box is displayed.

2. In this dialog box, select the **c07_ archi_ plumbing_tut1** option for Imperial (**M_c07_ archi_ plumbing_tut1** option for Metric) in the **Linked Name** column and then choose the **Unload** button; the **Unload Link** dialog box is displayed.

3. Choose the **OK** button; the dialog box is closed.

4. Choose **OK** in the **Manage Links** dialog box to close it.

5. Choose the **Default 3D View** tool from **View > Create > 3D View** drop-down; the 3D view of the system is displayed, as shown in Figure 7-23.

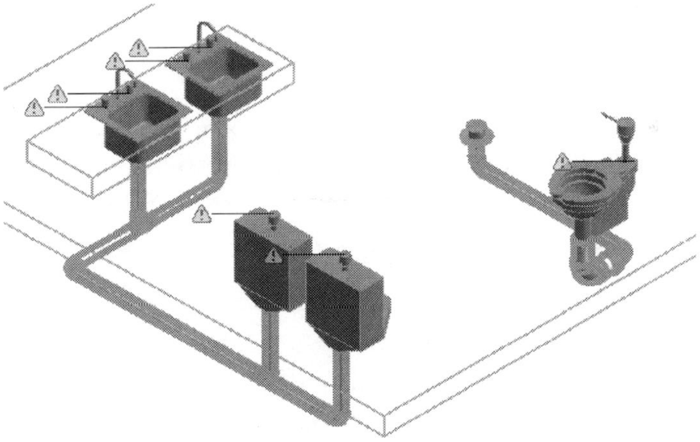

Figure 7-23 *The 3D view of the plumbing system*

Saving the Project

In this section, you need to save the project and the settings using the **Save As** tool.

1. Choose **Save As > Project** from the **File** menu; the **Save As** dialog box is displayed.

2. In this dialog box, browse to *C:\rmp_2025\ c07* folder. Next, in the **File name** edit box, enter the text **c07_Office-Space_tut1a** for Imperial or **M_c07_Office-Space_tut1a** for Metric and then choose the **Options** button; the **File Save Options** dialog box is displayed.

3. Now, choose the **OK** button; the **File Save Options** dialog box is closed and the **Save As** dialog box appears.

4. In the dialog box, choose the **Save** button; the current project file is saved with the specified name and the **Save As** dialog box is closed.

Closing the Project

1. Choose the **Close** option from the **File** menu; the file is closed and this completes the tutorial.

EXERCISE

Exercise 1 Community Center

Download the *c07_Community-Center_exer1.rvt* file for Imperial or *M_c07_Community-Center_exer1.rvt* file for Metric from *www.cadsofttech.com*. Create a sanitary system for the *Community-Center* project with the following parameters and project specifications. Figures 7-24 and 7-25 show the sanitary system.

(Expected time: 1hr)

1. Project View used for creating the plumbing system:
   Floor Plans > Mechanical >1ST FLOOR- PLUMBING
2. Families to be used:
   For Imperial **Sink-Vanity Round 19"x19"**
   Toilet-Commercial-Wall-3D 15" Seat Height
   Urinal-Wall-3D
   For Metric **M_Sink-Vanity Round 482 x 482 mm**
   Toilet-Commercial-Wall-3D 380mm Seat Height
   Urinal-Wall-3D
3. Pipe Types: PVC
4. Offset Values:
   Main Pipe **-1'(-305mm)**
   Branch Pipe **-1'(-305mm)**
5. File name to be assigned:
   For Imperial *c07_Community-Center_exer1.rvt*
   For Metric *M_c07_Community-Center_exer1.rvt*

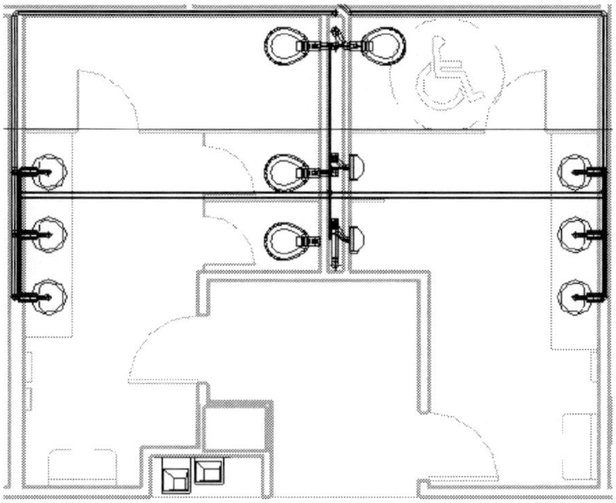

Figure 7-24 *Sanitary system*

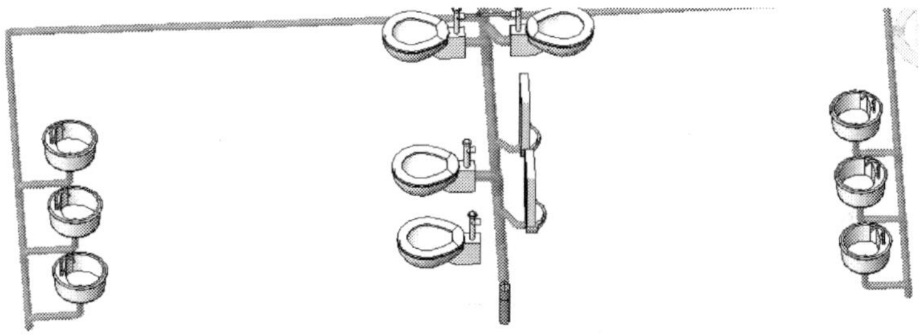

Figure 7-25 *The 3D view of a sanitary system*

This page is intentionally left blank

Chapter 8

Creating Fire Protection System

Learning Objectives

After completing this chapter, you will be able to:

- *Set up the fire protection system project*
- *Create space schedules*
- *Place sprinklers*

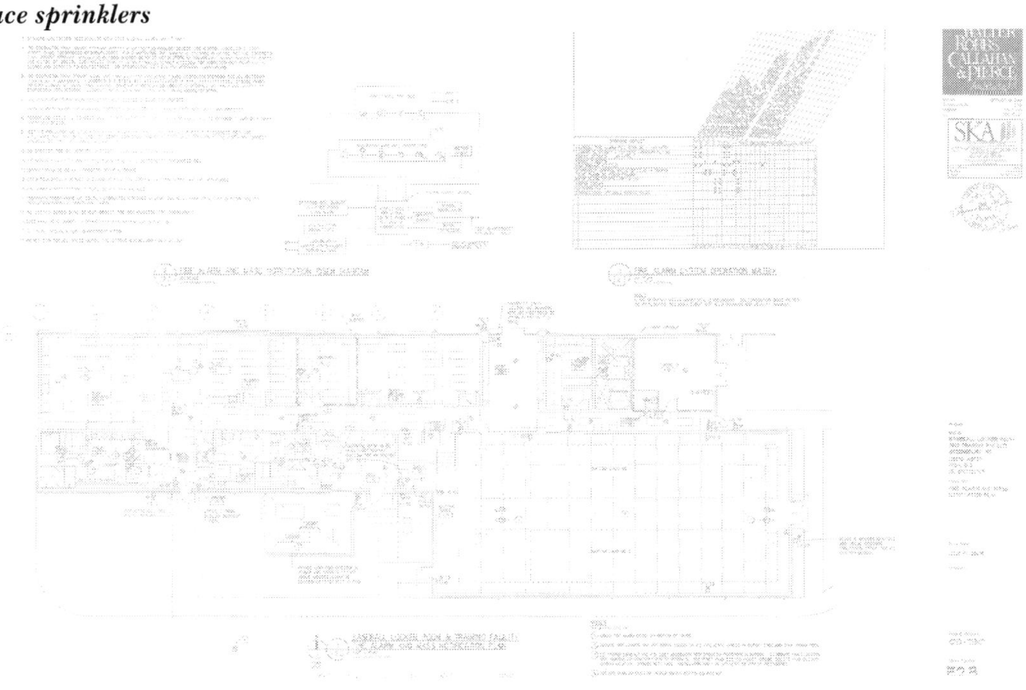

INTRODUCTION

Fire Protection or Fire Suppression is one of the major areas for a building design in a project. While designing a Fire Protection System for a project, the designer will use variety of methods and techniques. For example, for designing the schema of a fire protection plan, the designer should plan all the equipment required for the system. This will help in making the fire protection system more efficient and will also improve the coordination between other disciplines. A Fire Protection System consists of the following essential features: fire detection alarm, all types of fixed extinguishing systems, portable systems, private fire main and hydrants, pumping stations, smoke and heat evacuation system, centralized control system, auxiliary equipment, automotive and fire prevention system.

In this chapter, you will learn how to place fire protection equipment, create a wet fire protection system, create a dry protection system, and more.

FIRE PROTECTION SYSTEMS

A fire protection system comprises of sprinklers, pipes, and valves. In Revit, you can create a sprinkler system by placing sprinkler heads, such as upright and pendent as elements. These elements can be placed as hosted in the ceiling or as non-hosted elements. You can then connect the sprinklers with the pipes using the auto layout tools or by manually drawing the pipes. Further, you can check the fire protection system and its components for interferences with other components in a building. Figure 8-1 displays a 3d view of a fire protection system in an MEP project. The information stored within the system can be used for analysis or scheduling purposes. The various features and processes required for creating a fire protection system are discussed next.

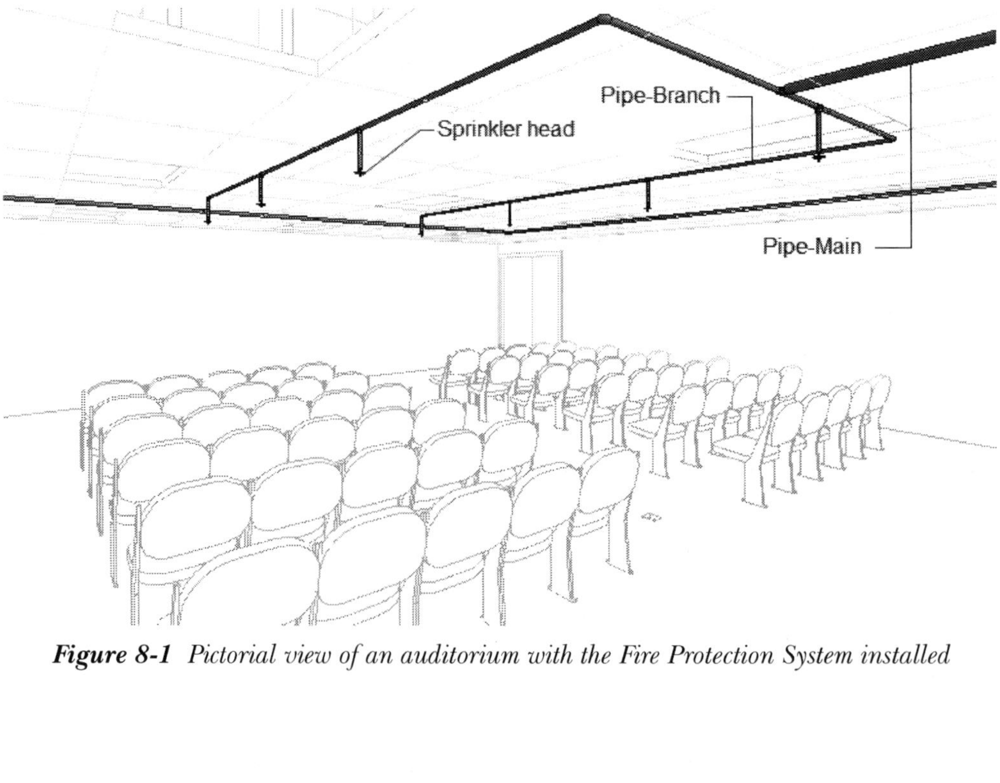

Figure 8-1 *Pictorial view of an auditorium with the Fire Protection System installed*

Sprinkler Libraries

A fire sprinkler, which is also called a sprinkler head, is an important component of a fire protection system. At the time of fire, the function of the sprinkler is to discharge water from its nozzle when a predetermined temperature has been exceeded. In an MEP project, you can use various sprinkler families for creating a fire protection system. To use the sprinkler families in a project, choose the **Load Family** tool from the **Load from Library** panel in the **Insert** tab; the **Load Family** dialog box will be displayed, as shown in Figure 8-2. In this dialog box, browse to **English-Imperial > Fire Protection > Sprinklers** (for metric, **English > US > Fire Protection > Sprinklers)** folder and select any of the sprinkler families that you want to insert. Next, choose the **Open** button; the selected sprinkler will be loaded in the project view. You can use this sprinkler in the project.

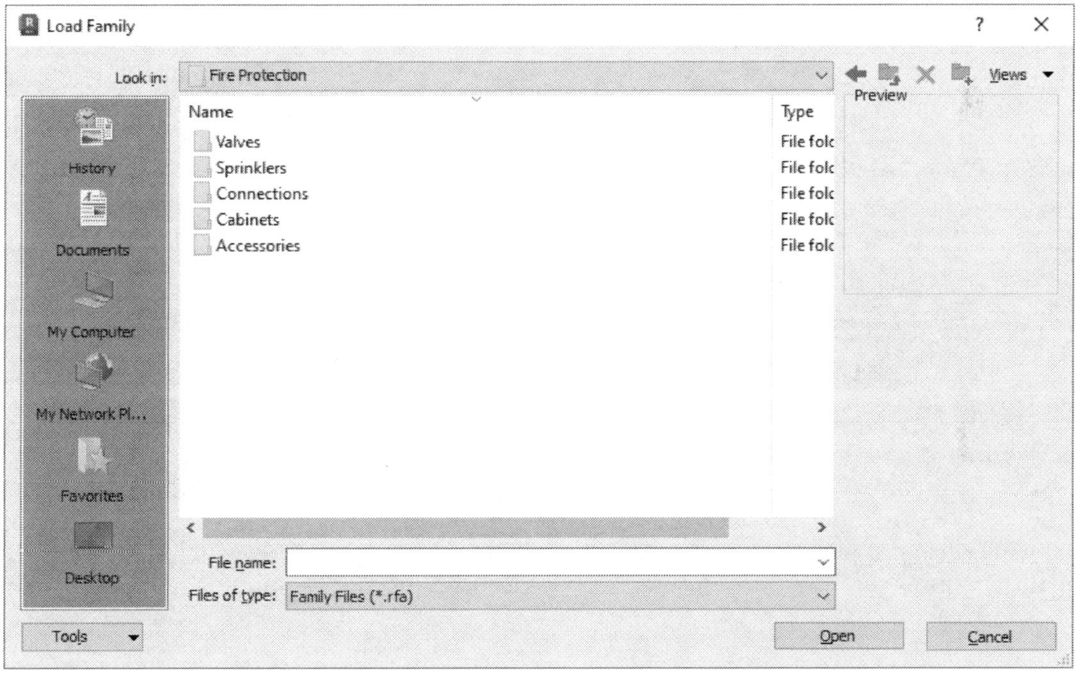

Figure 8-2 *The **Load Family** dialog box displaying various folders for fire protection equipment and accessories*

Piping Tools

In a fire protection system, the sprinklers are connected using pipes. You can connect the pipes manually to the sprinklers by using various tools available in the **Plumbing &Piping** panel of the **Systems** tab, as shown in Figure 8-3. Alternatively, you can connect the sprinklers with the pipe by using the auto-routing feature of Revit 2025.

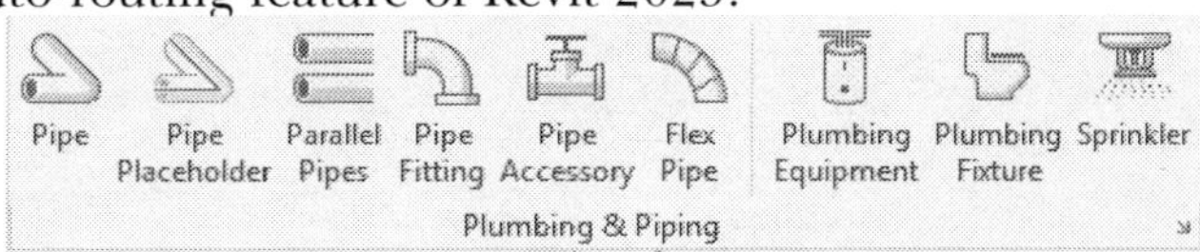

Figure 8-3 *The piping tools in the **Plumbing & Piping** panel*

Wet and Dry Fire Protection Systems

A wet pipe sprinkler system uses automatic sprinkler heads that immediately discharge water when exposed to heat from a fire. The system includes a network of pipes filled with pressurized water, connected through fire risers, automatic sprinklers, and an alarm check valve. In contrast, a dry pipe sprinkler system contains pressurized air or nitrogen instead of water. When a fire triggers a sprinkler, the air is released, allowing water to flow through the pipes. Dry systems are used in areas prone to freezing, such as unheated buildings or parking garages, to prevent water in the pipes from freezing. Sprinkler heads are specific to each system type and cannot be interchanged.

DESIGNING THE FIRE PROTECTION SYSTEM

In Revit, you can design both wet and dry fire protection systems. The designing of a fire protection system involves the following steps: setting up the fire protection system, creating the space schedule, placing sprinkler heads, and then connecting the sprinklers. To perform these steps, you need to create different views and pipe types, insert fittings, modify the pipes and fittings manually, create schedules, and size and tag the pipes. In the next sections, the procedure and the steps involved in designing a fire protection system are discussed next.

> **Tip**
> *While designing a fire protection system for a building, the following elements are considered critical: Water Supply in the building or site, occupancy of the building (light hazard, ordinary hazard, extra hazard), and building construction*

Setting Up a Fire Protection System

The setting up of a fire protection system, both wet and dry, involves various steps. First, you need to create a project view in which you can draw the layout of the fire protection system. Next, you need to create different pipe types and configure different conversion settings for the pipes that will be used to physically connect the sprinklers. Also, you need to sketch the supply pipe for the fire protection system. The steps involved in setting the fire protection system project are discussed next.

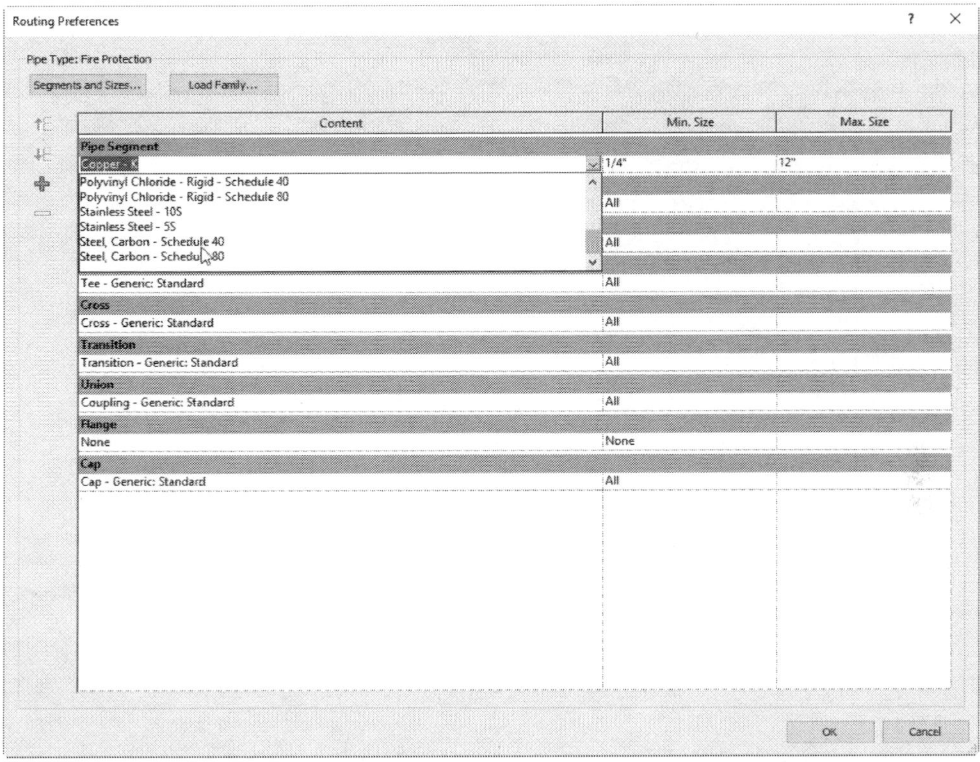

Figure 8-4 *Selecting the* ***Steel, Carbon - Schedule 40*** *option from the drop-down list in the* ***Routing Preferences*** *dialog box*

Creating the Space Schedule

Before placing the sprinklers in a project, it is necessary to estimate the quantity of sprinklers required for each space in a building. Figure 8-5 shows a space schedule for the fire protection system in an MEP project. By using the space schedule, you can place the sprinklers in the spaces as per the industry standards. To create a schedule for the estimation of the sprinkler, you can choose the **Schedule/Quantities** tool from the **Reports & Schedules** panel of the **Analyze** tab. The advantage of creating a schedule prior to adding sprinklers in the space is that you can keep track of the sprinklers added in the spaces as per the design requirement.

<Space Schedule-Fire Protection>

A	B	C	D	E
Number	Name	Level	Area	Sprinkler Quantity- Min. Required
101	CEO-Office	Level 1	156 SF	1.20
102	V.P. Marketing	Level 1	117 SF	0.90
103	Toilet-1	Level 1	70 SF	0.54
104	Rest Room	Level 1	97 SF	0.75
105	Cafeteria	Level 1	109 SF	0.84
106	Store	Level 1	84 SF	0.64
107	Purchase	Level 1	153 SF	1.18
108	International	Level 1	104 SF	0.80
109	H.R.	Level 1	130 SF	1.00
110	Central Area	Level 1	1015 SF	7.81
111	Quality	Level 1	160 SF	1.23
112	Open	Level 1	157 SF	1.21
113	Conference-2	Level 1	216 SF	1.66
114	Conference	Level 1	201 SF	1.55
115	Lounge	Level 1	844 SF	6.49
116	Meeting Room	Level 1	106 SF	0.81
117	Service	Level 1	84 SF	0.64
118	Shaft	Level 1	63 SF	0.49
119	Services	Level 1	75 SF	0.57
120	Accounts	Level 1	145 SF	1.11
121	Chase	Level 1	83 SF	0.64
122	Electrical	Level 1	139 SF	1.07
123	Server	Level 1	114 SF	0.87
124	Store-1	Level 1	58 SF	0.44
125	Service-2	Level 1	200 SF	1.54
126	Toilet-Mens	Level 1	56 SF	0.43
128	Toilet-c	Level 1	27 SF	0.20

Figure 8-5 *The schedule displaying the quantity of sprinklers required for different spaces in a project*

Tip

While designing a fire protection system for a building, the following coverage per sprinkler are used (maximum square foot or square metres): Light Hazard =225 sq. ft. (20.9 sq.m); Ordinary Hazard=130 sq. ft (12.077 sq. m); and Extra Hazard=100 sq.ft.(9.29 sq. m)

Placing Sprinkler Heads

A sprinkler head in a fire protection system discharges water when a fire is detected, typically due to rising temperatures. In MEP projects, sprinkler heads are either hosted or non-hosted. Hosted sprinklers, shown in **Figure 8-6**, are face-based and must be placed on surfaces like ceilings or walls, while non-hosted sprinklers are independent and can be placed freely with the correct offset. To place them, use the **Sprinkler tool** from the **Systems** tab in a suitable floor or ceiling plan view. Hosted sprinklers are best placed in the ceiling plan, and non-hosted ones in the floor plan. After loading the desired sprinkler family and adjusting settings in the **Properties palette**, place the sprinklers by clicking on the appropriate location.

Tip

While placing a standard sprinkler following are the maximum spacings recommended for different occupancy: Light and Ordinary Hazard=15'-0" (4.572 m) and Extra Hazard = 12'-0" (3.657m)

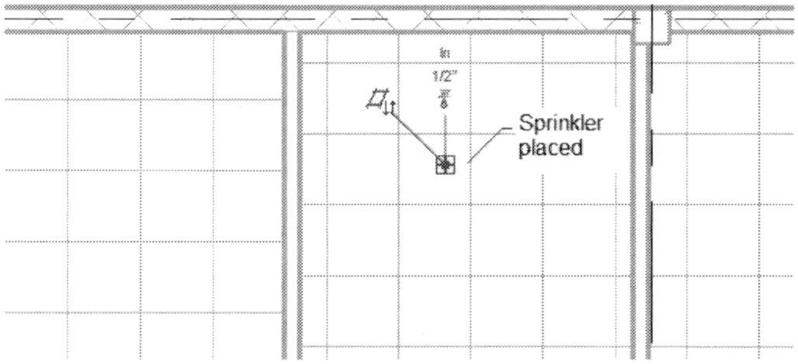

Figure 8-6 The sprinkler placed in the project view

Connecting the Sprinklers

After placing the sprinklers, they need to be connected through a piping system using either the auto-routing or manual connection method. To use the auto-routing method, select the sprinklers in the project view; this activates the **Modify | Sprinklers** tab. From there, choose the **Piping** tool in the **Create Systems** panel. This action opens the **Create Piping System** dialog box, as shown in **Figure 8-7**.

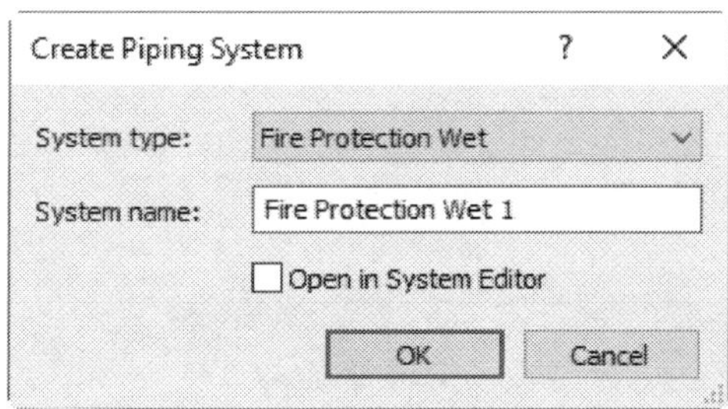

***Figure 8-7** The **Create Piping System** dialog box*

In this dialog box, select the type of system from the **System type** drop-down list and then enter a name for the system in the **System name** edit box. Now, choose the **OK** button; the **Create Piping System** dialog box will be closed and the **Modify | Piping Systems** contextual tab will be displayed, as shown in Figure 8-8.

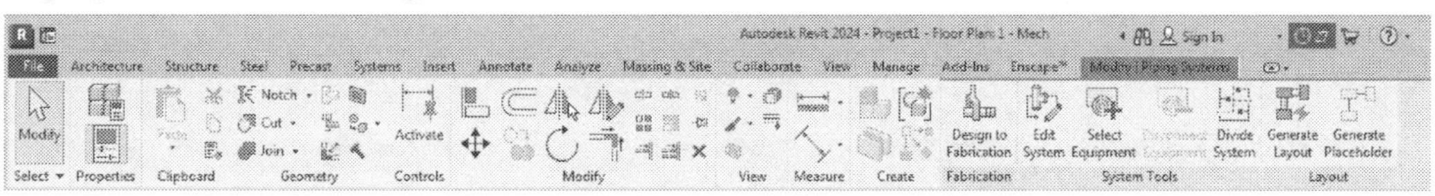

***Figure 8-8** The **Modify / Piping Systems** contextual tab*

To generate a pipe layout for sprinklers, go to the Layout panel and select the **Generate Layout** tool, which opens the contextual tab for layout configuration. In the **Options Bar**, choose a layout type from the **Solution Type** drop-down—**Network**, **Perimeter**, or **Intersections**, with **Network** selected by default. Use the **Previous Solution** and **Next Solution** buttons to preview different layout options. Click the **Settings** button next to Next Solution to open the **Pipe Conversion Settings** dialog box, where you can define branch and main pipe parameters. After setting preferences, click **OK** to close the dialog. To complete the setup, click the **Finish Layout** button to connect the sprinklers to the pipe network, as illustrated in **Figure 8-9**.

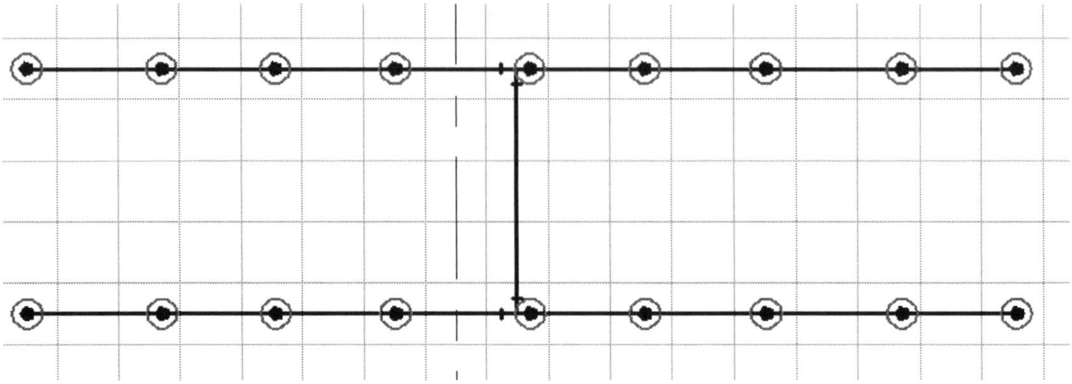

Figure 8-9 The sprinklers connected to the piping system

TUTORIAL

General instructions for downloading tutorial files:

1. Download the *c08_tutorial.zip* file for this tutorial from *https://www.cadsofttech.com*. The path of the folder is as follows: *Books > Civil/GIS BOOKS > Revit MEP > Revit MEP 2025 For Novices*

2. Now, save and extract the downloaded zip file at the following location: *C:\rmp_2025\c08_rmp_2025_tut*

Tutorial 1 Office Space- Fire Suppression

In this tutorial, you will create pipe types for dry and wet fire protection system. Also, you will create a space schedule to estimate the quantity of sprinklers required for each space in the model. **(Expected time: 45 min)**

1. File name to be used: *c08_office_fire_protections.rvt (M_ c08_office_fire_protections.rvt)*
2. File name to be assigned: *c08_Office-Space_tut1.rvt (M_c08_Office-Space_tut1.rvt)*

The following steps are required to complete this tutorial:

a. Open the *c08_office_fire_protections.rvt* (*M_ c08_office_fire_protections.rvt*) file.
b. Create pipe types.
c. Configure the pipe settings.
d. Create space schedule for estimating sprinklers.
e. Save the project using the **Save As** tool.
f. Close the project by using the **Close** tool.

Opening a Project

In this section, you will download the *c08_office_fire_protections.rvt (M_08_office_fire_protections.rvt)* file from *www.cadsofttech.com* and then open the file.

1. To open a file, choose **Open > Project** from **File** menu; the **Open** dialog box is displayed.

2. In the **Open** dialog box, browse to *C:\rmp_2025\c08_rmp_2025_tut* and then select the *c08_office_fire_protections.rvt* file (for Metric, select the *M_c08_office_fire_protections.rvt*). Now, choose the **Open** button; the selected file opens.

Note
The architectural model named c04_archi_spaces_rmp_2025.rvt linked in this tutorial file is located in the c04_rmp_2025_tut folder. It is recommended to maintain the relative path to the architectural model. However, if the link is lost, you need to reload the file using the ***Manage Links*** *dialog box which has already been discussed in the previous chapters.*

Creating Pipe Types

In this section, you will create new pipe types for the fire suppression system.

1. In the **Project Browser**, click on the "+" symbol on the left of the **Families** node to expand it.

2. In the expanded **Families** node, click on **Pipes > Pipe Types > Standard**; the **Standard** node is highlighted.

3. Next, right-click; a shortcut menu is displayed. From the shortcut menu, choose the **Duplicate** option; a duplicate node with the name **Standard 2** is created and is displayed under the **Standard** node.

4. Select the **Standard 2** node and then right-click; a shortcut menu is displayed. From the shortcut menu, choose the **Rename** option; the existing name is displayed in an edit box.

5. In the edit box, type **Wet Protection-Fire** and press ENTER; the new name entered is displayed, as shown in Figure 8-10.

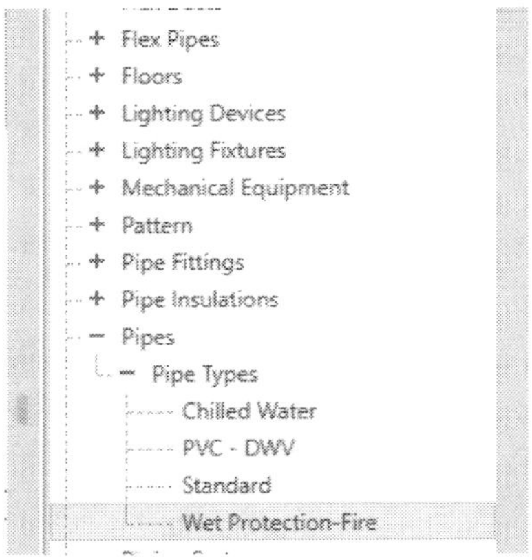

Figure 8-10 *The* ***Project Browser*** *displaying the* ***Wet Protection-Fire*** *type*

6. Next, ensure that the **Wet Protection-Fire** node is selected and then right-click; a shortcut menu is displayed. From the shortcut menu, choose the **Type Properties** option; the **Type Properties** dialog box is displayed, as shown in Figure 8-11.

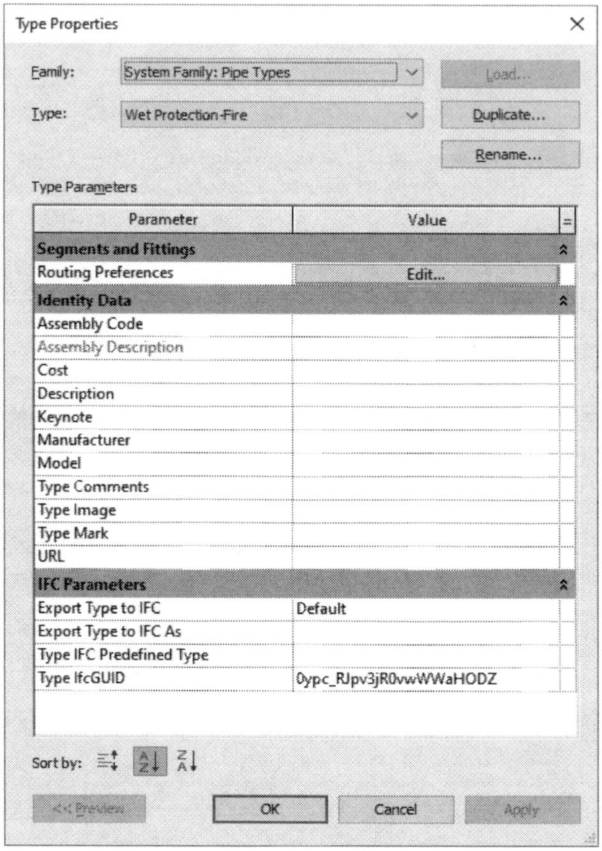

***Figure 8-11** The **Type Properties** dialog box for the wet protection type*

7. In this dialog box, click on the cell corresponding to the **Assembly Code** parameter and then choose the three dot button displayed inside it; the **Choose Assembly Code** dialog box is displayed.

8. In the displayed dialog box, select the **D-Services** > **D40-Fire Protection** option under the **Uniformat Classification** head, as shown in Figure 8-12.

*Figure 8-12 The **D40-Fire Protection** option selected*

9. Choose the **OK** button; the **Choose Assembly Code** dialog box is closed. Also, note that in the **Type Properties** dialog box, the **Assembly Code** parameter is assigned as **D40** and the **Assembly Description** parameter displays **Fire Protection** in its **Value** field.

10. In the **Type Properties** dialog box, choose the **Duplicate** button; the **Name** dialog box is displayed.

11. In the **Name** edit box of the **Name** dialog box, type the **Dry Protection-Fire** text and then choose the **OK** button; the **Name** dialog box is closed and the **Dry Protection-Fire** type is displayed in the **Type** drop-down list in the **Type Properties** dialog box.

12. Repeat steps 6 to 9 to assign the assembly code to the new type.

13. In the **Type Properties** dialog box, choose the **OK** button; the dialog box is closed. Notice that in the **Project Browser**, the **Dry-Protection-Fire** type is displayed under the **Pipe Types** head.

Configuring the Pipe Settings

In this section, you will configure the pipe types for the fire protection system.

1. Choose the **Mechanical Settings** button from the **Plumbing & Piping** panel of the **Systems** tab; the **Mechanical Settings** dialog box is displayed.

2. In the left pane of this dialog box, click on the **Conversion** node under the **Pipe Settings** node; various options related to the **Conversion** node are displayed in the right pane, as shown in Figure 8-13.

3. In the right pane, select the **Fire Protection Wet** option from the **System Classification** drop-down list.

4. In the **Main** area, click in the **Value** field corresponding to the **Pipe Type** parameter; a drop-down list is displayed.

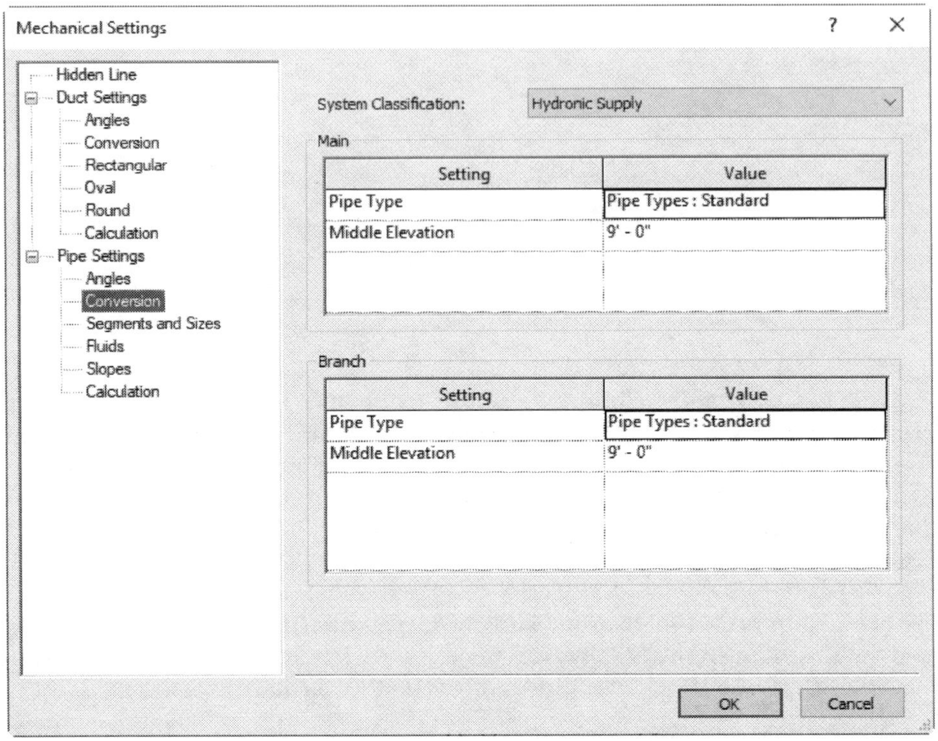

Figure 8-13 *Various options related to the* ***Conversion*** *node displayed in the* ***Mechanical Settings*** *dialog box*

5. From the drop-down list displayed, select the **Pipe Types: Wet Protection-Fire** option.

6. In the **Main** area, click in the **Value** field corresponding to the **Middle Elevation** parameter and replace the existing value of **9' - 0" (2750)** with **9' - 3" (2820)**.

7. In the **Branch** area, assign the same settings as specified in the **Main** area.

Note

The offset values specified for the main and branch pipes refers to the placement of the main and branch pipes with respect to the referenced level. Also, it is important to note that the branch offset allows you to automatically create branches that run above or below the main and other obstacles. This is useful for avoiding interference with pipes, duct, structural beams, or architectural components.

8. Now, repeat the procedure followed in steps 3 to 7 to assign the conversion settings for the **Dry Fire Protection** system. For creating the settings, use the following specifications:

   System Classification: **Fire Protection Dry**
   Pipe Type (Main and Branch areas): **Pipe Types : Dry Protection-Fire**
   Middle Elevation (Main and Branch area): **9'3"(2820 mm)**

9. After configuring the settings for the pipes, choose the **OK** button; the **Mechanical Settings** dialog box is closed.

Creating a Space Schedule for Estimating Sprinklers

In this section, you will create a space schedule and find out the quantity of the sprinklers in the spaces.

1. Choose the **Schedules/Quantities** tool from the **Reports & Schedules** panel of the **Analyze** tab; the **New Schedule** dialog box is displayed.

2. Click on the down arrow corresponding to the **Filter** list option and select the **Mechanical** check box from the drop-down list displayed. In the **Category** list box of the dialog box, click on the **Spaces** category. Now, in the **Name** edit box, enter a new name **Space Schedule-Fire Protection**. Also, ensure that in the **New Schedule** dialog box, the **Schedule building components** radio button is selected and the **New Construction** option is selected from the **Phase** drop-down list.

3. In the **New Schedule** dialog box, choose the **OK** button; the **Schedule Properties** dialog box is displayed with the **Fields** tab chosen, as shown in Figure 8-14.

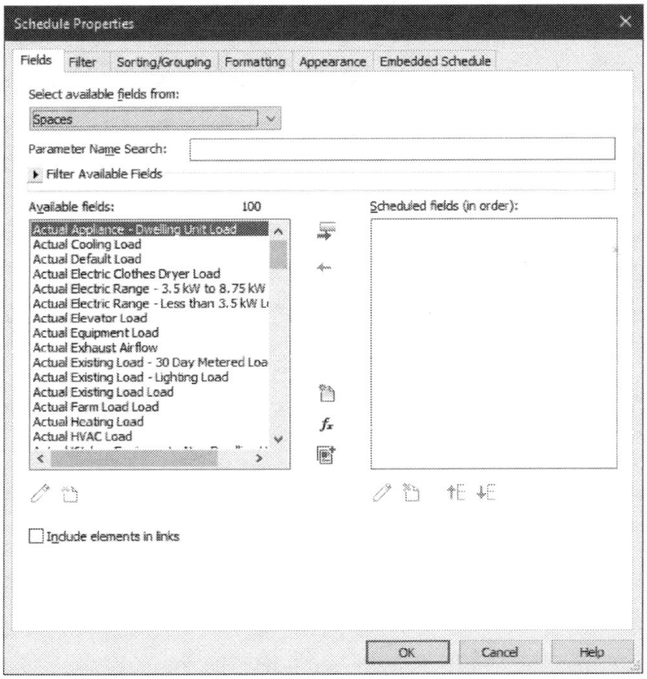

*Figure 8-14 The **Schedule Properties** dialog box with the **Fields** tab chosen*

4. In the **Available fields** area of the **Fields** tab, select the **Area** field. Now, press and hold the CTRL key and then select the following fields from the **Available fields** area: **Level**, **Name**, and **Number**.

5. Now, release the CTRL key and choose the **Add parameter(s)** button; the selected fields are added to the **Scheduled fields (in order)** area. Note that the added fields are selected and are highlighted in blue color, refer to Figure 8-15. Now, click on the **Number** field in the **Scheduled fields (in order)** area; other fields except the **Number** field get deselected.

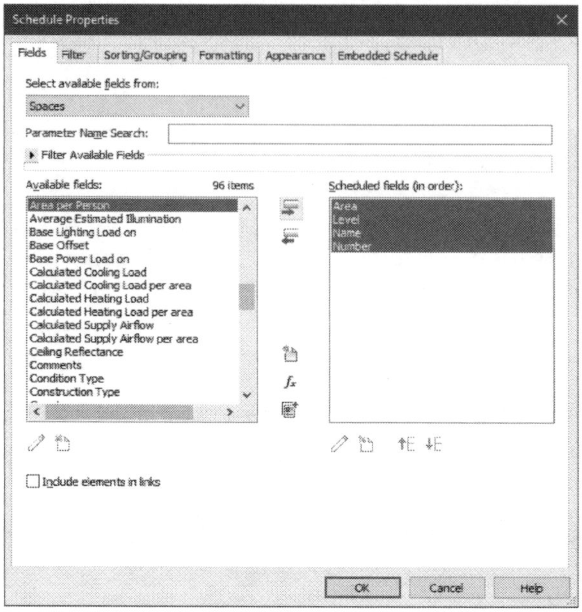

***Figure 8-15** The **Schedule Properties** dialog box displaying the added fields*

6. Now, choose the **Move parameter up** button thrice; the **Number** field moves up in the order above the **Area** field. Similarly, arrange the remaining fields in the order **Name**, **Level**, and **Area** (Top to bottom), refer to Figure 8-16.

7. Now, choose the **Add calculated parameter** button; the **Calculated Value** dialog box is displayed.

8. In the **Name** text box of the dialog box, enter **Sprinkler Quantity- Min. Required** and then ensure that the **Formula** radio button is selected.

9. Also, ensure that the **Common** and **Number** options are selected from the **Discipline** and **Type** drop-down lists, respectively.

10. Now, in the **Calculated Value** dialog box, choose the browse button next to the **Formula** edit box; the **Fields** dialog box is displayed.

11. In the displayed dialog box, select the **Area** field and then choose the **OK** button; the dialog box is closed and the **Area** field is added to the **Formula** edit box.

12. In the **Formula** edit box, type **/130** (for Metric unit type **/12**) after the **Area** text to complete the formula as Area/130 (for Metric unit type Area/12).

    The fire protection code requires one sprinkler for every 130 square feet (sprinkler required for every 12 square meter).

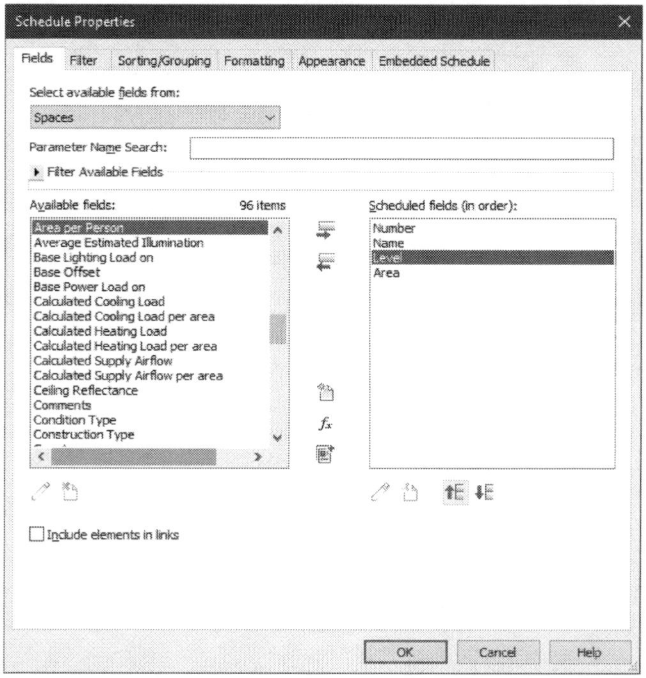

Figure 8-16 *The* ***Scheduled fields (in order)*** *area displaying the re-ordered fields*

Note

The formula that you enter in the ***Formula*** *edit box is case sensitive.*

13. Now, choose the **OK** button; the **Calculated Value** dialog box is closed and the **Sprinkler Quantity- Min. Required** parameter is added to the **Scheduled fields (in order)** area.

14. Choose the **Filter** tab from the **Schedule Properties** dialog box; the options in this tab are displayed.

15. Select the **Level** option from the **Filter by** drop-down list and also ensure that the **equals** option is selected in the drop-down list displayed next to it.

16. Next, select the **Level 1** option from the drop-down list located at the third position corresponding to the **Filter by** parameter, if it is not selected by default.

17. Now, ensure that the **(none)** option is selected in the **And** drop-down list.

18. In the **Schedule Properties** dialog box, choose the **Sorting/Grouping** tab and then select the **Number** option from the **Sort by** drop-down list. Also, ensure that the **Ascending** radio button located next to it is selected.

19. Now, ensure that the **(none)** option is selected in the **Then by** drop-down list and then select the **Grand totals** check box. Ensure that the **Title, count, and totals** option is selected in the drop-down list displayed next to it.

20. Choose the **Formatting** tab. Next, in the **Fields** area, select the **Sprinkler Quantity- Min. Required** field; the right pane displays the information related to the selected field.

21. In the right pane of the **Formatting** tab, choose the **Field Format** button; the **Format** dialog box is displayed.

22. In the **Format** dialog box, clear the **Use default settings** check box and then select the **Fixed** option from the **Units** drop-down list.

23. In the **Format** dialog box, select the **2 decimal places** option from the **Rounding** drop-down list.

24. Now, choose the **OK** button; the **Format** dialog box is closed. In the **Formatting** tab, select the **Calculate totals** option from the drop-down list located below the **Show conditional format on sheets** check box and then choose the **OK** button again; the **Schedule Properties** dialog box is closed; the schedule is displayed in the drawing area, as shown in Figure 8-17.

<Space Schedule-Fire Protection>

A	B	C	D	E
Number	Name	Level	Area	Sprinkler Quantity- Min. Required
101	CEO-Office	Level 1	156 SF	1.20
102	V.P. Marketing	Level 1	117 SF	0.90
103	Toilet-1	Level 1	70 SF	0.54
104	Rest Room	Level 1	97 SF	0.75
105	Cafeteria	Level 1	109 SF	0.84
106	Store	Level 1	84 SF	0.64
107	Purchase	Level 1	153 SF	1.18
108	International	Level 1	104 SF	0.80
109	H.R.	Level 1	130 SF	1.00
110	Central Area	Level 1	1015 SF	7.81
111	Quality	Level 1	160 SF	1.23
112	Open	Level 1	157 SF	1.21
113	Conference-2	Level 1	216 SF	1.66
114	Conference	Level 1	201 SF	1.55
115	Lounge	Level 1	844 SF	6.49
116	Meeting Room	Level 1	106 SF	0.81
117	Service	Level 1	84 SF	0.64
118	Shaft	Level 1	63 SF	0.49
119	Services	Level 1	75 SF	0.57
120	Accounts	Level 1	145 SF	1.11

Figure 8-17 *The schedule of the fire protection displayed*

Saving the Project

In this section, you need to save the project and settings using the **Save As** tool.

1. To save the project with the specified settings, choose **Save As > Project** from **File Menu**. As you are saving the project for the first time, the **Save As** dialog box is displayed.

2. In this dialog box, browse to *C:\rmp_2025\c08_rmp_2025_tut* and then in the **File name** edit box, enter **c08_Office-Space_tut1** (for Metric **M_c08_Office-Space_tut1**).

3. In the displayed dialog box, choose the **Save** button to save the current project file with the specified name and to close the **Save As** dialog box.

Closing the Project

1. To close the project, choose the **Close** option from the **File** menu.

EXERCISE

Exercise Fire Protection System

Download the *c08_Conference-Center_exer1.rvt* file from *https://www.cadsofttech.com*. The path of the file is as follows: *Books > Civil/GIS BOOKS > Revit MEP > Revit MEP 2025 For Novices*
Open the *c08_Conference-Center_exer1.rvt* file and create a fire protection system in the conference room for the *Conference-Center* project. For location of the room, refer to Figure 8-18. While creating the fire protection system, you will add sprinklers and then create a wet pipe system, as shown in Figure 8-19.

1. Project view to be used :
   Ceiling Plans > Mechanical > 1 - Ceiling Mech

2. Family type to be used:
   Sprinklers
   For Imperial Sprinkler-Pendent_Semi-Recessed-Hosted : 1/2" Pendent
   For Metric M_Sprinkler - Pendent - Semi-Recessed - Hosted : 15mm Pendent

3. File name to be assigned: *c08_Conference-Center_exer1a.rvt*

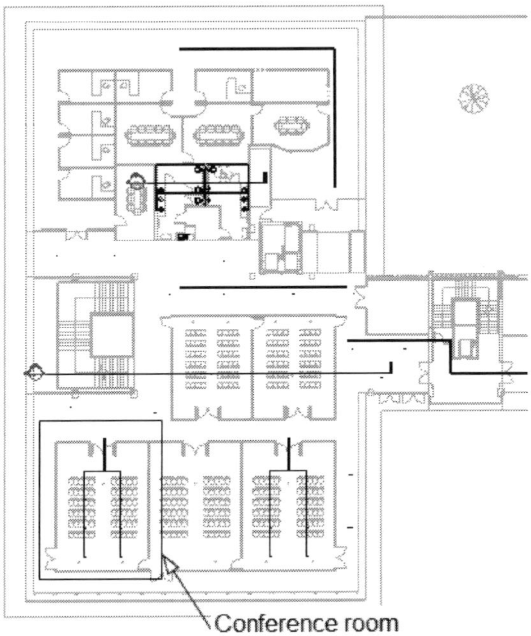

Figure 8-18 *The conference room for which fire protection system will be created*

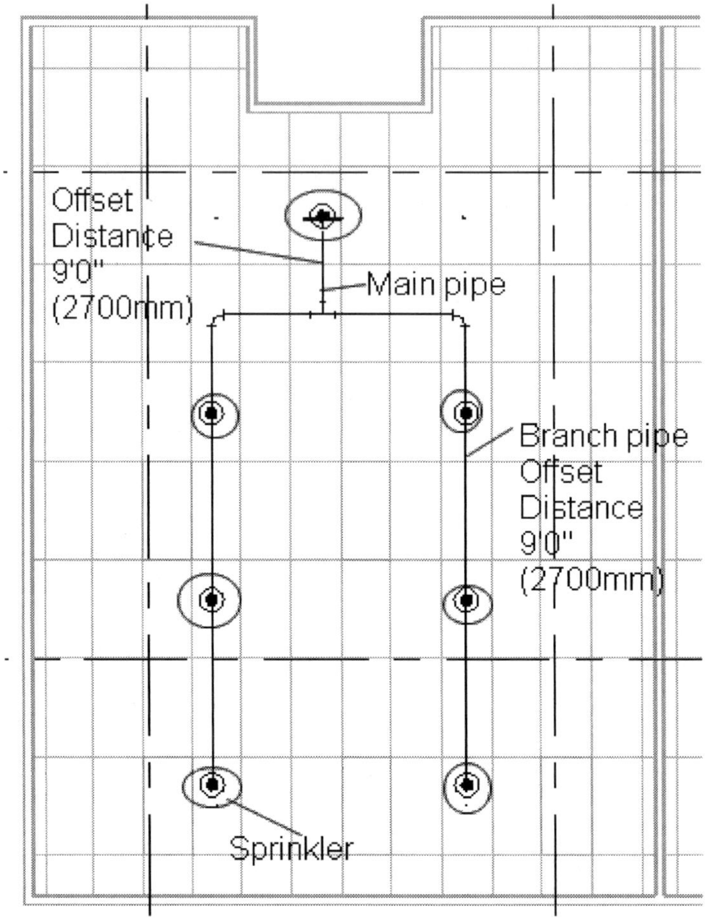

Figure 8-19 *The fire protection system created for the conference room*

Chapter 9

Creating Construction Documents

Learning Objectives

After completing this chapter, you will be able to:

- *Add dimensions*
- *Add text notes*
- *Add tags to elements*
- *Use callout views*
- *Create drafting views*
- *Duplicate views*
- *Add sheets to a project*

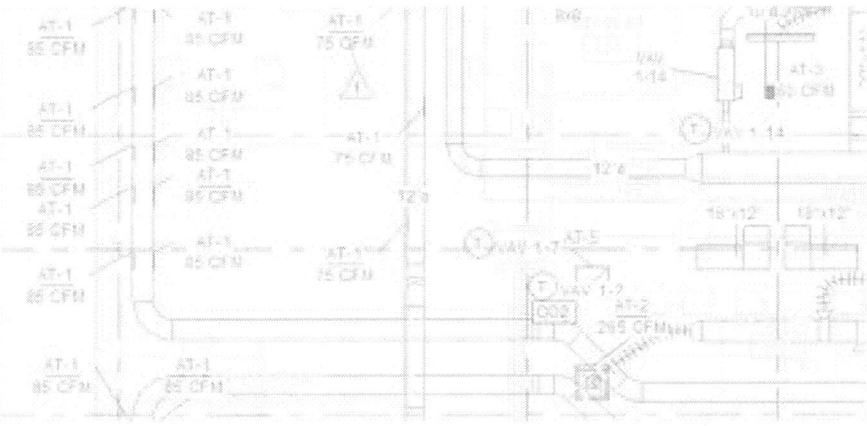

INTRODUCTION

In previous chapters, you learned to create building envelopes, design systems, and perform space analysis. This chapter focuses on documenting a project in Revit by creating working drawings using various tools. You'll learn how to add dimensions, callouts, tags, and text annotations to elements like ducts and pipelines. The chapter also covers adding, importing, and exporting views, placing them on sheets, and creating schedules and legends for comprehensive project documentation.

BIM CONCEPTS

In Revit MEP, creating comprehensive construction documents is essential for ensuring seamless project execution and collaboration. Use the ***View Templates****,* ***Schedules****, and* ***Sheets*** *tools to standardize and organize your documentation efficiently. Integrate all MEP systems with accurate parameters and annotations, adhering to BIM workflows to enhance coordination among stakeholders. Following the principles of ISO 19650, ensure your documents align with standardized naming conventions, classification systems, and a Common Data Environment (CDE) to maintain consistency, traceability, and compliance.*

DIMENSIONING

Dimensions play a crucial role in project documentation, especially in MEP design, by translating visual layouts into precise, real-world measurements. They define key parameters such as length, width, height, and angles, ensuring accurate construction on-site. Dimensions are as vital as visual representations, helping avoid misinterpretations and inconsistencies. In MEP projects, they reflect the actual size of elements and guide all stakeholders—like engineers, contractors, and estimators—throughout the construction process. The chosen unit of measurement also indicates the required level of precision for execution.

Types of Dimensions

In Revit, dimensions can be either temporary or permanent. Temporary dimensions appear when creating or selecting an element but are not visible in project views. Permanent dimensions, however, are view-specific and display the size or distance between elements like ducts, pipes, or conduits. You can also dimension arc lengths, angles, and spot elevations. Dimension styles such as size and color can be adjusted through the **Type Properties** dialog box. Both dimension types use the project's default units, but only permanent dimensions remain visible in the specific view where they were added.

Using Temporary Dimensions

In Revit, temporary dimensions appear dynamically when drawing, placing, or selecting elements, helping to position components accurately in relation to surrounding elements. These dimensions are not view-specific and show up in any view where elements are being manipulated. Temporary dimensions guide placement by snapping to the nearest perpendicular element based on predefined snap settings, as shown in **Figure 9-1**, where a return air diffuser is added. You can customize reference points for these dimensions using the **Temporary Dimension Properties** dialog box, accessed via **Manage > Settings > Additional Settings > Annotations** (see **Figure 9-2**). This dialog allows you to choose reference options for walls-such as Centerlines, Center of Core, Faces, or Faces of Core- and for doors and windows, either Centerlines or Openings, ensuring accurate and efficient placement during drafting.

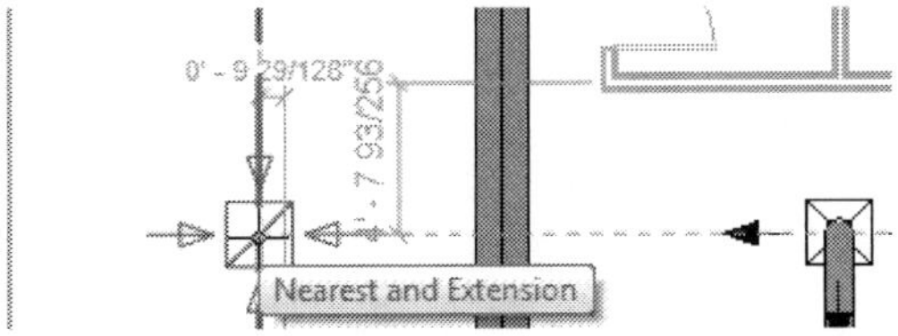

Figure 9-1 *Temporary dimensions displayed while placing a return air diffuser*

While using temporary dimensions in Revit, you can set the reference point of the element being placed by adjusting the settings in the **Temporary Dimension Properties** dialog box. To access it, go to **Manage > Settings > Additional Settings > Annotations**, and select **Temporary Dimensions**, which opens the dialog box shown in **Figure 9-2**. This dialog is divided into two sections: **Walls** and **Doors and Windows**. In the **Walls** section, you can choose from four reference options—**Centerlines**, **Center of Core**, **Faces**, and **Faces of Core**. For **Doors and Windows**, you can set the reference to either **Centerlines** or **Openings**. These settings allow you to control how temporary dimensions reference walls, doors, and windows during element placement.

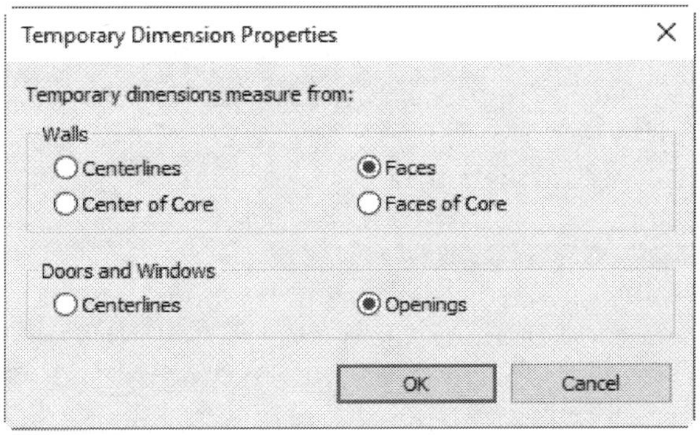

Figure 9-2 *The* ***Temporary Dimension Properties*** *dialog box*

Entities in a Dimension

Before using the dimensioning tools, it is important to understand various entities in a dimension. Figure 9-3 shows the terms used for various entities in a dimension. In the next section, these entities are discussed briefly.

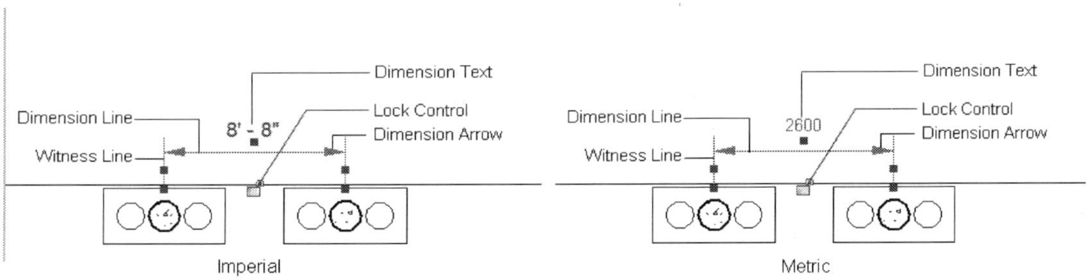

Figure 9-3 *Various entities in a permanent dimension*

Adding Permanent Dimensions

Permanent dimensions are added for a specific measurement. In Revit, you can access various dimension tools from the **Dimension** panel in the **Annotate** tab, as shown in Figure 9-4. You can choose appropriate dimension type and dimension tool to add dimensions to an element. The dimensioning tools are discussed next.

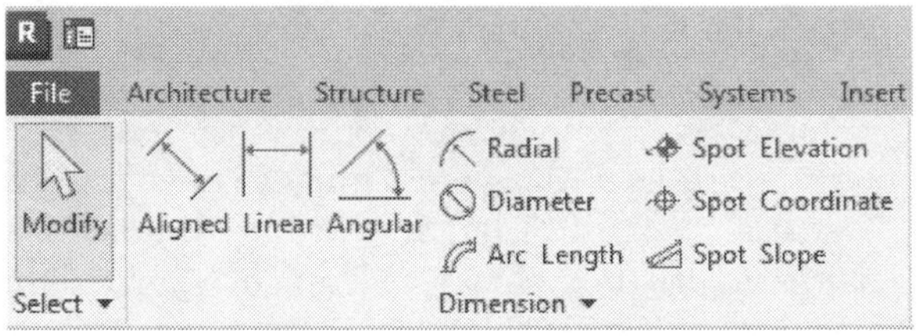

Figure 9-4 *Various dimensioning tools in the* ***Dimension*** *panel of the* ***Annotate*** *tab*

> **Tip**
> *While working in a BIM project, dimensions are the primary data points to communicate the design intent. You can add dimension to plans, elevations and sections, detail and drafting views, as well as sheets. Dimensions are also used to create parameters for the objects in BIM.*

Modifying Dimension Parameters

You can customize the appearance of a dimension by modifying parameters such as text font and the gap between the witness line and the element. To do this, select a dimension type like linear, aligned, angular, radial, or diameter, and the **Modify | Dimensions** tab will appear. For spot elevation, spot coordinates, or spot slope dimensions, a specific contextual tab will be shown instead. Within the tab, you can use the **Label** drop-down to assign a parameter to the selected dimension.

> **Tip**
> *Assigning a parameter to the selected dimension makes the task of the Revit modeler easy. Once you assign a parameter to the dimension, it can be then controlled from the* ***Global Parameters*** *dialog box. This dialog box can be invoked by typing GL or choosing the* ***Global Parameters*** *tool from the* ***Settings*** *panel of the* ***Manage*** *tab.*

To create a parameter for a selected dimension, first ensure the dimension is selected. Then, use the **Create Parameter** tool from the **Label Dimension** panel on the **Modify | Dimensions** tab. In the **Global Parameter Properties** dialog box, enter a name for the parameter and click **OK**. The new parameter will appear in the Label drop-down list and will be assigned to the selected dimension. A pencil icon will appear next to the dimension text, indicating that a global parameter has been successfully applied.

When you select any of the dimensions in a project, its properties will be displayed in the **Properties** palette. Figure 9-5 shows the **Properties** palette of the selected linear dimension.

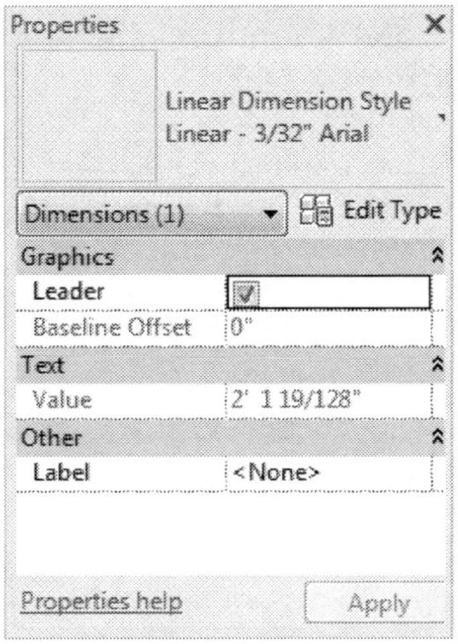

Figure 9-5 *The* ***Properties*** *palette displaying various properties of the linear dimension type*

In the Properties palette, various instance parameters can be modified depending on the selected dimension type in the drawing area. To change type properties, click the **Edit Type** button, which opens the **Type Properties** dialog box (Figure 9-6). Here, you can create a new dimension style tailored to your project by selecting the **Duplicate** button. The dialog box also provides access to multiple type parameters, which will be discussed in the following section.

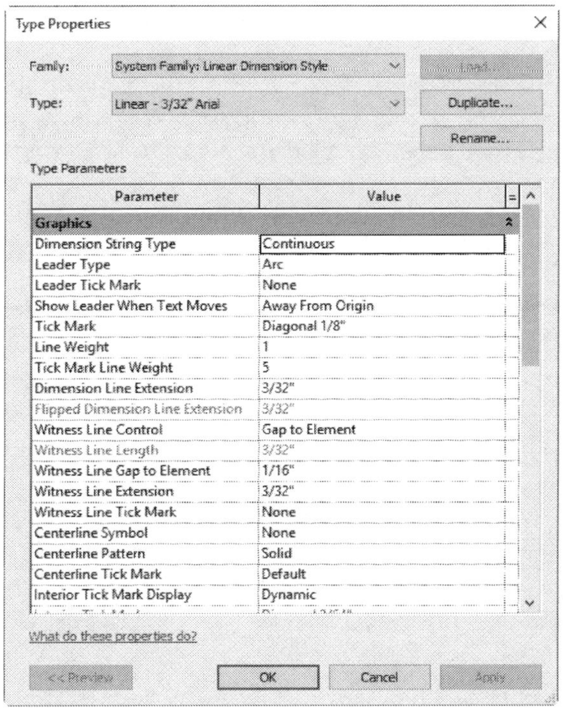

Figure 9-6 *The* ***Type Properties*** *dialog box*

Figure 1 shows various parameters used to customize leader and dimension styles. The **Leader Type** parameter lets you choose between an Arc or Line for the leader line. The **Leader Tick Mark** defines the type of tick mark used, while **Show Leader When Text Moves** allows you to set when the leader is displayed. The **Tick Mark** parameter specifies the mark type, and **Line Weight** sets the thickness of the dimension line (from 1 to 16). **Tick Mark Line Weight** adjusts the thickness of the tick mark line, and **Dimension Line Extension** controls how far the dimension line extends beyond the witness line. Figure 9-6 highlights the **Equality Text** parameter under the "Other" section, allowing you to replace the default EQ label with custom text. Additionally, parameters such as **Text Size** and **Text Offset** can be modified for dimension text. Click **Apply** to save changes.

Locking Permanent Dimensions

When you add a permanent dimension, a lock control symbol is displayed. It also gets displayed when you select a permanent dimension. This symbol can be used to lock or unlock the dimension of an element. When the symbol is unlocked, you can modify the dimension along with the element. When you lock a dimension, the element corresponding to it is also locked along with it. This means you cannot modify its dimension value. You can, however, move the element along with dimension. Once the dimension is locked, you must unlock it to change its value.

Tip

In a Common Data Environment (CDE), locking a dimension may be a good practice to avoid any unauthorized change(s) made by other users. However, if a parameter is assigned to a dimension, it cannot be locked and therefore the locking symbol will not appear with the dimension.

Converting Temporary Dimensions into Permanent Dimensions

As discussed, in the earlier sections of this chapter, temporary dimensions are displayed while creating various elements such as grids, ducts, panels, pipes, and other MEP system elements. They can be used to specify the size and locationf of an element in the structural system. You can move or resize an element by clicking on the temporary dimension and entering a new value. When you create an element, a conversion control symbol is also displayed along with the temporary dimension, as shown in Figure 9-7. When you click on this symbol, the temporary dimension is converted into a permanent dimension, as shown in Figure 9-8.

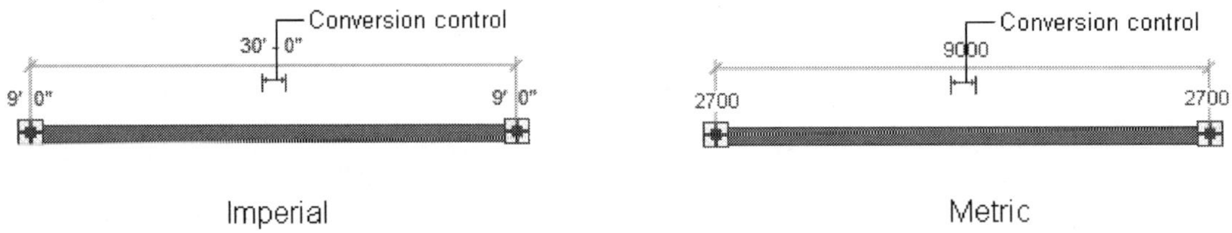

Figure 9-7 *The conversion control symbol displayed in the selected temporary dimension*

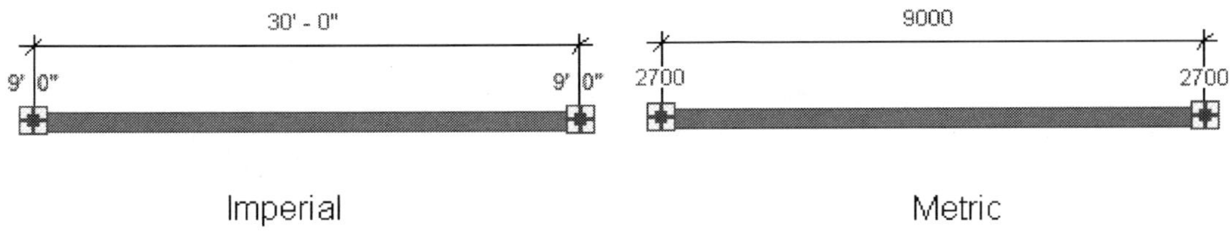

Figure 9-8 Temporary dimension converted into permanent dimension

TEXT NOTES

In Revit, text notes are important part of a project detail. They not only help in adding the specification of various elements but also in conveying the specific design intent. Revit provides a variety of options to add text notes to different system detail views by using the **Text** tool. Various options to add and modify a text in a model view are discussed next.

Adding Text Notes

Ribbon:	Annotate > Text > Text

In Revit, text notes are view-specific entities. You can add different text notes to different views. Generally, text notes are added in the plan view, section view, callout view, or in drafting view. The selection of a view in which the text notes are added depends on the project requirement. To add a text note to a desired view, invoke the **Text** tool from the **Text** panel; the **Modify | Place Text** tab will be displayed, as shown in Figure 9-9.

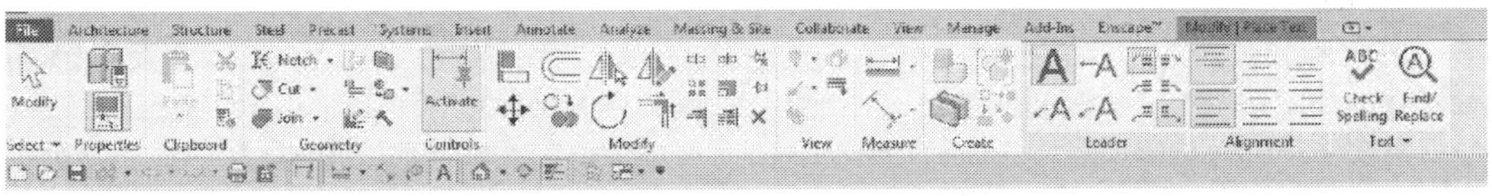

*Figure 9-9 Various options in the **Modify / Place Text** tab*

In this tab, tools from the **Leader** panel allow you to add leaders with text notes. By default, the No Leader tool is selected, so no leader line is attached when inserting a text note. To add a leader, you can choose the One Segment tool for a straight single-segment line, the **Two Segments** tool for a double-segment straight line, or the **Curved** tool for a curved leader line from the **Format** panel. These options are shown in Figure 9-10.

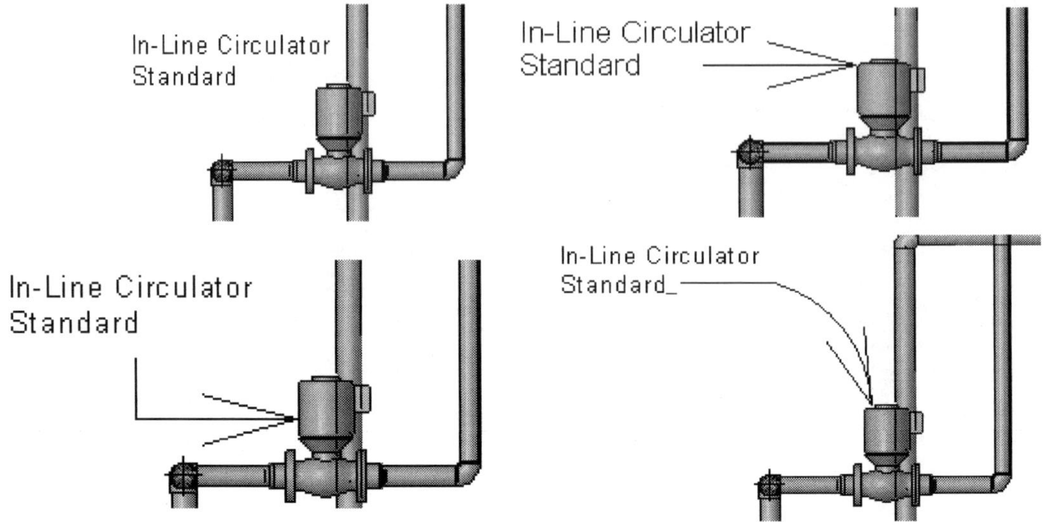

***Figure 9-10** Illustration showing various options of a leader in a text note*

After selecting the desired leader line display option, you can use tools from the Leader panel to position the leader line relative to the text note. For example, you can place it at the top, middle, or bottom of the left or right side of the text note using tools like **Leader at Top Left** or **Leader at Bottom Right**. By default, **Leader at Top Left** and **Leader at Bottom Right** are selected. Next, you can set text alignment using **Align Left**, **Align Center**, or **Align Right** tools from the Paragraph panel, with **Align Left** chosen by default. After setting these options, click near the desired location to add a text note—if a leader is enabled, it will appear from the point clicked. Once the leader is set, click again to place a text box and access the **Edit Text** contextual tab. Here, use formatting options in the **Font** and **Paragraph** panels, such as **Bold**, **Italics**, **Underline**, **Subscript**, **Superscript**, and **List** tools. You can also adjust indents and access the **Text Editor Options** menu to toggle border and background settings while editing. After typing the text, click **Close** in the Edit Text panel to exit, then select **Modify** from the Select panel. Finally, use the **Check Spelling** tool in the Tools panel to verify spelling, as shown in **Figure 9-11**.

***Figure 9-11** The **Check Spelling** dialog box*

Editing Text Notes

You can customize a text note either before or after placing it. When selected, control dots appear—blue dots on each side for resizing, and a rotation control at the top-right to rotate the text box. Note that rotating the box doesn't affect the leader arrowhead, which stays in place while the leader tail adjusts (Figure 9-12). You can also drag the leader head or elbow and reposition the entire text box using the drag controls. Upon selection, the **Modify | Text Notes** tab appears, offering tools like **Copy**, **Mirror**, **Rotate**, and **Array**. Text alignment can be changed using options in the **Format** panel, including adding or removing straight leaders. To modify the text, click inside the box, edit it directly, or change font and size in the **Properties** palette. Horizontal alignment and arc leaders can also be adjusted here. For deeper customization, use the **Edit Type** button in the Properties palette to access the **Type Properties** dialog box (Figure 9-12).

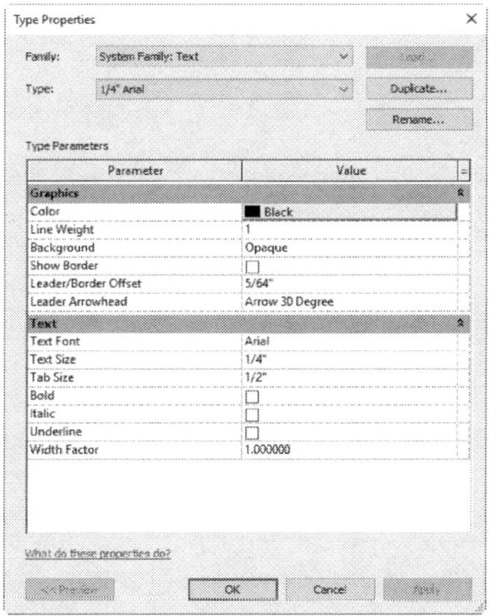

***Figure 9-12** The **Type Properties** dialog box for the selected text*

ADDING TAGS

Tags are essential in MEP systems for identifying and organizing elements within Revit drawings. As shown in **Figure 1**, you can easily add tags to various components like ducts, pipes, and diffusers. Tags help create schedules by linking each tagged element with relevant information in tabular form. Revit includes built-in tools to add and edit tags, and many are pre-loaded in the default template. When placing elements, Revit automatically tags certain ones, while others require manual loading from the library. Tags are view-specific, meaning they appear only in the view where they're added. As illustrated in **Figure 9-13**, their visibility can be controlled using the **Visibility/Graphics** tool under the **View** tab. The **Annotation Categories** tab in the dialog box lists all tag categories, allowing users to enable or disable them as needed.

Tagging Elements by Category

Ribbon: Annotate > Tag > Tag by Category

Tags are essential in MEP systems for identifying and organizing elements within Revit drawings. As shown in **Figure 9-13**, you can easily add tags to various components like ducts, pipes, and diffusers. Tags help create schedules by linking each tagged element with relevant information in tabular form. Revit includes built-in tools to add and edit tags, and many are pre-loaded in the default template. When placing elements, Revit automatically tags certain ones, while others require manual loading from the library. Tags are view-specific, meaning they appear only in the view where they're added. As illustrated in **Figure 9-13**, their visibility can be controlled using the **Visibility/Graphics** tool under the **View** tab. The **Annotation Categories** tab in the dialog box lists all tag categories, allowing users to enable or disable them as needed.

Tip

Tagging by category enables a Revit modeler to quickly retrieve the information from the selected component. As a result, while loading a tag for the category, the Revit modeler needs to ensure that the required tag family is associated with the corresponding category.

Tagging All Elements in a View

You can tag all elements visible in the current view by using the **Tag All** tool. On invoking this tool from the **Tag** panel, the **Tag All Not Tagged** dialog box will be displayed, as shown in Figure 9-13. This dialog box has two radio buttons: **All objects in current view** and **Only selected objects in current view**. The **Only selected objects in current view** radio button gets activated only if you select an object before invoking the **Tag All Not Tagged** dialog box. The selection of the **All objects in current view** radio button ensures that all objects present in the current view are tagged, provided their categories are selected in the **Category** column of the **Tag All Not Tagged** dialog box. The table displayed below the radio buttons shows the listing of category tags of all elements present in the current view and their corresponding loaded tags.

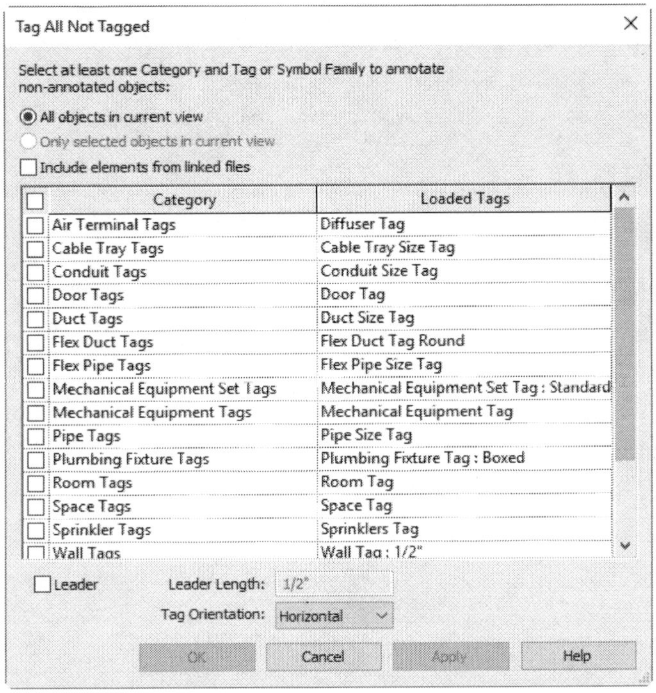

Figure 9-13 *The* ***Tag All Not Tagged*** *dialog box*

From the table displayed in the **Tag All Not Tagged** dialog box, select the check box corresponding to the desired category. The **Leader** check box can be used to add a leader to them. You can also select the orientation and length of the leader in this dialog box. Choose the **Apply** button to apply the settings to the current view. For example, you can tag all air diffusers in a floor plan view by using the **Tag All** tool, as shown in Figure 9-14.

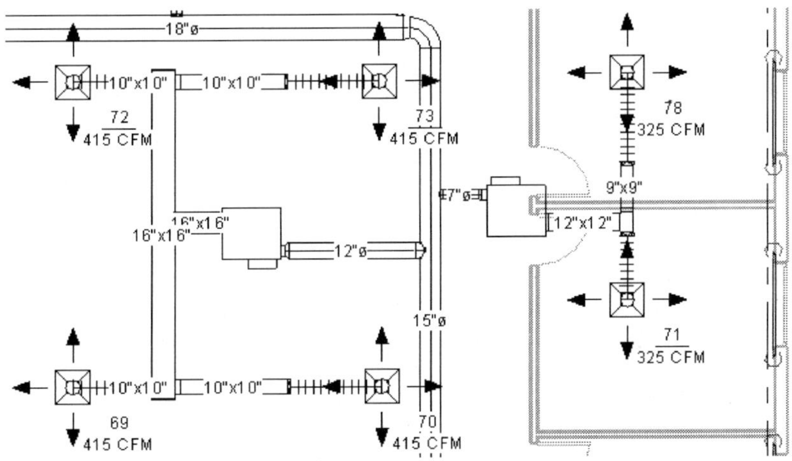

Figure 9-14 *All air diffusers tagged using the* ***Tag All*** *tool*

> **Tip**
> *If you want to use the **Tag All Not Tagged** tool, you should use it only at the final stage of construction documentation to avoid overcrowding of data at the design stage. Also, before using this tool, you must ensure that all components to be tagged are assigned a tag value.*

CALLOUT VIEWS

A callout view, as shown in **Figure 9-15**, is an enlarged view used to provide more detail on specific parts of a project model. Engineers frequently use callout views to examine areas of a design more closely, such as showing the connection between a duct and a diffuser. You can create callout views for plan, section, or elevation views, and the callout tag added in a view links to the callout view. The parent view, where the tag is placed, is essential; if deleted, the callout view is also deleted. The callout tag consists of a bubble, head, leader line, and reference labels. The bubble highlights the area to be enlarged, the head represents the callout view with a detail number and sheet number, and the leader line connects the head to the bubble. The reference label indicates the referenced callout views.

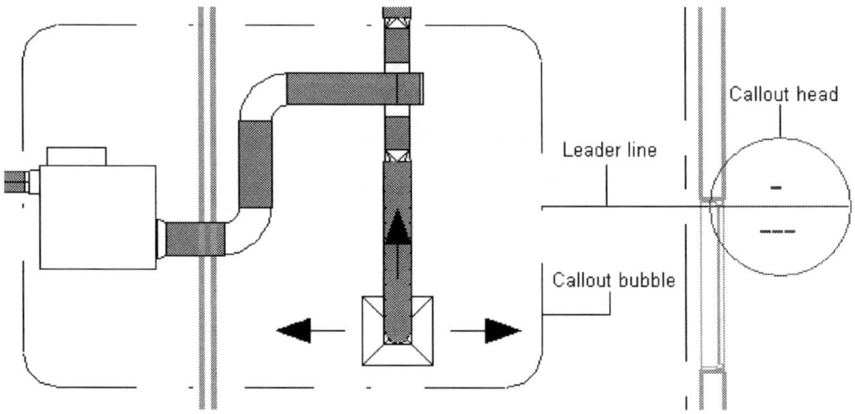

Figure 9-15 *The display of a callout tag*

Creating a Callout Using the Rectangle Tool

Ribbon: View > Create > Callout drop-down > Rectangle Tool

To create a rectangular callout view, choose the **Rectangle** tool from **View > Create > Callout** drop-down; the **Modify | Callout** contextual tab will be displayed. The **Properties** palette will display the properties related to the callout view to be created. In the **Properties** palette, the **Type Selector** drop-down list displays the type of callout view to be generated. In this drop-down list, you can select any of these two types: **Detail View : Detail** and **Floor Plan : Floor Plan** (for current view). Select the **Detail View : Detail** type if you want to provide detailed information about a specific part in a model. Select the **Floor Plan : Floor Plan** (for current view) type if you want to provide more information about a part of the current view. The **Scale** pop-up in the **View Control Bar** can be used to set the view scale. In the **Options Bar**, the **Reference Other View** check box is cleared by default. You can select this check box to create a reference callout. On selecting this check box, the drop-down list next to it is activated. In this drop-down list, you can select an option to specify a view that the callout will refer to. After selecting the desired type and option, move the cursor to the top left corner of the area that you want to enlarge and then drag it toward the lower right corner to create a callout bubble, refer to Figure 9-15. Release the left mouse button when the required area is enclosed in it.

Creating a Callout Using the Sketch Tool

Ribbon:	View > Create > Callout drop-down > Sketch

In Revit, you can create a customized callout view making use of various draw tools. To do so, invoke the **Sketch** tool from **View > Create > Callout** drop-down; the **Modify| Edit Profile** contextual tab will be displayed. The **Draw** panel of this tab contains various sketching tools such as **Line**, **Rectangle**, **Pick Line**, and so on. Choose any tool from this panel to sketch the desired callout view.

Displaying Callout View

When you create a callout in the existing view, a new callout view is added to the **Project Browser** under the parent category. For example, if you create a callout view in the section view, the callout view will be added under the **Sections** heading in the **Project Browser**. In the **Project Browser**, double-click on the name of the callout view to be displayed; the corresponding callout view will be displayed in the drawing window. Alternatively, highlight the callout bubble and right-click to display the shortcut menu. Next, choose the **Go To View** option from the shortcut menu to display the callout view.

Modifying the Properties of a Callout View

You can modify the appearance of a callout bubble in the parent view by using the bubble controls displayed on selecting a callout view. You can modify the extents of a callout view by using its drag controls. The rotation control can be used to rotate the callout bubble along with its leader and tag. The leader elbow control can be dragged to the desired location.

The instance properties of a callout view are different for different types of callouts. To modify the instance properties of a view type callout, select it from the drawing; the instance properties of the selected callout will be displayed in the **Properties** palette, as shown in Figure 9-16. The different instance properties displayed in the **Properties** palette are discussed next.

In the **Properties** palette, you can enter the name to be assigned to a new callout view in the **Value** field of the **View Name** instance parameter under the **Identity Data** head. Similarly, you can enter the title to be given on the sheet in the value field of the **Title on Sheet** parameter.

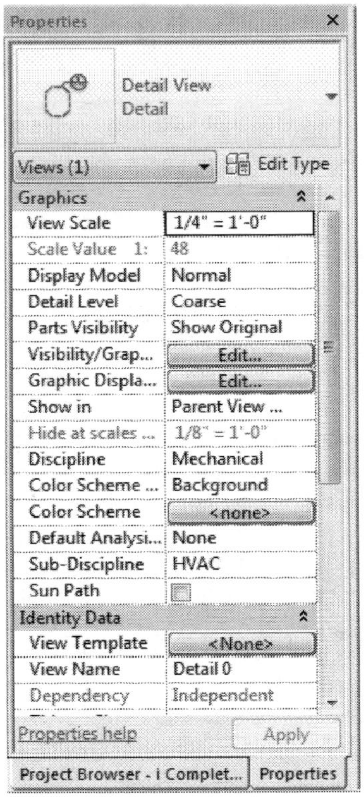

*Figure 9-16 The instance properties of the callout view in the **Properties** palette*

The **Display Model** parameter under the **Graphics** head is used to set the display type for the view of a building model. By default, this parameter is set to **Normal**. You can set the value of the **Display Model** parameter to **Halftone** to display the model element in the current view as a faded image. You can set the value of this parameter to the **Do not display** option to hide the model element in the current view.

The **Underlay** parameter under the **Graphics** head can be used to set the display of model elements, which are displayed in other levels and are invisible in the current view. By default, the **Underlay** parameter is set to **None**. You can select a level from the drop-down list corresponding to this parameter to make the elements of the selected level visible in the current view. The elements that are displayed as underlay appear dim.

The amount of details to be displayed in a callout view can be controlled by using the **Detail Level** parameter under the **Graphics** head. To display details, you can select any of the following three options from the drop-down list corresponding to this parameter: **Coarse**, **Medium**, and **Fine**. By default, the **Medium** option is selected in the drop-down list. As a result, model elements are displayed with fewer details in the current view. You can select the **Fine** option from this drop-down list to display the layers of various materials used in a building model. On selecting the **Fine** option, additional lines are displayed in the callout view. These lines describe composite materials.

The **Edit** button in the Value column for the **Visibility /Graphics Overrides** parameter can be used to control the visibility of different models and annotation elements in the callout view. The **View Scale** instance parameter is used to set the scale for a callout view. You can use the drop-down list in the Value column to select a scale for the callout view.

Creating Details in a Callout View

Ribbon:	Annotate > Detail > Detail Line

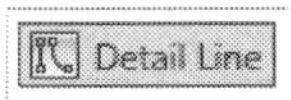

Revit provides various tools to create details in a callout view. You can sketch lines using the **Detail Line** tool and also add detail components provided in Revit's library.

The **Detail Lines** tool is used to create lines for a detail view. Detail lines are view-specific and appear only in the view in which they are created. You can use the callout view of a building model and trace detail lines over the image using their varying thickness.

To create detail lines for a detail view, choose the **Detail Line** tool from the **Detail** panel of the **Annotate** tab; the **Modify | Place Detail Lines** tab will be displayed. Select the type of detail lines from the **Line Style** drop-down list in the **Line Style** panel of this tab. You can select the appropriate detail line based on its usage. For example, wide lines can be used to show insulated pipe, whereas thin lines can be selected to represent normal pipe.

You can draw these lines by using the sketching tools displayed in the **Draw** panel of the **Modify | Place Detail Lines** tab. To add detail lines, you can trace over the underlay elements. On doing so, you will notice that the cursor snaps at various elements on the underlay lines. The entire detail can be sketched using a variety of line thicknesses to achieve the desired graphical representation. You can also add dimensions and symbols to details. However, after completing a detail, you can hide the underlay callout view. To do so, you need to select the callout view from the drawing and then select the **Do not Display** option from the drop-down list corresponding to the **Display Model** parameter in the **Properties** palette.

Note

After adding the detail lines to the callout view, you can add the filled region to the drafted details by invoking the ***Filled Region*** *tool from* ***Annotate > Detail > Region*** *drop-down.*

Tip

In an MEP BIM project, the callout views can be used extensively to create shop and fabrication drawings for HVAC and piping. You can use the callout views for a specific area in the drawing to show details of fixings and fittings for piping and plumbing.

DRAFTING DETAILS

Drafted details are created when you want to access the details that are not referenced to the existing project views. These details are not linked to a building model and therefore, they do not update with it. To create a drafted detail, first create a drafting view and then use the drafting tools provided in Revit to sketch the details. You can also import in-built details from Revit's detail library and use them.

After the drafted details are created, they can be used as reference details. The methods used for creating a drafting view and drafting details are discussed in the next section.

Creating a Drafting View

Ribbon:	View > Create > Drafting View

To create a drafting view, invoke the **Drafting View** tool from the **Create** panel of the **View** tab; the **New Drafting View** dialog box will be displayed, as shown in Figure 9-17. In this dialog box, enter a name for the drafting view in the **Name** edit box. Next, select the scale for the detailing from the **Scale** drop-down list. To specify a user-defined scale, select the **Custom** option from the **Scale** drop-down list and then specify a value in the **Scale** value **1** edit box. Next, choose the **OK** button; the drafting view will be created and added under the **Drafting Views** subhead of the **Mechanical** head in the **Project Browser**.

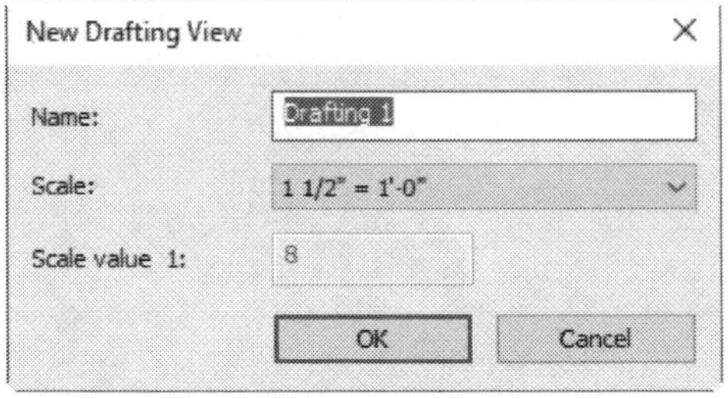

Figure 9-17 *The* ***New Drafting View*** *dialog box*

Note

In Revit, drafting views are created as independent views. As a result, when you create drafting views, project views are not displayed in the drawing window.

Drafting a Detail

The method of drafting a detail in a drafting view is similar to the one used in a callout view. In a callout view, lines are traced over an enlarged project view, whereas in a drafting view, the project view is not available and a new detail is drafted as if you are drafting on a blank sheet.

You can draft a detail by using the tools in the **Detail** panel of the **Annotate** tab. The **Detail Line** tool enables you to draw lines with varying thickness. The **Detail Component** tool in the **Component** drop-down is used to add in-built detail components from Revit's detail library. The **Insulation** tool enables you to add a graphical insulation symbol to the drafted detail. You can create graphical patterns to represent various building materials using the **Region** tool. The usage of these tools is similar to the tools in a callout view. You can also add dimensions, text notes, and break lines to complete a drafted detail.

DUPLICATING VIEWS

You can create multiple copies of a project view in Revit. These multiple views are called duplicate views. All duplicate views are dependent on the original view from which they are created. Therefore, these views are updated automatically if any changes are made in the original view. Duplicate views are useful when you want to use or place the same view on more than one sheet. These views are also useful in case of huge projects when the overall project view is too large to fit into a single sheet. In such cases, you can crop the parent view to make small segments by using the crop regions, create dependent views from them, and then place the cropped dependent views on the sheet. In Revit, you can create three types of duplicate views. The tools used for creating these duplicate views are: **Duplicate View**, **Duplicate with Detailing**, and **Duplicate as Dependent**. You can use the **Duplicate View** tool to create a view that contains only the model geometry from the current view. The **Duplicate with Detailing** tool can be used to create a view that contains view-specific elements from the current view. You can use the **Duplicate as**

Dependent tool to create a view that is dependent on the current view. The method of creating duplicate view as a dependent view is discussed next.

Creating Duplicate View as Dependent View

Ribbon: View > Create > Duplicate View drop-down > Duplicate as Dependent

You can create a dependent view from a plan, elevation, section, or callout view. To create a dependent view, open the view from which you want to create duplicate views and then choose the **Duplicate as Dependent** tool from **View > Create > Duplicate View** drop-down. Alternatively, select the name of the view in the **Project Browser** and right-click; a shortcut menu will be displayed. Choose **Duplicate View > Duplicate as Dependent** from the shortcut menu; the dependent view will display a crop region boundary and will be added in the **Project Browser** as a dependent view under the primary view.

Tip
*Crop regions help to exclude unwanted content from a crop view, thereby reducing the view size. In Revit, there are two types of crop regions that allow you to crop views for the model and annotation categories. You can control the visibility of a crop region by using the **Crop View** or **Do not Crop View** tool from the **View Control Bar**.*

Next, select the crop region boundary to display drag controls. Using the drag controls, resize the primary view and then crop it to include only the required portion of the view in the dependent view. Next, select the name of the dependent view in the **Project Browser** and right-click; a shortcut menu will be displayed. Choose **Rename** from the shortcut menu; the **Rename View** dialog box will be displayed. In this dialog box, enter a name for the dependent view and choose the **OK** button; the duplicate view will be renamed. Again, open the primary view and choose the **Show Crop Region** button from the **View Control Bar** to display the crop region boundary. Similarly, create another dependent view as explained above. In this way, you can crop the primary view and create multiple dependent views. To navigate to the primary view from the dependent view, select the crop region boundary of the dependent view and right-click; a shortcut menu will be displayed. Choose **Go to Primary View** from the shortcut menu; the primary view will be displayed on the screen. You can also navigate to the dependent views from the primary view. To do so, choose the **Show Crop Region** button from the **View Control Bar** if the crop regions are not displayed in the primary view. On doing so, the crop regions for all dependent views will be displayed. Select the crop region of the required dependent view and then right-click; a shortcut menu will be displayed. Choose **Go to View** from the shortcut menu; the corresponding dependent view will open. You can display the primary or dependent views by double-clicking on their respective view names in the **Project Browser**.

Note
*The procedure to create duplicate views by using the **Duplicate View** and **Duplicate with Detailing** tools is similar to that of creating duplicate view/views as dependent views.*

SHEETS

Ribbon: View > Sheet Composition > Sheet

In Autodesk Revit, a sheet or a drawing sheet is a document set that is used for the final working drawings in a site. A drawing sheet contains sheet views that consist of multiple drawing views or schedules added in a project. Sheets are defined by a border and a title block. To create a drawing sheet, you first need to add a sheet view to the project and then add the required project views to the added sheet views.

Adding a Drawing Sheet to a Project

To create a drawing sheet for a project, first you need to decide its title block. In Revit, title blocks are loaded as families. A title block is used to convey project information and drawing sheet title. Project information includes client name, project name, project title, project number, date of issue of drawing, drawing sheet number, scale, and so on. The project information can be added to a project by using the **Project Information** tool in the **Settings** panel of the **Manage** tab. The type of title block will decide the size of the sheet. In Revit, there are some predefined title blocks that can be used in a project. To load these title blocks so that they can be used in the sheet, choose the **Load Family** tool from the **Load from Library** panel in the **Insert** tab; the **Load Family** dialog box will be displayed, as shown in Figure 9-18. In this dialog box, browse to **English-Imperial > Title blocks** (for Metric **English > Title blocks**). From this location, you can select the required title block for the project.

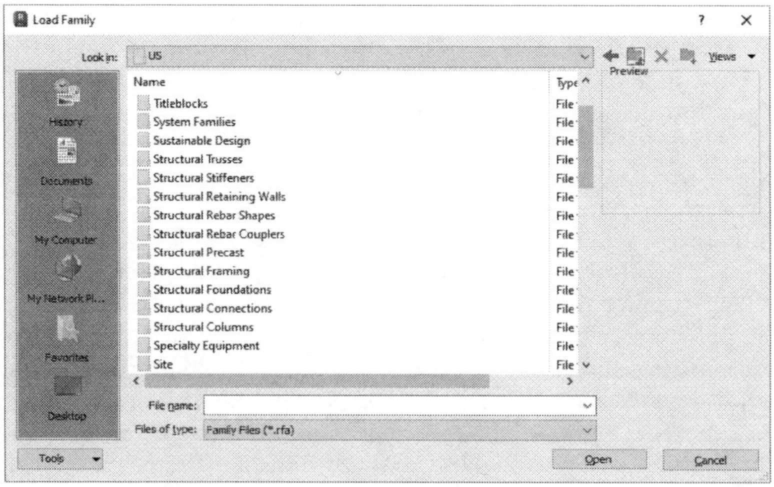

*Figure 9-18 The **Load Family** dialog box*

Note

The sizes of title blocks available in the library are based on ANSI standards for sheet size. These title blocks are available in both vertical (portrait) and horizontal (landscape) alignments.

After selecting a file for the title block from the **Load Family** dialog box, choose the **Open** button; the selected title block(s) will be loaded into the project file. Next, you need to add a drawing sheet to the project. To do so, choose the **Sheet** tool from the **Sheet Composition** panel of the **View** tab; the **New Sheet** dialog box will be displayed. This dialog box has two areas: **Select**

titleblocks and **Select placeholder sheets**. Select the required title block for the drawing sheet from the list of title blocks and choose the **OK** button; the Autodesk Revit will create a sheet view and display the borders and the title block in the drawing window, as shown in Figure 9-19. As you add sheets to the project, their names and numbers will be displayed in the **Sheets** head in the **Project Browser**.

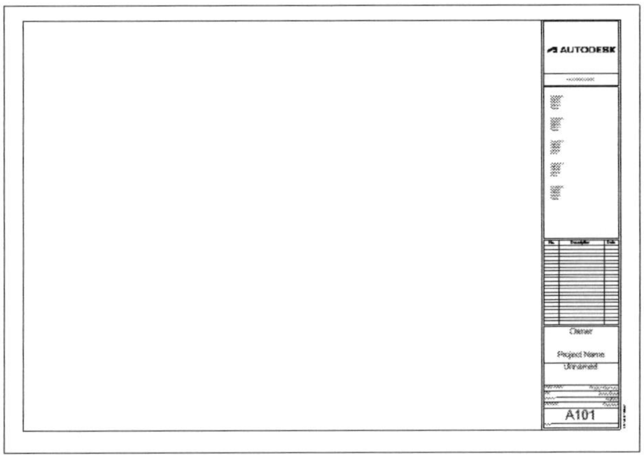

Figure 9-19 *Sheet view displayed in the drawing window*

Adding Views to a Drawing Sheet

After adding a drawing sheet to a project, you need to add project views to it. To do so, choose the **View** tool from the **Sheet Composition** panel of the **View** tab; the **Views** dialog box will be displayed, as shown in Figure 9-20.

This dialog box displays the list of project views available in a project. Select the project view that you want to add to the sheet and choose the **Add View to Sheet** button; the selected view will appear as a viewport represented by a rectangle attached to the cursor. Move the cursor to the drawing area in the sheet and click at the preferred location to add the selected view. Next, in the **Properties** palette, change the **View Scale** parameter by selecting a suitable scale from the drop-down list in its value field and choose **Apply**; the added view will adjust to the specified scale in the sheet, as shown in Figure 9-21.

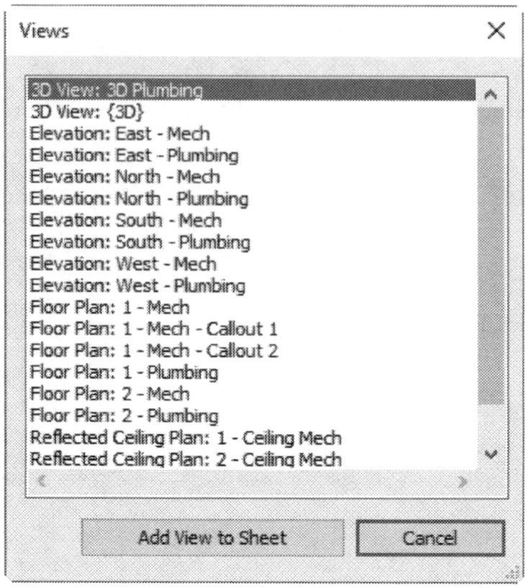

***Figure 9-20** The **Views** dialog box*

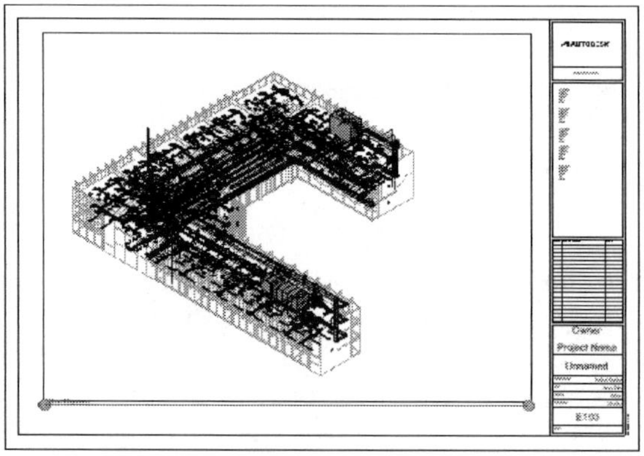

***Figure 9-21** The specified view added to the sheet*

The viewport of the inserted view will display a label line at the bottom left corner. The label line shows the view name, view scale, and view number. You can modify the properties associated with the viewport in the **Properties** palette. You can also activate the project view and then pan it to adjust the view area that you need to display in the viewport. Various options to modify the properties of the views in the viewport and to pan the view areas are discussed in the next section.

Modifying a Building Model in Sheets

In sheets, sometimes you may need to make quick modifications in the elements of building model. To do so, you need to activate a view and make the desired modifications in it. To activate a view from a drawing sheet, select a viewport. Next, choose the **Activate View** button from the **Viewport** panel in the **Modify | Viewports** contextual tab; the view will be activated. In the activated view, you can work on the building model as in any other project view. The elements in the building model can be edited using the editing tools. When the building model is edited, the parametric change engine of Autodesk Revit modifies other project views immediately. As a result, all project views update automatically. For example, modifications made to the location of air diffuser in an activated viewport of the drawing sheet are automatically reflected in the corresponding floor plans, sections, and other associated views. After making necessary modifications in the model, ensure that the viewport is selected and then choose the **Deactivate View** tool from **View > Sheet Composition > Viewports** drop-down to return to the drawing sheet view. Note that new views can also be added to the project when the viewport is activated. You can also create a new plan, elevation, section, callout, detail view, and so on while working on the activated view. The activated view can then be deactivated. Also, the newly created views can be added to the drawing sheet.

Adding Schedules to a Drawing Sheet

You can add schedules created in a project to a drawing sheet. To do so, drag the schedules to be added from the **Project Browser** and drop them into the drawing sheet; the preview image of the schedule will be attached to the cursor. Click at the desired location in the drawing to place the schedule. The selected schedule will appear with blue triangles for each column and a break line on the right side, as shown in Figure 9-22. You can modify the appearance and properties of a schedule. The blue triangles representing the controls for the column width can be dragged to modify the width of each column in the schedule. The break line on the right border represents the split control. It is used to split the schedule table into multiple sections. When the split control is used, the schedule is split into two sections. These sections are placed adjacent to each other. In this way, the schedule can be split into a number of sections that fit into the drawing sheet.

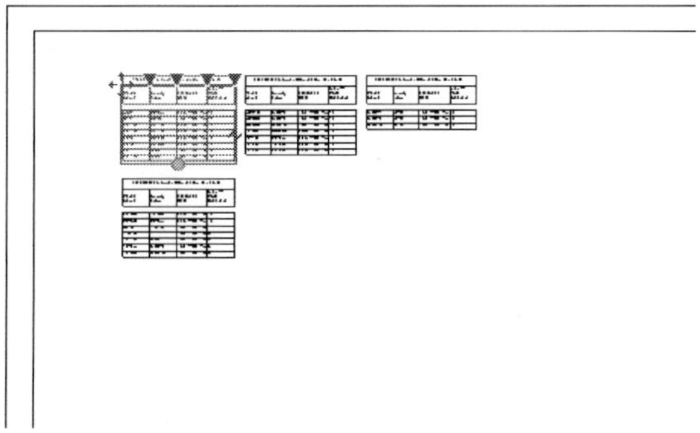

Figure 9-22 *The selected schedule with its controls*

The sections of a schedule can be moved by using the move control available at the center of the schedule section. They can also be resized by dragging the blue dot created at the schedule's end. When the schedule section is resized, the additional or reduced rows are automatically adjusted between the sections. However, the last section cannot be resized, as it contains the remaining rows. The split sections of a schedule can be rejoined. As they are sequential, a section can be joined only to its previous or next section. To join a split section, drag a section by using the move control over the previous or the next section; the two schedule sections will merge into a single schedule section.

Adding Guide Grids to a Sheet

Guide grids help maintain consistent view placement across sheets in Revit. As shown in , you can add a guide grid using the **Guide Grid** tool from the **Sheet Composition** panel in the **View** tab. When selected, the **Assign Guide Grid** dialog box appears , where you can name the grid. Once added, the grid appears on the sheet with default spacing. You can then select it to open the **Modify | Guide Grid** tab and adjust its properties in the **Properties Palette** . Here, you can change the spacing and rename the grid as needed. It's recommended to create a few guide grids and apply them consistently across sheets to maintain alignment.

TUTORIAL

General instructions for downloading the model file and performing the tutorials:

1. Download the *c09_rmp_2025_tut.zip* file from *www.cadsofttech.com*. The path of the file is as follows: *Books > Civil/GIS BOOKS > Revit MEP > Revit MEP 2025 For Novices*

2. Next, browse to *C:\rmp_2025* and create a new folder with the name *c09_rmp_2025_tut*. Next, save and extract the file in this folder.

Tutorial 1 Office Space-Tags

In this tutorial, you will add dimensions to a project file in different views. Further, you will annotate spaces and rooms using various annotation tools. Use the following project parameters: **(Expected time: 40 min)**

The following steps are required to complete this tutorial:

a. Open the project file.
   For Imperial *c09_Office-Space_tut1.rvt*
   For Metric *M_c09_Office-Space_tut1.rvt*
b. Dimension grid lines.
c. Add tags to the Duct and Mechanical Equipment.
d. Save the project using the **Save As** tool.
e. Close the project using the **Close** tool.

Opening the Project File

In this section, you will open the downloaded project file in Revit.

1. Choose **Open > Project** from the **File** menu; the **Open** dialog box is displayed.

2. In this dialog box, browse to *C:\rmp_2025\c09_rmp_2025_tut* location and choose the project file.
   For Imperial *c09_Office-Space_tut1.rvt*
   For Metric *M_c09_Office-Space_tut1.rvt*

3. Next, choose the **Open** button in this dialog box; the selected project file opens in the drawing window.

Dimensioning the Grid Line

In this section, you will add dimensions to the grid lines displayed in the **Basement** plan view.

1. Double-click on the **B1 BASEMENT** node under the **Floor Plans > Architectural** head in the **Project Browser** to open the corresponding floor plan view, if it is not selected by default.

2. Choose the **View** tab and then invoke the **Visibility/Graphics** tool from the **Graphics** panel of this tab; the **Visibility/Graphic Overrides for Floor Plan: B1 BASEMENT** dialog box is displayed.

3. In this dialog box, choose the **Annotation Categories** tab and then select the **Dimensions** check box displayed under the **Visibility** column, if it is not selected by default.

4. Choose the **Apply** button to save the changes and then choose the **OK** button; the **Visibility/ Graphic Overrides for Floor Plan: B1 BASEMENT** dialog box is closed.

5. Choose the **Annotate** tab and then choose the **Aligned** tool from the **Dimension** panel.

6 In the **Type Selector** drop-down list of the **Properties** palette, ensure that the **Linear Dimension Style : Linear - 3/32" Arial** option (for imperial system) or the **Linear Dimension Style : Diagonal - 2.5mm Arial** option (for Metric system) is selected. Next, choose the **Edit Type** button; the **Type Properties** dialog box is displayed.

7. In this dialog box, choose the **Duplicate** button; the **Name** dialog box is displayed.

8. In the **Name** edit box of the dialog box, enter **Small Dot- 3/32" Arial** (for Metric enter **Small Dot- 2.5mm Arial)** and then choose the **OK** button; the **Name** dialog box is closed.

9. In the **Type Properties** dialog box, click in the value field of the **Tick Mark** parameter; a drop-down list is displayed. Select the **Arrow Filled 15 Degree** option from this drop-down list.

10. In the **Type Properties** dialog box, click in the value field of the **Centerline Pattern** parameter; a drop-down list is displayed. Select the **Center** option from the drop-down list.

11. Choose the **OK** button in the **Type Properties** dialog box; the dialog box closes. Also, notice that in the drawing area, the dimension symbol appears with the cursor.

12. Place the cursor and click over grid 1a; the grid gets highlighted and the dimension line is attached to the grid.

13. Place the cursor on grid 1 and click when grid 1 gets highlighted. Notice that a dimension is displayed between grid 1a and grid 1, as shown in Figure 9-23.

14. Move the cursor over grid 2 and click when it gets highlighted.

15. Repeat the procedure followed in step 14 and place the dimensions till grid 9. After you have clicked over the grid 9, move the cursor up and place it above the grid bubbles, as shown in Figure 9-24.

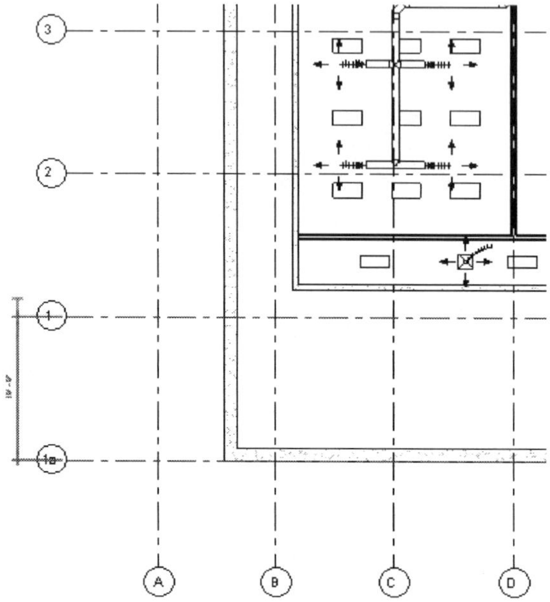

Figure 9-23 *The dimension created between grid 1a and grid 1*

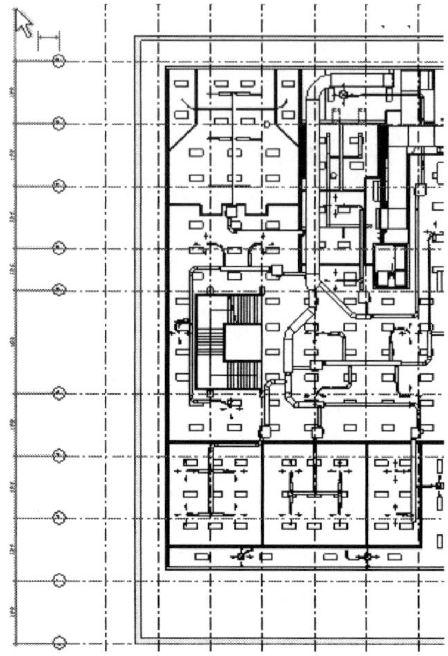

Figure 9-24 *Dimensions placed at a location above grid bubbles*

16. Now, click at the desired location, the dimensions are added. Note that you can place the dimensions based on your requirement.

17. Next, repeat the procedures followed in steps 12 to 16 and add dimensions for all vertical grids from A to T. The dimensions are placed above the structural plan view, refer to Figure 9-25.

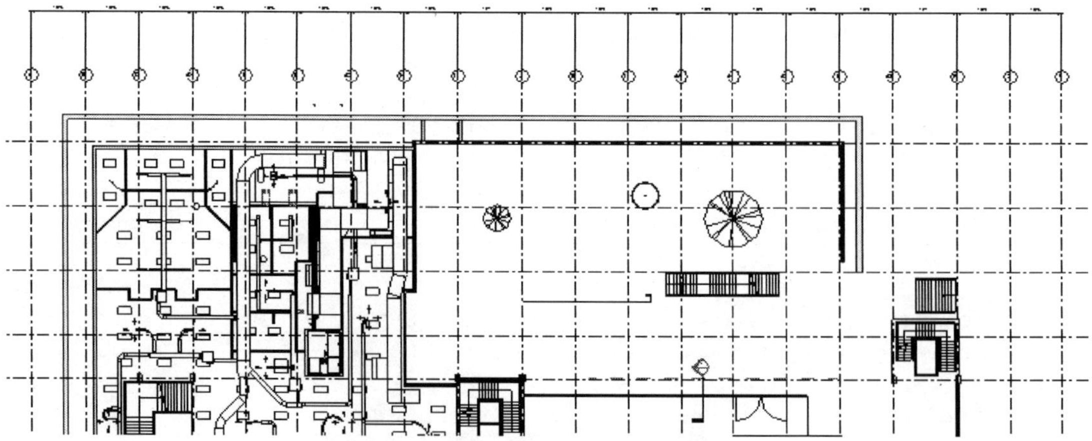

Figure 9-25 *The dimensions placed at a location above the grid bubbles*

18. Next, press ESC twice to exit the tool.

Adding Tags to Duct and Mechanical Equipment

In this section, you will add tags to ducts and mechanical equipment in the **B1_HVAC DUCTWORK** Mechanical plan view.

1. Double-click on the **B1_HVAC DUCTWORK** node under **Floor Plans > Mechanical** head; the view is changed.
2. Next, choose the **Annotate** tab and then invoke the **Tag All** tool; the **Tag All Not Tagged** dialog box is displayed.
3. Select the **Duct Tags** check box in the **Category** column and click in the cell corresponding to it; a drop-down list is displayed. Select the **Duct label Rectangular** option from the drop-down list. (For Metric **M_Duct Size Tags loaded tags**).
4. Next, select the check boxes corresponding to the **Mechanical Equipment Tags** and **Room Tag** categories (For Metric **M_Mechanical Equipment Tags** and **M_Room Tags**) in the **Category** column, as shown in Figure 9-27.

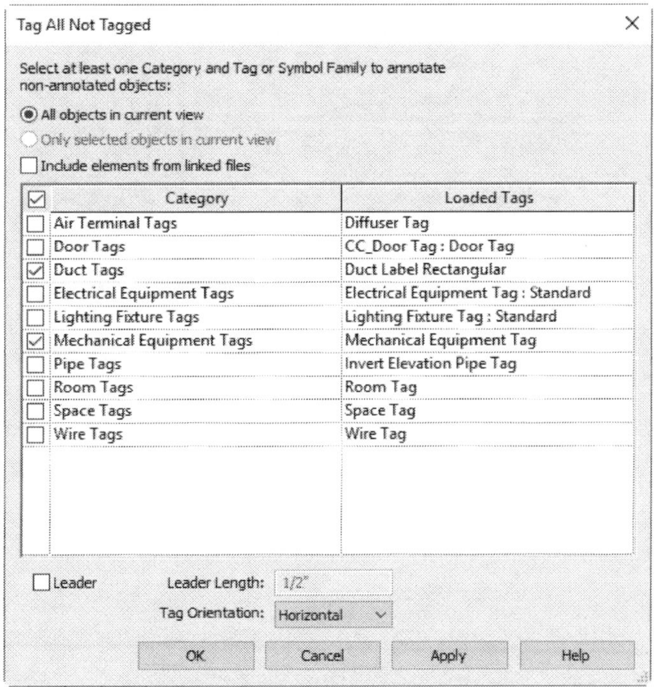

***Figure 9-27** The **Tag All Not Tagged** dialog box with selected categories*

5. Choose the **Apply** button; the **Category Visiblity is Disable** message box is displayed prompting you to to turn on the visibility of the room in the current view. Choose the **OK** button; the message box is closed. Next, choose the **OK** button; the **Tag All Not Tagged** dialog box closes and the selected tag categories are displayed in the **B1_HVAC DUCTWORK** Mechanical plan view, as shown in Figure 9-28.

6. Next, choose the **Tag by Category** tool from the **Tag** panel and place the cursor over the diffuser; the diffuser is highlighted.

7. Click on the highlighted diffuser when the preview of the tag appears; the diffuser tag is added to the selected diffuser. Repeat this step to add tags to all diffusers in the conference rooms, refer to Figure 9-29.

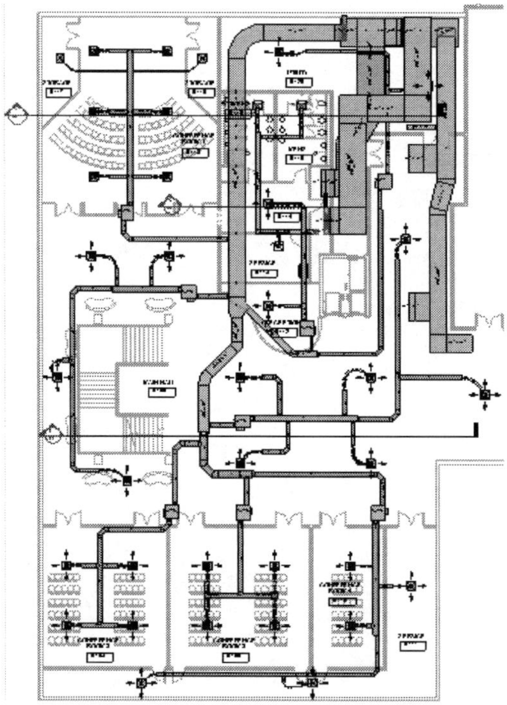

Figure 9-28 *The Mechanical plan view displaying the added tags*

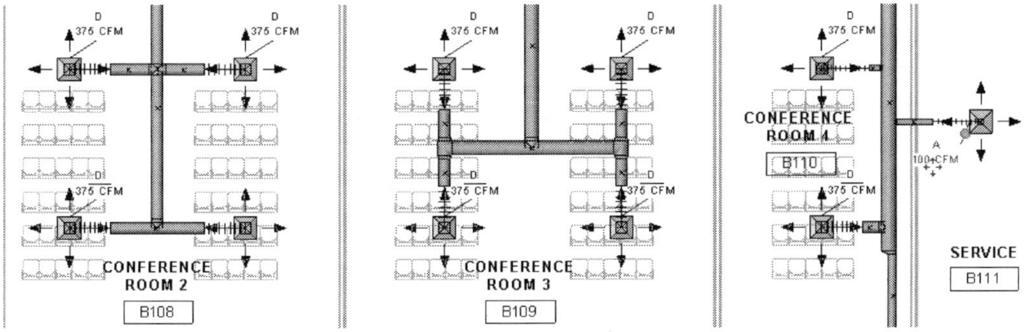

Figure 9-29 *The Mechanical floor plan view displaying the added diffuser tags*

8. Choose the **Modify** button in the **Select** panel to exit the selection.

Saving the Project

In this section, you will save the project file by using the **Save As** tool.

1. To save the project, choose **Save As > Project** from the **File** menu; the **Save As** dialog box is displayed.
2. In this dialog box, browse to *C:\rmp_2025\c09_rmp_2025_tut* and then enter **c09_Office-Space_tut1_tag.rvt** for Imperial or **M_c09_Office-Space_tut1_tag.rvt** for Metric in the **File name** edit box.
3. Now, choose the **Save** button; the **Save As** dialog box closes and the project file is saved.

Closing the Project

1. To close the project, choose the **Close** option from the **File** menu.

EXERCISE

Exercise 1 Conference Hall

Download the *c09_Conference-hall_exer1.rvt* file from *https://www.cadsofttech.com*. The path of the file is as follows: *Books > Civil/GIS BOOKS > Revit MEP > Revit MEP 2025 For Novices*. Now add callout view and add tags to ducts and diffusers at the **1ST FLOOR-HVAC DUCT** mechanical floor plan view of the downloaded file. Also. add space tags to the **1ST FLOOR-HVAC DUCT** mechanical floor plan view. Refer to Figures 9-30 and 9-31 to complete the exercise.

(Expected time: 1 hr)

1. Add callout view to the **1ST FLOOR-HVAC DUCT** mechanical plan view.
2. Add tags to round ducts in the **1ST FLOOR-HVAC DUCT** mechanical plan view.
3. Add tag to air diffusers in the **1ST FLOOR-HVAC DUCT** mechanical plan view.
4. Add space tag to **1ST FLOOR-HVAC DUCT** mechanical plan view.
5. File name to be assigned:

For Imperial *c09_Conference-hall_exer1a.rvt*
For Metric *M_c09_Conference-hall_exer1a.rvt*

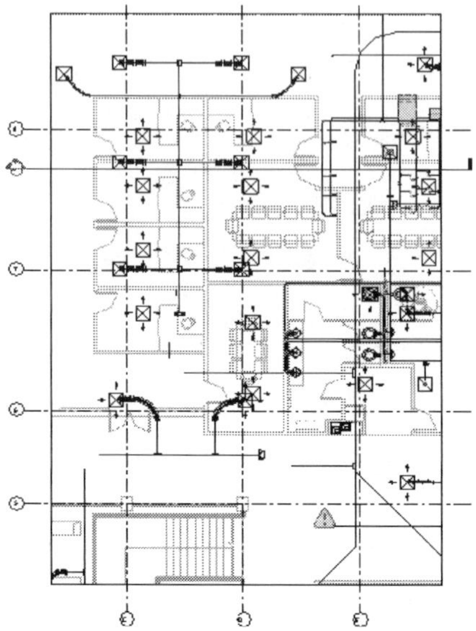

Figure 9-30 *The tags added to round ducts and air diffusers*

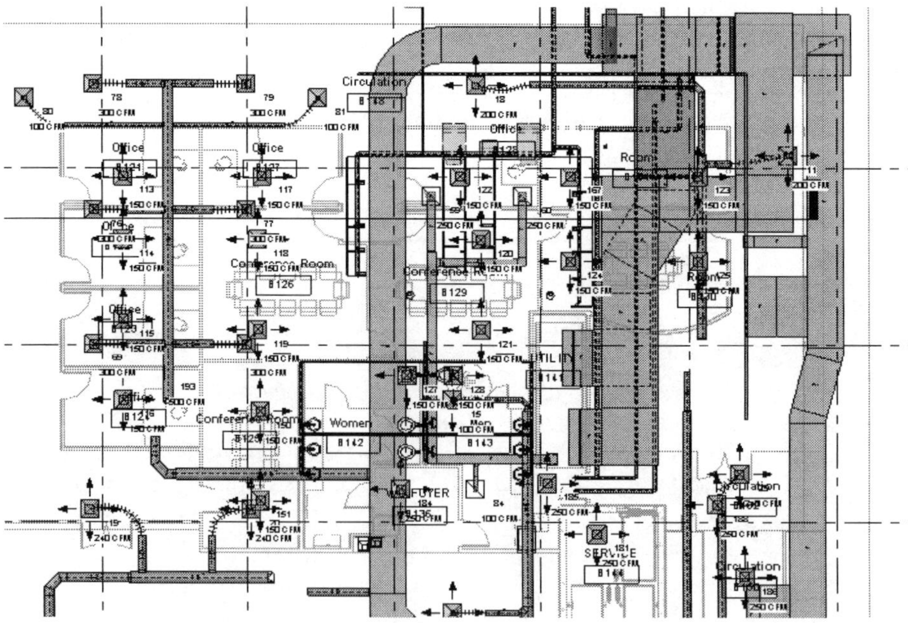

Figure 9-31 *The space tags added to the* ***1ST FLOOR-HVAC DUCTS*** *mechanical plan view*

Chapter 10

Creating Families and Worksharing

Learning Objectives

After completing this chapter, you will be able to:

- *Create massing in Conceptual Mass Environment*
- *Create massing in Family Editor Environment*
- *Create massing in Project Environment*
- *Understand worksharing*
- *Understand Worksharing Monitor*

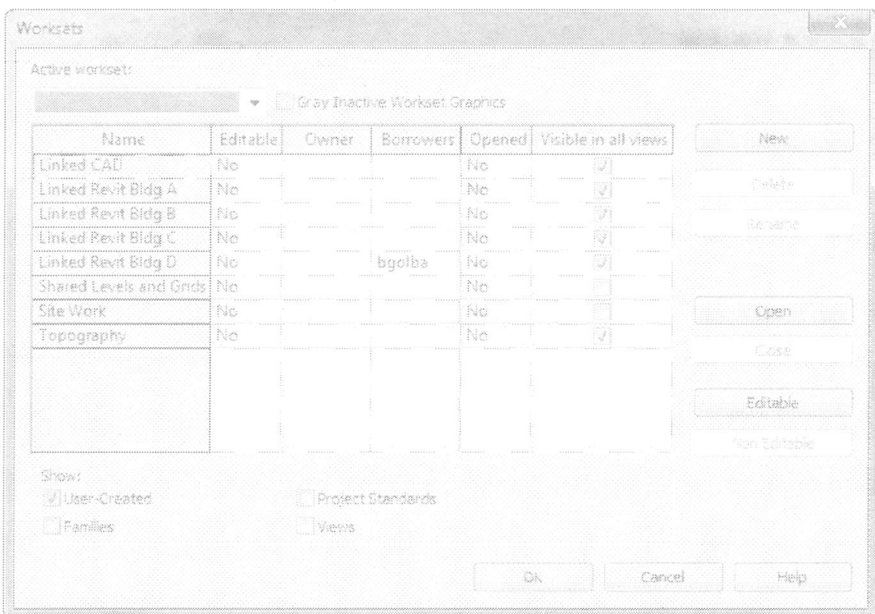

INTRODUCTION TO MASSING

In earlier chapters, you learned to use different system elements and components to create an MEP model. These system elements are parametrically associated and enable you to generate an MEP model based on specific design requirements such as duct types, cable trays, plumbing fittings, width, spaces, and so on. Each of the elements must be assigned specific properties to achieve the desired element parameters, thereby making system model accurate. Needless to say, this is a fairly time-consuming procedure. Autodesk Revit provides you an easy alternative to create system elements, known as massing.

UNDERSTANDING MASSING CONCEPTS

At the conceptualization stage of a project, massing tools in Revit allow you to explore and present spaces, volumes, and shapes in a simplified, three-dimensional form. These tools help convey the basic structure of a building without detailed components. As shown in **Figure 10-1**, you can create and combine various geometric shapes to represent elements like a water closet in a plumbing fixture. This approach is similar to modeling with foam blocks—shapes can be added, cut, or merged to form a complete system, offering flexibility and clarity during the early design phase.

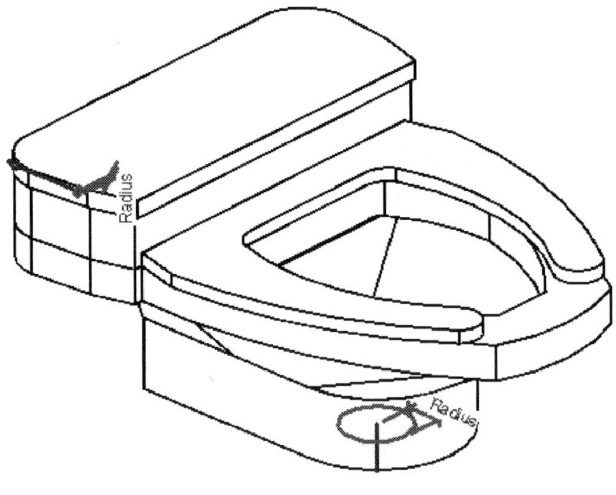

Figure 10-1 *Water closet created using the massing geometry*

As shown in **Figure 1**, Autodesk Revit supports design continuity by allowing the same model to evolve from concept to completion. During the early stages, users can control the visibility of geometry between massing (building volumes) and shell (building elements), and extract information like total area from the massing model (**Figure 2**). However, as illustrated in **Figure 3**, massing tools are limited to conceptual design using simple geometric forms. While predefined massing families from Revit libraries can be used, it's important to avoid using them for detailed elements like columns, cavity walls, or footings, as Revit only translates massing into basic shell geometry.

CREATING THE MASSING GEOMETRY

In Autodesk Revit, massing geometry can be created in three environments: Family Editor, Conceptual Design, and Project. To create it in the Family Editor, go to **File > New > Family**, which opens the **New Family - Select Template File** dialog box. As shown in **Figure 10-2**, select the **Generic Model.rft** template from the **English-Imperial** folder, then click **Open** to launch the Family Editor environment. Here, you can use various tools from the ribbon to design the massing geometry.

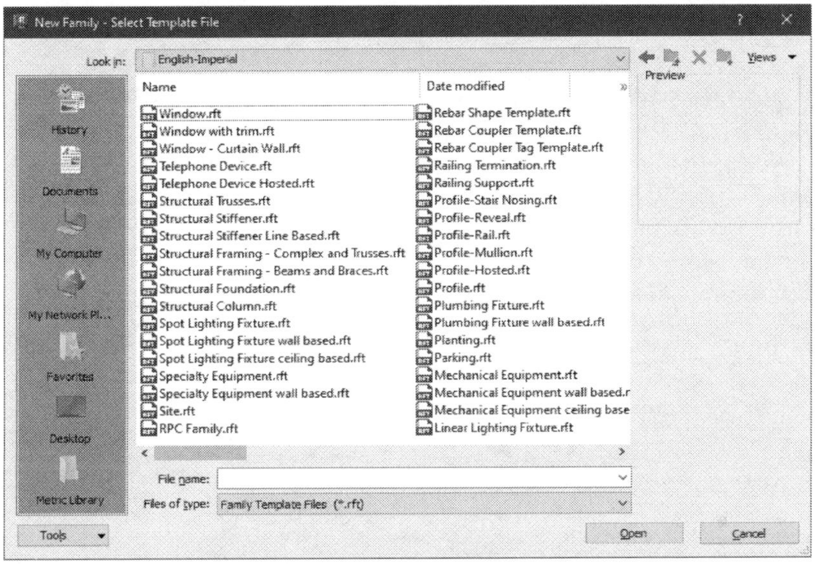

Figure 10-2 *Selecting a file from the* ***New Family- Select Template File*** *dialog box*

The Conceptual Design environment in Revit is used for creating conceptual masses with advanced modeling tools. To begin, go to **File > New > Conceptual Mass**, which opens the **New Conceptual Mass - Select Template File** dialog box (**Figure 1**). From there, select the **Mass** template located in the **Conceptual Mass** folder and click **Open** to launch a new file in the Conceptual Design environment. In this file, you can use various tools from the ribbon to create massing geometry, as shown in **Figure 10-3**.

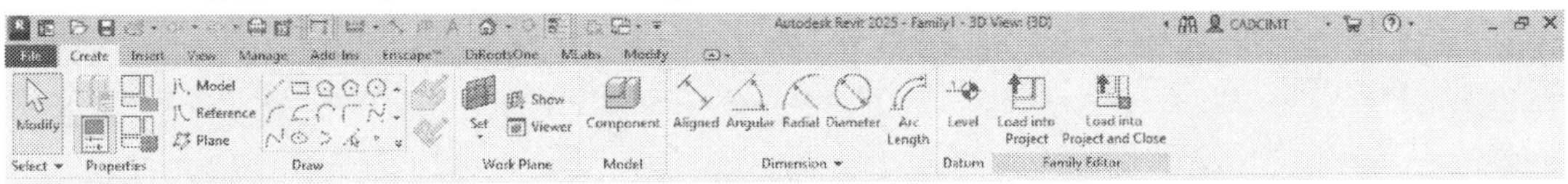

Figure 10-3 *Different tools in the ribbon of the Conceptual Design Environment*

The Project environment is the most common environment used in a project design. To start creating a mass in this environment, open a new file or an existing file by choosing **New > Project** or **Open > Project** from the **File** menu. After opening a new file or an existing file, choose the **Massing & Site** tab, as shown in Figure 10-4. In this tab, the massing tools can be accessed from the **Conceptual Mass** panel. The **Conceptual Mass** panel contains the following tools for massing: **In-Place Mass**, **Show Mass**, and **Place Mass**.

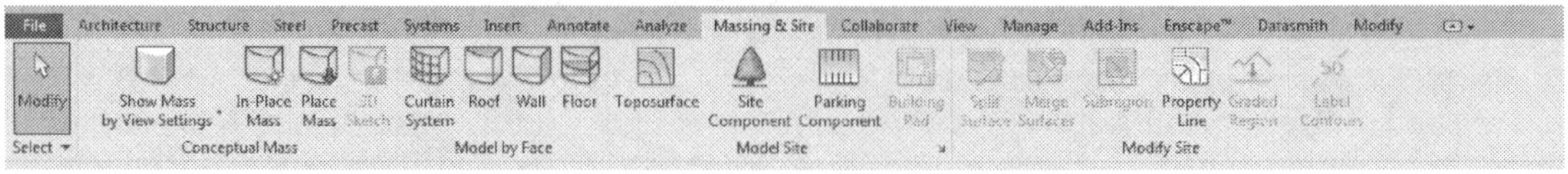

Figure 10-4 *Various tools in the **Massing & Site** tab*

The tools in the **Model by Face** panel are used to convert the conceptual mass created into real building elements such as walls, floors, roofs, and curtain systems. These tools are also called the Building Maker tools.

Creating a Massing Geometry in the Family Editor

As shown in **Figure 1**, to create massing geometry in Revit, you begin by selecting **New > Family** from the File menu. In the **New Family - Select Template File** dialog box, choose **Generic Model.rft** from the English_I folder and click **Open** to launch a new file. In the Family Editor environment, the **Create** tab offers tools for creating both solid and void massing geometries. Solid forms can be created using tools like **Extrusion**, **Blend**, **Revolve**, **Sweep**, or **Swept Blend** from the **Forms** panel (**Figure 10-5**). For void forms, similar tools are available in the **Void Forms** drop-down. Designers can use these tools individually or in combination, depending on the complexity and shape of the desired massing geometry.

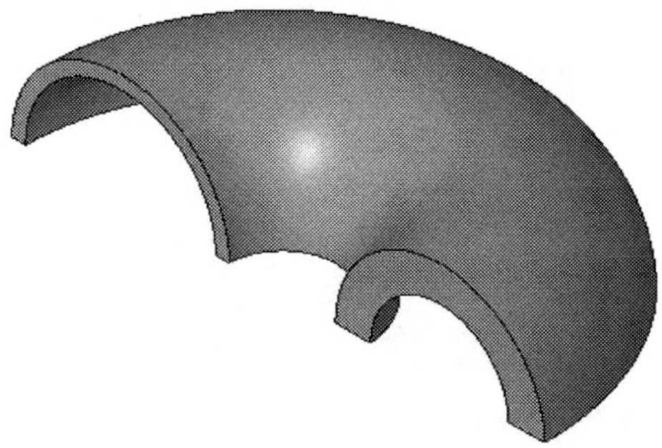

Figure 10-5 *Massing geometry created by using the **Swept Blend** tool*

Editing a Massing Geometry in the Family Editor

In Autodesk Revit, you can easily edit a massing geometry in the family editor environment. It can be edited using the drag controls or by editing the massing parameters.

Creating Cuts in a Massing Geometry Using the Family Editor

You can cut a massing geometry by creating a void form in it. This void form is cut or subtracted from the massing geometry it intersects. You can create void forms by using the tools displayed in the **Void Forms** drop-down in the **Forms** panel of the **Create** tab, as shown in Figure 10-6. The **Void Forms** drop-down displays five tools: **Void Extrusion**, **Void Blend**, **Void Revolve**, **Void Swept Blend**, and **Void Sweep**. You can choose an appropriate tool to generate the shape and volume of the void form. The method of creating a void form using these tools is similar to that of creating a solid form.

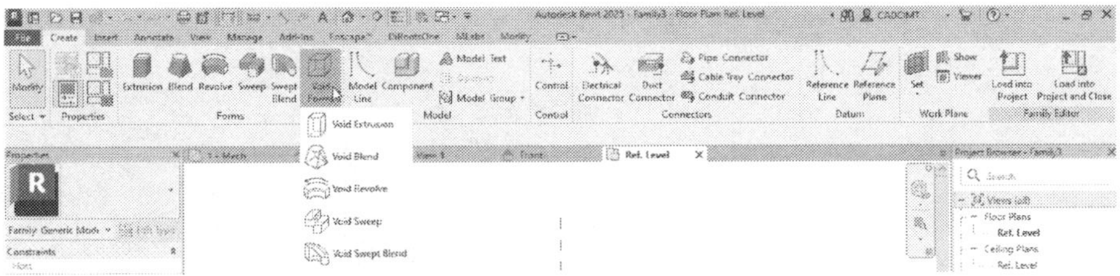

Figure 10-6 *Tools in the* ***Void Forms*** *drop-down*

When any of the tools is invoked from the **Void Forms** drop-down is invoked, the corresponding contextual tab is displayed. For example, if you invoke the **Void Extrusion** tool, the **Modify | Create Void Extrusion** tab will be displayed. You can use the options in this tab to sketch the profile for the extruded void geometry. After sketching the profile of the extruded void geometry, choose the **Finish Edit Mode** button from the **Mode** panel of the **Modify | Create Void Extrusion** tab; the **Modify | Void Extrusion** tab will be displayed. From this tab, you can use various editing options to modify the extruded void geometry. Next, click in the drawing area or press **ESC**; the void form will automatically cut its shape and volume from the intersecting massing geometry.

For example, Figure 10-7 shows the profile of a duct sketched using the **Void Extrusion** tool. Now, when you choose the **Finish Edit Mode** button from the **Modify | Create Void Extrusion** tab, the cutting geometry, a shaft in this case, is generated and automatically cut from the larger massing geometry. Figure 10-8 shows the resulting massing geometry.

Figure 10-7 *The profile for creating a void form*

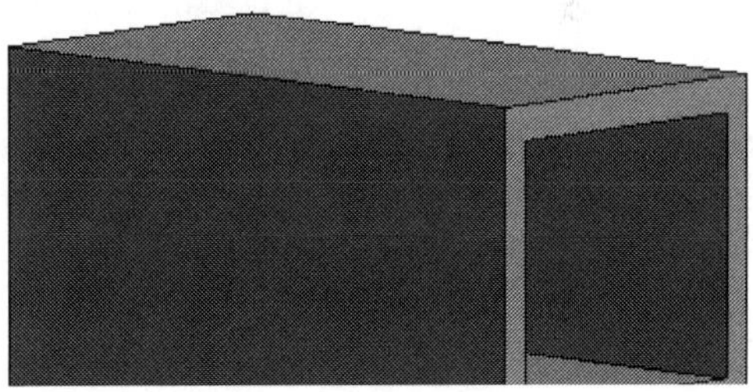

Figure 10-8 *The void form created using the*

Similarly, you can also use the **Void Revolve**, **Void Sweep**, **Void Blend**, and **Void Swept Blend** tools from the **Void Forms** drop-down to create the void form. For example, to create a pipe bend, you can use the **Void Sweep** tool to sketch a semicircular profile and to specify its path, as shown in Figure 10-9. The resulting void form will be cut from the cuboid to create an arched opening through the building block, as shown in Figure 10-10.

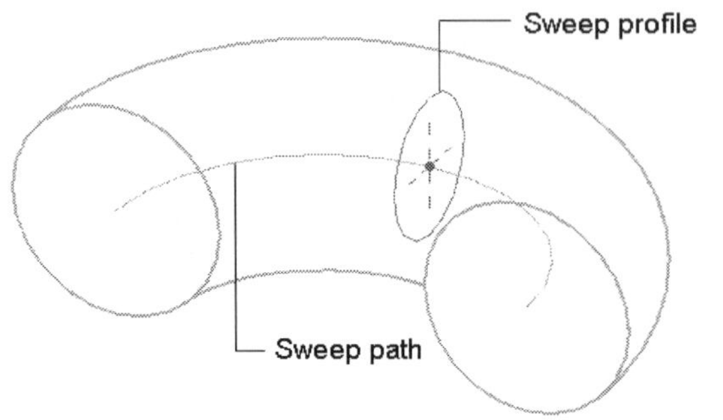

Figure 10-9 Sketching the profile for creating a void form

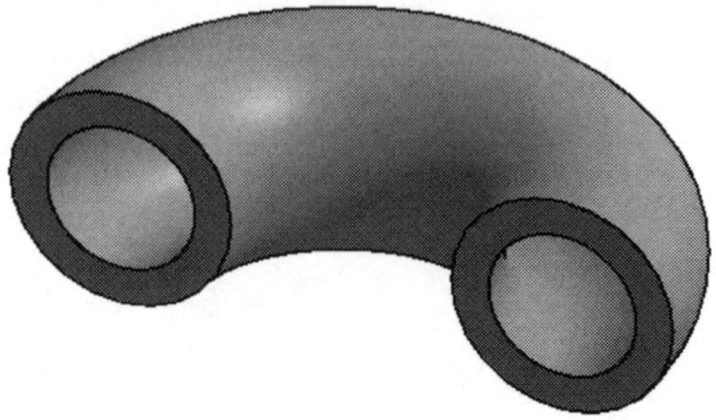

Figure 10-10 The generated void form and the resulting massing geometry

You can also use various editing tools such as **Copy**, **Mirror**, **Array**, and so on to create multiple copies of the void form profile of a single massing geometry. For example, a plumbing fixture has a semicircular vertical cut as the void form on one of its sides. You can select the 2D profile of the void form and use the **Copy** tool to create its duplicate.

Loading Massing Geometry into a Project

After creating massing geometries in the Family Editor, you can load them into the Project environment. To do so, choose the **Load into Project** tool from the **Family Editor** panel of the **Modify** tab; the current project file will appear, if one project file is opened on the screen. Note that if more than one project file is opened in the current session, the **Load into Projects** dialog box will be displayed, as shown in Figure 10-11. In this dialog box, you can select the check box(es) to select the project(s) in which the created mass will be loaded. After selecting the required check box(es), choose **OK**; the mass will be loaded into the selected project file(s) corresponding to the check box(es) selected. In this project file, the **Modify | Place Component** tab is chosen by default and you will notice that the mass created in the Family Editor appears in the drawing area along with the cursor.

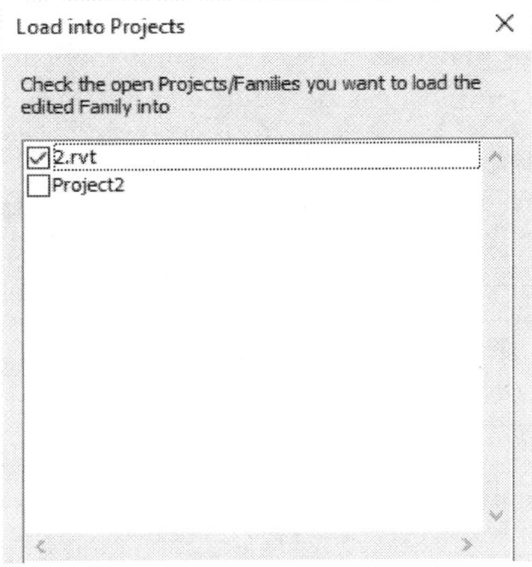

*Figure 10-11 The **Load into Projects** dialog box*

Placing the Massing Geometry in a Project

In Project environment of Autodesk Revit, you have the option to add a predefined massing geometry to the project using the **Place Mass** tool. To do so, invoke this tool from the **Conceptual Mass** panel of the **Massing & Site** tab; the **Massing - Show Mass Enabled** window will be displayed. Choose the **Close** button from the window; the **Revit** message box will be displayed. Choose the **Yes** button in this message box; the **Load Family** dialog box will be displayed. In this dialog box, browse to the **Mass** sub folder in the **English-Imperial** folder and then select a massing family. After selecting the massing family, choose the **Open** button; the **Load Family** dialog box will be closed and the **Modify | Place Mass** tab will be displayed. Also, the selected family will be loaded into the project. Now, click in the drawing area to place the loaded massing geometry.

In Revit, you can either align a mass geometry with the selected faces of the components created or place it on the defined work plane based on your requirement. To place a mass geometry on a selected face, choose the **Place on Face** tool from the **Placement** panel of the **Modify | Place Mass** tab and then place the cursor on the face of the component with which you want to align the massing geometry. You will notice that the mass geometry is aligned with the face on which the cursor is placed. Now, you can click to place the mass on the plane. Similarly, while placing a mass in a drawing, you can place it at a desired level. To do so, choose the **Place on Work Plane** tool from the **Placement** panel of the **Modify | Place Mass** tab; the options for placing the mass will be displayed in the **Options Bar**. In the **Options Bar**, select the required level from the **Placement Plane** drop-down list, as shown in Figure 10-12. Click in the drawing area; a mass will be created at the selected level. Before aligning the mass geometry with the desired face or placing it in a work plane, you can also rotate it about an axis. To do so, select the **Rotate after**

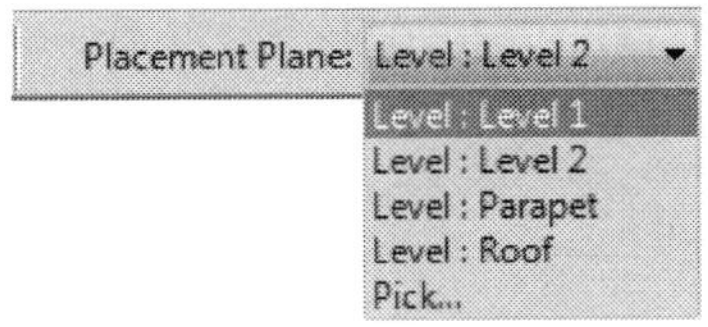

*Figure 10-12 The **Placement Plane** drop-down list*

placement check box from the **Options Bar**. After placing the mass geometry in the drawing, you can edit its geometrical properties such as height, width, radius, and so on, depending upon the type of geometries created. To do so, modify the instance parameters displayed in the **Properties** palette. In the **Dimensions** category of the **Properties** palette, you can select the desired dimension instance parameter and change its value in respective value field.

Creating the In-Place Mass in a Project

In the Project environment, you have the option to create an in-place massing geometry by using the **In-Place Mass** tool. Invoke this tool from the **Conceptual Mass** panel of the **Massing & Site** tab; the **Name** dialog box will be displayed. In this dialog box, enter the name of the mass in the **Name** edit box and choose **OK**; the **Create** tab will be displayed. You can use the options in this tab to sketch the massing profile and convert it into a solid or void form. To sketch the profile for the mass, choose the **Model** tool from the **Draw** panel; the **Modify | Place Lines** tab will be displayed. You can use various sketching tools available in the **Draw** panel of this tab. While sketching, you can also use the **Reference Plane** tool from the **Draw** panel to draw references for the sketch. After sketching the profile, you can use any of the tools displayed in the **Create Form** drop-down in the **Form** panel. The **Create Form** drop-down displays two tools: **Solid Form** and **Void Form**. You can use the **Solid Form** tool to create a solid form and the **Void Form** tool to create a void form. To use any of the tools from the **Create Form** drop-down, you need to select the sketched profile from the drawing and then invoke the **Solid Form** or **Void Form** tool; the **Modify | Form** tab will be displayed. Using the options in this tab, you can specify the settings to change the instance property of the mass created, divide the surfaces of the mass, and can modify the geometrical elements of the mass. You can change the instance properties of the massing geometry like material and visibility from the **Properties** palette. After changing the instance properties of the selected massing geometry, you can change the geometrical property of the surface by adding edges and profiles. The tools for changing the geometrical properties are: **X-Ray**, **Add Edge**, and **Add Profile**. You can invoke these tools from the **Form Element** panel of the **Modify | Form** (solid form) tab.

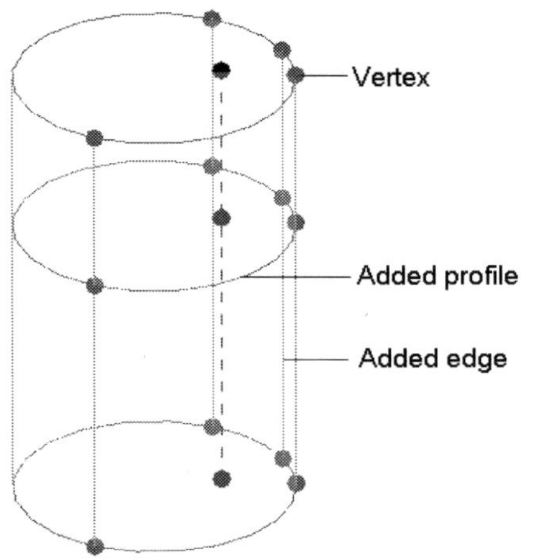

*Figure 10-13 Massing geometry with added edge and profile in the **X-Ray** mode*

The **X-Ray** tool in the **Element** panel of the **Modify | Form** tab is used to display the geometry skeleton like vertices and edges of the mass. The **Add Edge** tool can be used to add edges to the form of the mass and the **Add Profile** tool is used to add profile to the surface of the mass. Figure 10-13 shows the added edge and profile in the **X-Ray** mode.

MASSING IN CONCEPTUAL DESIGN ENVIRONMENT

Conceptual designing is the very first phase of a design process. The primary objective of a conceptual design is to create a representation of the idea for creating the system mass of a project. For creating a building project, the conceptual design is very important for architects, engineers, and designers as it helps in finding out the final representation of the design intent. Thus, it enables them to create a more specific sets of plans.

Interface of the Conceptual Design Environment

To open the interface of the Conceptual Design Environment, choose **New > Conceptual Mass** from the **File** menu. On doing so, the **New Conceptual Mass-Select Template File** dialog box will be displayed. In this dialog box, select the *Metric Mass.rft* file in the **Conceptual Mass** folder and then choose the **Open** button; the interface of the Conceptual Design Environment along with the settings of the *Mass.rft* template file will be displayed, as shown in Figure 10-14. In the interface of the Conceptual Massing Environment, you can use various tools in the ribbon to create the conceptual mass.

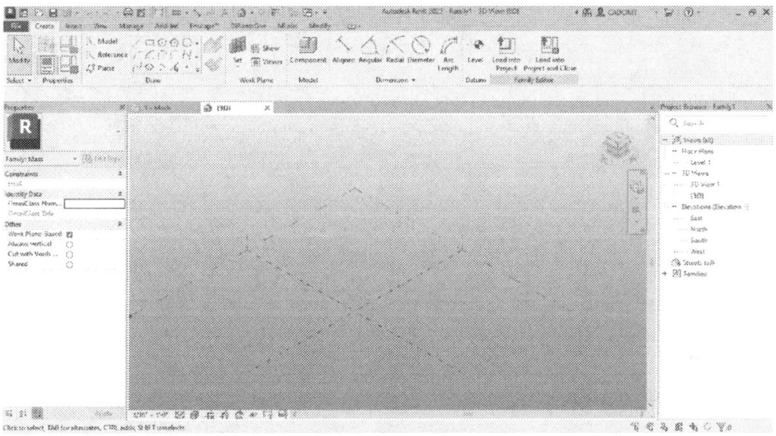

Figure 10-14 The Conceptual Design Environment interface

Creating Masses in the Conceptual Design Environment

In the Conceptual Design Environment, you can create the conceptual mass by using various tools from the ribbon. The ribbon of the Conceptual Design Environment, as shown in Figure 10-15, consists of the following tabs: **Create**, **Insert**, **View**, **Manage**, and **Modify**.

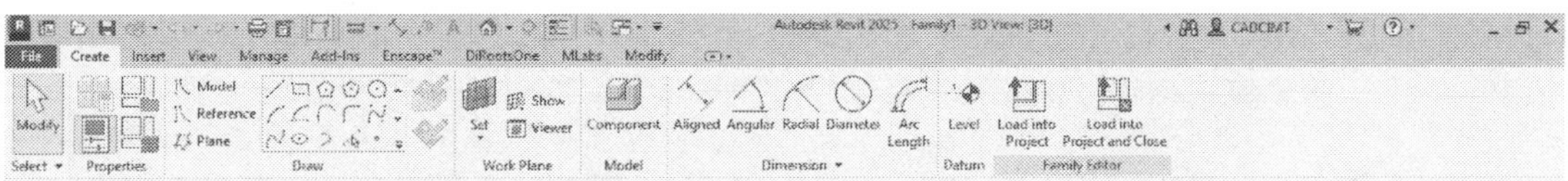

Figure 10-15 The ribbon of the Conceptual Design Environment

The **Create** tab is chosen by default. From the **Create** tab, you can access the tools required to create the Conceptual Mass. From the **Workplane** panel of the **Create** tab, you can choose any of the tools such as **Set**, **Show**, and **Viewer** that can be used for setting, showing, and viewing workplane to create the conceptual mass, respectively. You can choose the **Level** tool in the **Datum** panel to add levels that will assist in the creation of the conceptual mass. You can add levels to 3D and elevation views in the Conceptual Design Environment. The **Draw** panel of the **Create** tab contain tools that can be used to sketch the conceptual mass. In this panel, you can use the **Model**, **Reference**, and **Plane** tools to sketch the model lines, reference lines, and reference planes to create the conceptual mass. When you choose the **Model** or **Reference** tool from the **Draw** panel, the **Draw on Face** and the **Draw on Work Plane** tools will be activated. You can use any of these tools to specify the plane in which conceptual mass will be created. Also, on choosing the **Model** or **Reference** tool from the **Draw** panel, a contextual tab will be displayed along with various options in the **Options Bar**. This tab contains options that help in sketching the model or the reference line. In the **Family Editor** panel of the **Create** tab, you can choose the **Load into Project** tool to load the conceptual mass into the project file. From the **Dimension** panel of the **Create** tab, you can choose various tools to dimension the sketch of the conceptual mass. In the ribbon of the Conceptual Design Environment, the tools in the other tabs such as **Insert**, **View**, **Manage**, and **Modify** are same as those in the Project Environment. In the next sections, the method and the tools involved in the process of drawing and creating the conceptual mass are discussed.

Tip

*You can divide any face of the form. To do so, select the face and then choose the **Divide Surface** tool from the **Divide** panel of the **Modify / Form** tab; the **Modify / Divided Surface** tab will be displayed. You can use the tools from the displayed tab and the options in the **Options Bar** to divide the selected face.*

WORKSHARING CONCEPTS

Worksharing is a method of distributing work among people involved in a project to accomplish it within the stipulated period of time. In Worksharing, each person involved in the project is assigned a task that has to be accomplished through proper planning and coordination among the team members.

Worksharing Using Workset Tools

In Autodesk Revit, you work in an integrated single file model. This implies that all the construction related information and documents are available in a single model file and this forms the basis of the worksharing process.

The process of worksharing can be implemented using the **Worksets** tool. Worksets are a collection of elements such as ducts, VAV box, AHU, diffusers and more that define the domain and the ownership of the user in a project. For example, a project may contain building elements like beams, columns, mechanical equipments, plumbing fixtures, electrical elements, architectural elements, and so on. In a worksharing environment, you can assign each element to a workset, which defines its ownership to the user, thereby restricting other users to change or modify it in the project without prior permission. Therefore, only one user can edit a workset at a time.

When you start a new Revit project, the worksharing is not activated by default because the software assumes that you are working in a single user environment. When you transform your project from a single user to a multi-tasking environment, you need to understand the concept of central file system and the utility of using the worksets. A central file is a master file of the project and stores the entire building information. This file is connected and shared by all other files representing individual users and it monitors the progress of the entire project. The files that are shared and connected to the central file are called local files. These files are a copy of the central file and are saved locally at the individual users workstation. These files can also be saved in a specified network location, thereby differentiating them from the central file. These local files act as an interfacing mechanism for working on the central file.

Process of Worksharing

With the start of a new project, you start creating the building geometry with basic detailing. But as the project progresses, you need to share the project with multiple users and speed up the process of work. To invite multiple users to work in your project, you can save the file that you have worked as a central file at a particular network location and then create copies of it in the local hard drives of individual team members. By doing so, you divide the work into different parts. Everybody will start working with the copied central file from his or her individual workstation. Before starting with the file, each member working in the project is instructed to rename the copied central file at their individual hard drive location and then open it in Revit to proceed further. The flow diagram in Figure 10-16 explains the process of worksharing in brief.

Elements in each separate file are tied to the ownership rule linked to the central file. This restricts individual team members to edit an element in his or her local file, which is owned by someone in local files. In the worksharing process, the central file behaves as the controller of all the elements that are shared by members.

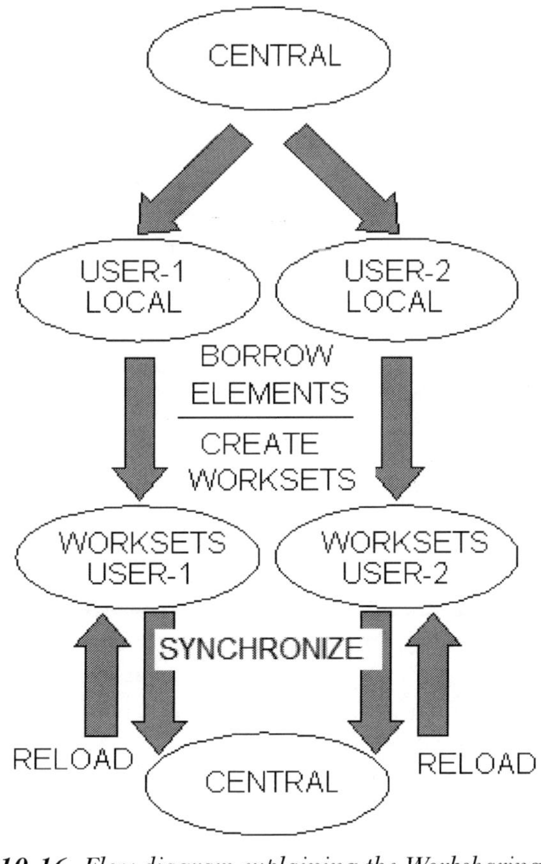

Figure 10-16 *Flow diagram explaining the Worksharing process*

Note

*If the model is not saved then enable the worksharing process by choosing the **Collaborate** tool from the **Collaborate** tab. On doing so, the **Collaborate- Revit Model Not Saved** dialog box will be displayed, as shown in Figure 10-17. This dialog box will prompt you to either save the model and continue the process cancel the process. Select the **Save the model and continue** option from the dialog box to perform the collaboration process.*

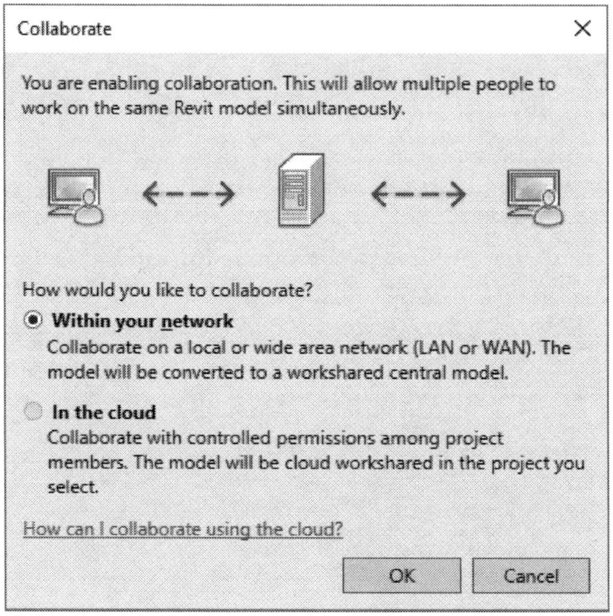

***Figure 10-17** The **Collaborate** dialog box*

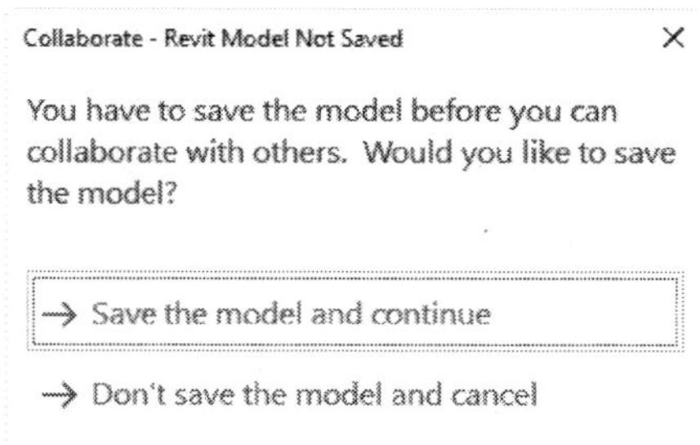

***Figure 10-18** The **Collaborate - Revit Model Not Saved** dialog box*

In the **Worksets** dialog box, there are six different columns: **Name**, **Editable**, **Owner**, **Borrowers**, **Opened**, and **Visible in all views**, as shown in Figure 10-19. The **Name** column displays the worksets that are defined or predefined in the project. The **Editable** column shows the status of the modification of the elements and components of the workset in the project. The **Owner** column specifies the ownership of the particular workset in the project.

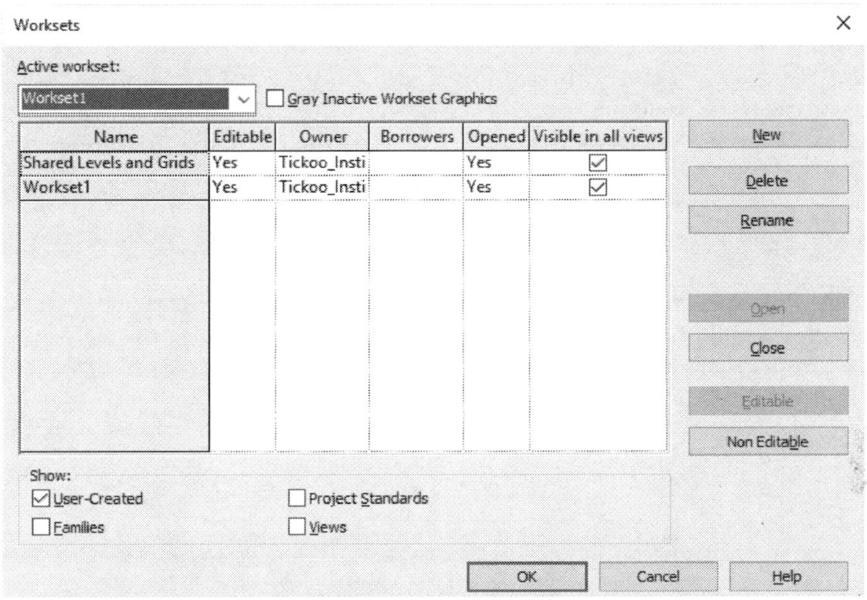

***Figure 10-19** The **Worksets** dialog box displaying different columns*

The ownership of the workset refers to the name of the user who had created it or who currently owns it. This name refers to the user of the master file or the user of the local file. By default, Revit denotes the name of the user, which is determined by the log-on name of the operating system. In Revit, before enabling the workset, you can change the username. To do so, choose the **Options** button from the **File** menu; the **Options** dialog box will be displayed. In this dialog box, choose the **General** tab and specify a new name in the **Username** edit box.

Note

Note that the currently signed Autodesk ID will be assigned as the username for the project.

The **Borrowers** column in the **Worksets** dialog box displays the name of the user currently using the elements of the worksets that do not belong to him or her. In this process, the owner or any other user can only use the borrowed worksets, if the borrower allows to do so. The owner can be a borrower of its own workset if he or she makes the workset as non-editable. To do so, choose the **Non-Editable** button in the **Worksets** dialog box The workset can be made non-editable if its project is synchronized to central.

The **Opened** column is used to select a workset definition. There are two options in this column: **Yes** and **No**. These options can be used to control the visibility of the elements in a group. By default, all worksets in the **Worksets** dialog box are assigned the **Yes** option. It means all the elements and components of these worksets are visible in the file. But, if you choose the **No** option, the visibility of elements and components of worksets will be disabled.

The **Visible in View** column allows you to show workset in all views. This column carries check boxes corresponding to the **Shared Level and Grids** and **Workset 1 Name** parameters. These check boxes are selected by default. On clearing these check boxes, the workset will not be displayed in any view.

You can select various options from the **Active workset** drop-down list located at the top in the **Worksets** dialog box. This drop-down list is used to select those worksets that are user defined and have the **Yes** status displayed in the **Opened** column of the **Worksets** dialog box. Select the appropriate workset available in the drop-down list to make it the current workset. If you select the **Gray Inactive Workset Graphics** check box, the elements that do not belong to the active workset will turn gray in the drawing area and the active elements will be displayed in black. If you clear this check box, all elements in the drawing area, whether they are active or inactive, will turn black.

To create a new workset definition, choose the **New** button in the **Worksets** dialog box; the **New Workset** dialog box will be displayed, as shown in Figure 10-20. Enter the name of the workset in the **Enter new workset name** edit box and choose the **OK** button to return to the **Worksets** dialog box.

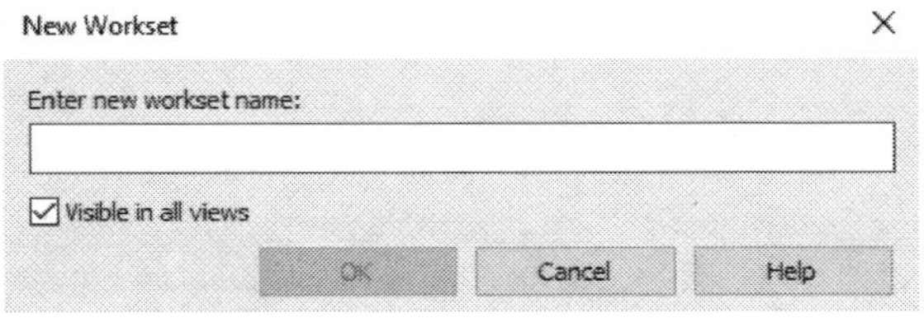

Figure 10-20 *The **New Workset** dialog box*

The **Delete** button in the **Worksets** dialog box is used to completely remove the selected workset. Note that you can delete only that workset definition which is editable. When you choose the **Delete** button, the **Delete Workset** dialog box will be displayed, as shown in Figure 10-45. This dialog box can be used to delete the elements belonging to the deleted workset or to move these elements to a different workset definition. If you select the **Deleted** radio button in this dialog box, elements that belong to the workset being deleted, will be deleted. Similarly, if you want to move these elements to a different workset, select the **Moved to** radio button and then select the appropriate workset from the **Moved to** drop-down list. Choose the **OK** button to return to the **Worksets** dialog box. You can rename the editable worksets. To do so, select the workset in the **Name** column and choose the **Rename** button; the **Rename** dialog box will be displayed. Specify a name in the **New** edit box to change the name of the selected workset. The **Open** and **Close** buttons in the **Worksets** dialog box are used to change the status of the selected workset in the **Opened** column to **Yes** and **No**, respectively. Similarly, the **Editable** and **Non Editable** buttons can also be used to change the status of the selected workset in the **Editable** column to **Yes** and **No**, respectively. While using the workset for the first time, if you choose the **Non Editable** button before saving the file to the central file, the **Revit** warning message box will be displayed. This message box warns you to save your file to the central file.

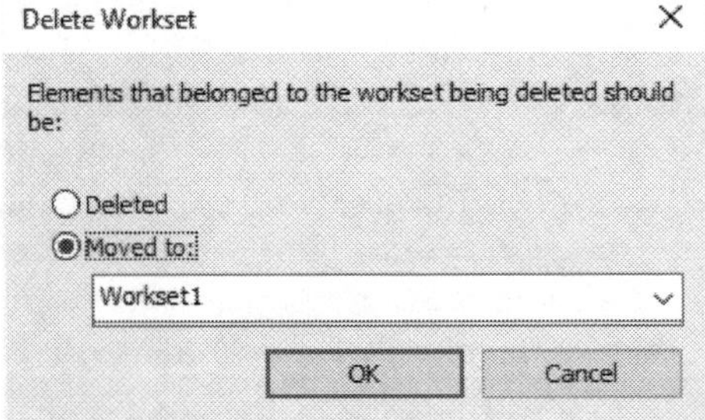

Figure 10-21 *The **Delete Workset** dialog box*

Saving Methodology in Worksharing

While working in a worksharing environment, it is important to learn the methods of saving your files. There are two methods of saving a file and sharing your files with others: saving a local file and synchronizing with the central file. You can save your file as a local file by choosing **Save** from the **File** menu. The **Save** option is generally used to save the local file. Using this method, you can secure the work that you have done without affecting the worksharing process. The other method is to synchronize the file with the central file. The **Synchronize with Central** option is used to publish the work in the central file. The moment you synchronize your file to central, you acquire the changes made by other team members to their local models, provided they had also saved their files to the central file at that particular interval.

In Revit, you can synchronize a local file with the central file by choosing the **Synchronize Now** tool or the **Synchronize and Modify Settings** tool from the **Synchronize** panel. If you choose the **Synchronize Now** tool, the current project file will be saved to the central file. On choosing the **Synchronize and Modify Settings** tool, the **Synchronize with Central** dialog box will be displayed, as shown in Figure 10-22. This dialog box allows you to customize the way to save your file to central. In the **Synchronize with Central** dialog box, the **Central Model Location** text box shows the path to the central file in the network. You can then use the **Browse** button to change the path, if required. After synchronizing the required file with central, you can use various check boxes to relinquish borrowed elements and various worksets, if required. The **Comment** text box in the **Synchronize with Central** dialog box allows you to publish the message regarding the changes you made during a particular session to the team member.

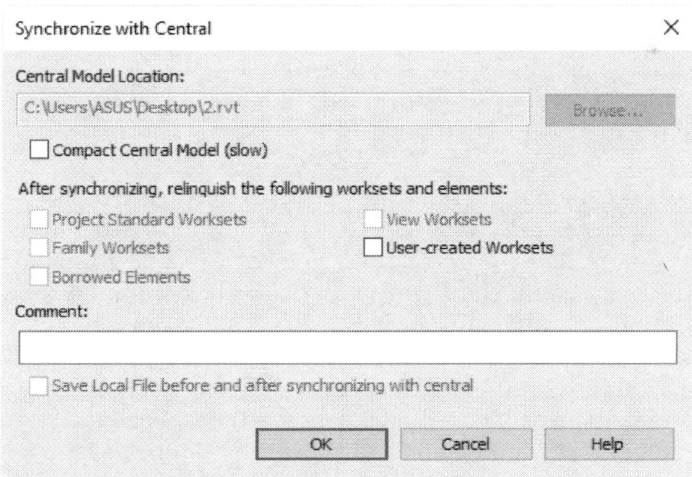

***Figure 10-22** The **Synchronize with Central** dialog box*

Element Ownership Concepts

In a worksharing environment, you need to know about the status of ownership of a particular element or workset in the project. To check the status of ownership of an element, select it and check whether you can modify or reposition it. If you own the element, it will be highlighted in blue and can be modified and repositioned. In case, you do not own the element that you have selected, it will still be highlighted in red, and a blue icon is attached to it indicating it cannot be modified.

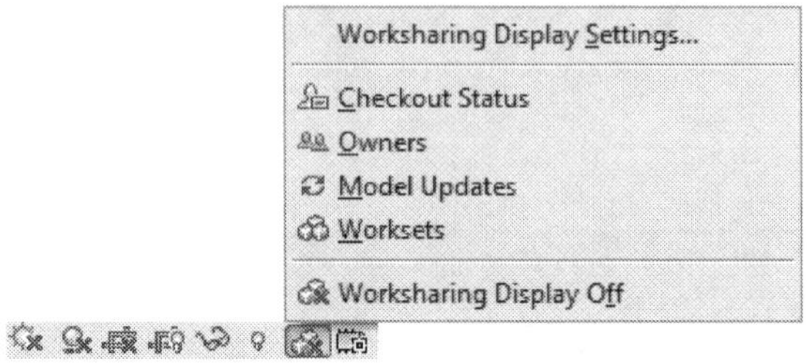

***Figure 10-23** Options in the flyout displayed upon choosing the **Worksharing Display Off** option*

In Autodesk Revit, you can use color to visualize the status of workshared elements. To do so, choose the **Worksharing Display Off** option from the **View Control Bar** and then select various options from the flyout displayed, refer to Figure 10-23. You can use these options to customize the display modes to visually distinguish the following in a view: borrowed elements, current element owners, model updates, and element checkout status.

In Revit, it is easy to retain the ownership status of a particular element or a component. There are several methods to get the ownership of a particular element or a workset and these are discussed next. One of the common methods of owning an element or a component is by trying to move or edit it. In this case, Autodesk Revit facilitates you to own it, provided it is not owned by anyone else in the worksharing environment. You can also get the ownership of an element by clicking the left mouse button over the blue icon that is displayed along with it. Next, to modify or edit an element, select it and click on the right mouse button and then choose the **Make Elements Editable** option from the shortcut menu displayed. In context to ownership, you need to know other concepts like Borrowing Elements, Placing Requests, and Granting Requests. These concepts are discussed next.

Sometimes, you or your team member may need to take the ownership of some of the elements within a particular workset rather than the entire workset. In such a situation, you can borrow elements from other users. To do so, select the element to be borrowed and click on the blue icon displayed with the element; the **Revit 2025** dialog box will be displayed with a warning message box regarding the ownership of the element. This message box also informs you to place request to the owner to use it in your file. Choose the **Place Request** button in the dialog box to request the owner to grant permission for using the element. On doing so, the **Check Editability Grants** dialog box will be invoked. Choose the **Check Now** button from the dialog box to check the status of the request made and choose the **Continue** button to resume the work.

Autodesk Revit also provides you with the flexibility to borrow components from worksets that are not editable by a team member. This is possible by getting permission from the user of the component, and then the borrowed elements or components can be edited, saved to central, and relinquished. In this release, the process of granting or denying permission to borrow an element or a component for editing it has been made more fast and interactive.

Monitoring the Worksharing Process

In Revit, you can monitor the worksharing process by using the **Worksharing Monitor for Revit 2025** program. This program is available for Revit subscription customers only. This is installed along with the regular Revit 2025 program.

The Worksharing Monitor for Revit 2025 program is useful for Revit projects that involves file-based worksharing environment, in which multiple people work on one project. The advantages of using the Worksharing Monitor for Revit 2025 program over conventional method of monitoring are:

- You can track team members involved in the project. By doing so, the BIM Manager of the project can identify the progress of the project and the people who are responsible for it

.

- You can track whether the current file is up-to-date or not. By knowing this, the owner of the local file can synchronize it with the central file or reload the latest changes in the central file.

- Using worksharing you can find out the the current status of the synchronization with central operation. This will enable the user to know whether the synchronization with central process has finished or not.

- You can track the status of request that has been placed to the owner of a particular element that is needed to be edited. By knowing this, the user can take decision to proceed with the work or not.

Note

The Worksharing Monitor for Revit 2025 program is strongly recommended for the projects that involve people from many disciplines. It is not recommended for standalone Revit projects.

To start the monitoring process for a workshared project, choose the **Start** button at the lower left corner of the screen (default position) in the taskbar (Windows 10) and then choose **All Autodesk > Worksharing Monitor for Revit 2025**, the worksharing monitoring process gets started by displaying the **Worksharing Monitor for Autodesk Revit 2025** window, as displayed in Figure 10-24.

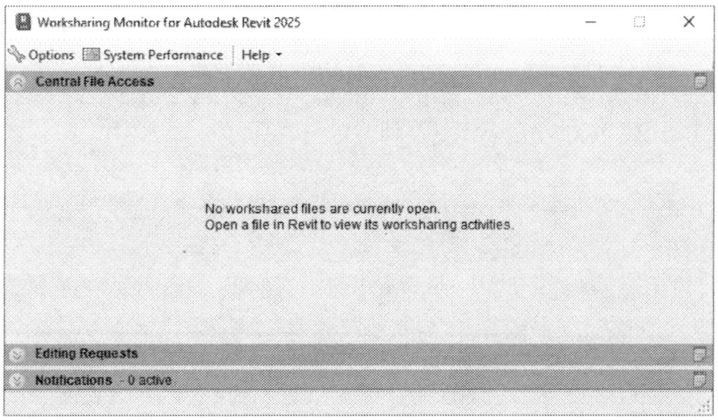

***Figure 10-24** The **Worksharing Monitor for Autodesk Revit 2024** window*

In the **Worksharing Monitor for Autodesk Revit 2025** window, there are three panes: **Central File Access**, **Editing Request**, and **Notification**. Click on the double-arrow button next to the **Central File Access** text to expand the pane, if it is not expanded, refer to Figure 10-48. The

Central File Access pane displays information about various files associated with the project along with the users who are responsible for it.

The **Central File Access** pane displays a table with the following column heads: **Current User**, **Action**, **Progress**, and **Step/Time Remaining**. The **Current User** column displays the name of the users who are working in the project.

The **Action** column displays information about what users are doing in the project. The following actions are displayed under the **Action** column depending on the situation of the project: **Working in local, Opening central file**, **Working in central**, **Synchronizing with central**, and **Reloading latest**. The **Working in local** action will be displayed if the user is working in a local copy of the project file. If the user is opening the central file to work on it, the **Opening central file** action will be displayed. Similarly, the **Working in central action** will be displayed if the user is working directly in the central file for the project. Alternatively, the **Synchronizing with central action** will be displayed if the user is saving changes to the central file. The **Reloading latest action** will be displayed if the user is updating the local copy of the project file with the most recent version of the central file.

Note

*In the **Worksharing Monitor for Autodesk Revit 2025** window the user name will be displayed in gray indicating that the user is working in the project but is not currently using Worksharing Monitor. As a result, this user will not receive information about editing requests through Worksharing Monitor.*

The **Progress** column in the **Worksharing Monitor** displays the progress bar for the corresponding action taken by the user. **The Step/Time Remaining** column display the time remaining for the progress of the worksharing process to complete.

In the **Worksharing Monitor for Autodesk Revit 2025**, the **Editing Requests** pane will display information about various requests that have been issued permission from the users to work on part of a project that they own. Figure 10-25 shows various requests in the **Editing Requests** pane of the **Worksharing Monitor for Autodesk Revit 2025** window. The **Notifications** pane of the **Worksharing Monitor for Autodesk 2025** window displays warnings and notices, providing information about issues that interfere in the progress of a Revit project.

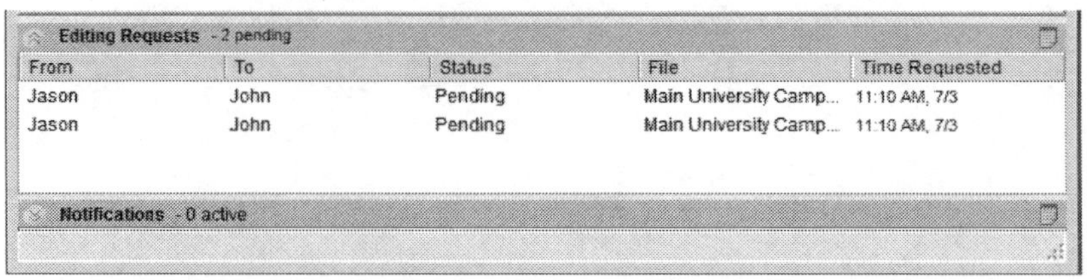

Figure 10-25 *The options in the **Editing Requests** pane*

In the **Worksharing Monitor for Autodesk 2025** window you can specify settings related to usage of the options of the panes. To do so, choose the **Options** button; the **Options** dialog box will be displayed, as shown in Figure 10-26.

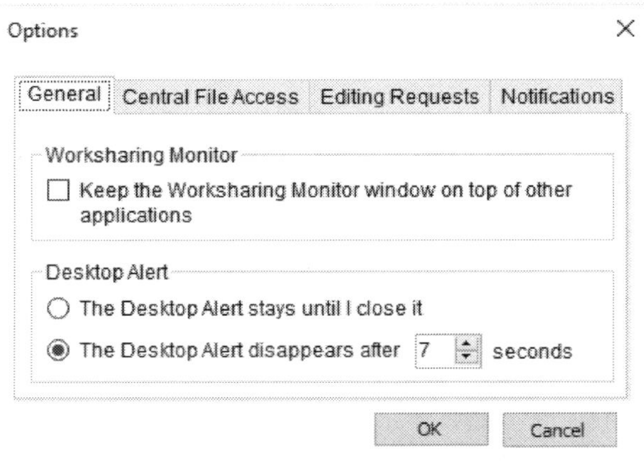

Figure 10-26 *The **Options** dialog box*

In this dialog box, the **General** tab is chosen by default. In the **Worksharing Monitor** area, you can select the **Keep the Worksharing Monitor window on the top of other applications** check box to keep the **Worksharing Monitor for Autodesk Revit 2025** window on the top of other application. Specify how long the desktop alerts will be displayed in the **Worksharing Monitor for Autodesk Revit 2025** window by selecting the required radio button in the **Desktop Alert** area. You can choose the **Central File Access** tab of the **Options** dialog box to display various options related to central file access issues. The options in this tab can be used to review the options that are available in the **Central File Access** pane of the **Worksharing Monitor for Autodesk Revit 2025** window. You can select any option in this tab to indicate that you want to receive desktop alerts about central file issues.

In the **Options** dialog box, you can choose the **Editing Requests** tab. On doing so, various options related to this tab will be displayed. You can select an option in this tab to indicate that you want to receive desktop alerts related to worksharing. The desktop alerts will be displayed when another user sends you an editing request, or when your editing requests are granted or denied.

In the **Options** dialog box, choose the **Notifications** tab to display various options related to notification issued during the worksharing process. In this tab, you can select an option to indicate that you want to receive desktop alerts when Worksharing Monitor issues a notification or a problem is resolved.

After specifying various options in the **Options** dialog box, you can choose the **OK** button to close it. In the **Worksharing Monitor for Autodesk Revit 2025** window, you can choose the **System Performance** button to review and monitor the performance of hardware system you are currently using. On choosing the **System Performance** button, the **System Performance** window will be displayed, as shown in Figure 10-27. In this window, you can use various options to monitor various hardware resources of a system such as physical memory, virtual memory, CPU load, and disk space. In this window, you can select the **Keep System Performance on top** check box to keep the **System Performance Monitor** window on top of other applications. Also, you can change the transparency of the **System Performance Monitor** window by using the **Window transparency** spinner. You can set the transparency between 0% and 80%.

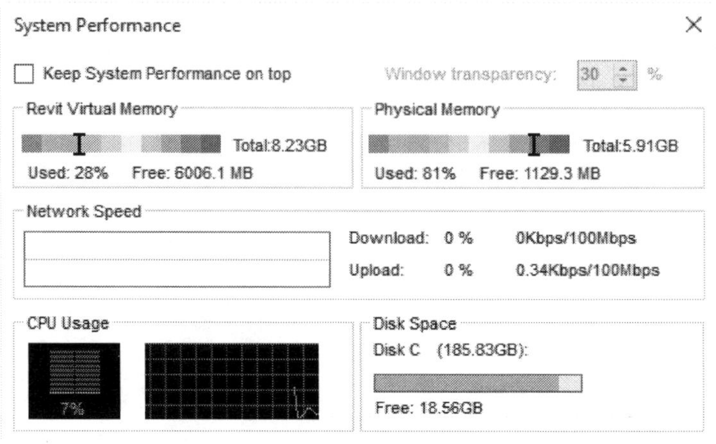

***Figure 10-27** The **System Performance** window*

TUTORIAL

Tutorial 1 Dual Duct VAV box

In this tutorial, you will create a Dual Duct VAV box in the conceptual mass family environment and then set various parameters to the VAV box created.

(Expected time: 1hr 30 min)

1. File name to be assigned:
   For Imperial *Dual_Duct_VAV*
   For Metric *M_Dual_Duct_VAV*

The following steps are required to complete this tutorial:

a. Open the Family Editor environment for massing.
b. Create the VAV box body.
c. Create the VAV box outlet sleeve.
d. Create the VAV box inlet sleeve.
e. Create the VAV box control box.
f. Attach the outlet duct connector.
g. Attach the inlet duct connector.
h. Save the project.
i. Close the project using the **Close** tool.

Opening the Family Editor Environment for Massing

In this section, you will open the Family Editor environment.

1. Choose **New > Family** from the **File** menu; the **New Family- Select Template File** dialog box is displayed.

2. In this dialog box, select the folder **English_I** for Imperial unit system or **English** for the Metric unit system from the **Look in:** drop-down list. Next, select the template file:
   Imperial system: *Mechanical Equipment.rft*
   Metric system: *Metric Mechanical Equipment.rft*
3. After selecting the template, choose the **Open** button; the reference planes are displayed in the working area.

Creating the VAV box Body

In this section, you will create the body of the VAV box.

1. Make sure that you are in the **Ref. Level** view, if not, then double-click on the **Ref. Level** sub node under the **Floor Plans** node in the **Project Browser**. Next, choose the **Reference Plane** tool from the **Datum** panel of the **Create** tab; the **Modify | Place Reference Plane** contextual tab is displayed.

2. In this contextual tab, choose the **Pick Lines** tool from the **Draw** panel. Next, enter the value **2'** (**610 mm**) in the **Offset** edit box in the **Options Bar** and press ENTER.

3. Now, pick the vertical reference plane twice to create the two reference planes at a distance of **2'** (**610 mm**), one on the right and the other on the left of reference plane.

4. Next, enter the value **1'** (**305 mm**) in the **Offset** edit box in the **Options Bar** and press ENTER. Pick the horizontal reference plane twice to create two reference planes at a distance of **1'** (**305 mm**), one above and the other below the existing reference plane.

5. Choose the **Annotate** tab and then choose the **Aligned** tool from the **Dimension** panel.

6. Next, add dimensions to the created reference planes, as shown in Figure 10-28. Choose the **Modify** button to exit the dimension tool.

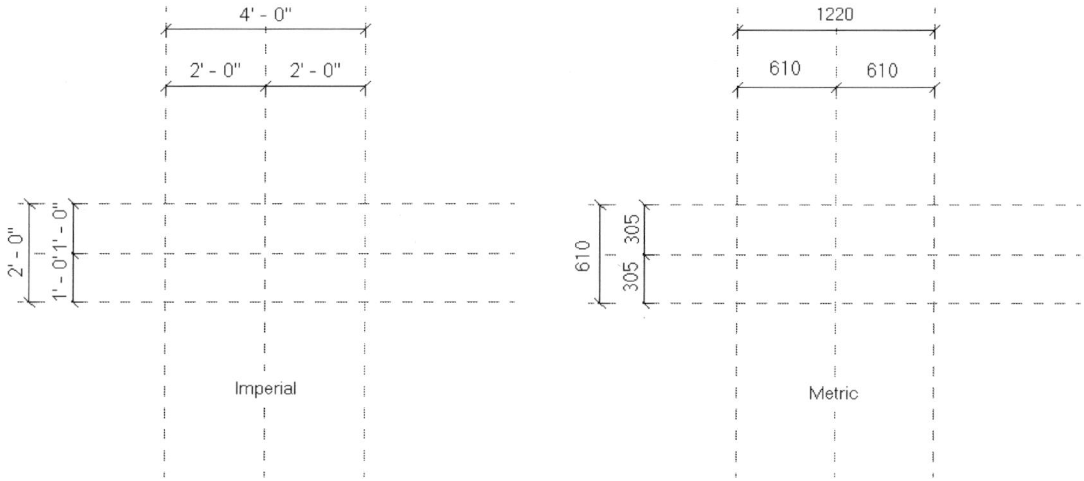

Figure 10-28 *Dimensions added to the created reference planes*

7. Select the **2'** (**610mm**) dimension line for the vertical reference plane; a blue **EQ** symbol appears. Click on the symbol; the dimensions are replaced by permanent **EQ** symbols.

> **Tip**
> *The* ***EQ*** *symbol ensures that whenever there is a change in the position of one of the reference planes with equality constraint, the other reference plane moves symmetrically.*

8. Next, select the **1'** (**305mm**) dimension line for the horizontal reference plane, and assign the equality constraint to it, as done in step 7, refer to Figure 10-29.

9. Select the **2'** (**610mm**) dimension line; the **Modify Dimensions** contextual tab is displayed. In this tab, choose the **Create Parameter** button from the **Label dimension** panel; the **Parameter Properties** dialog box is displayed, as shown in Figure 10-30.

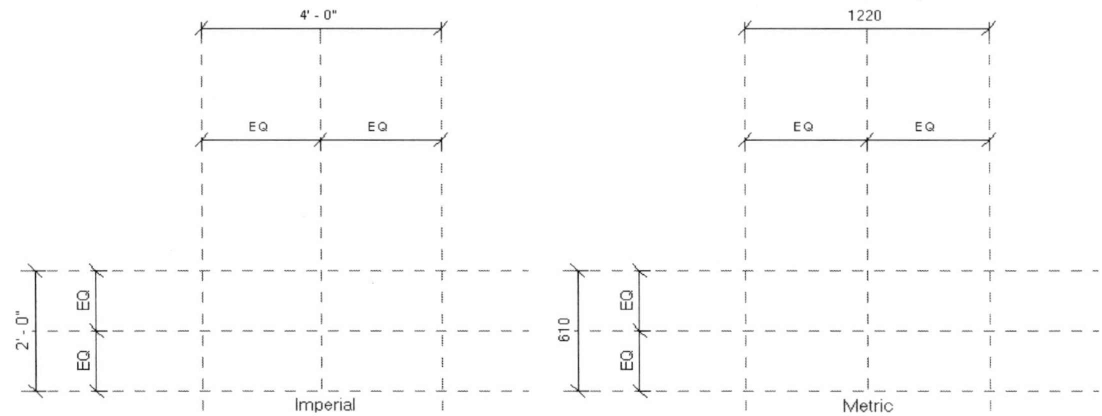

Figure 10-29 *Assigning equality constraint to the reference plane*

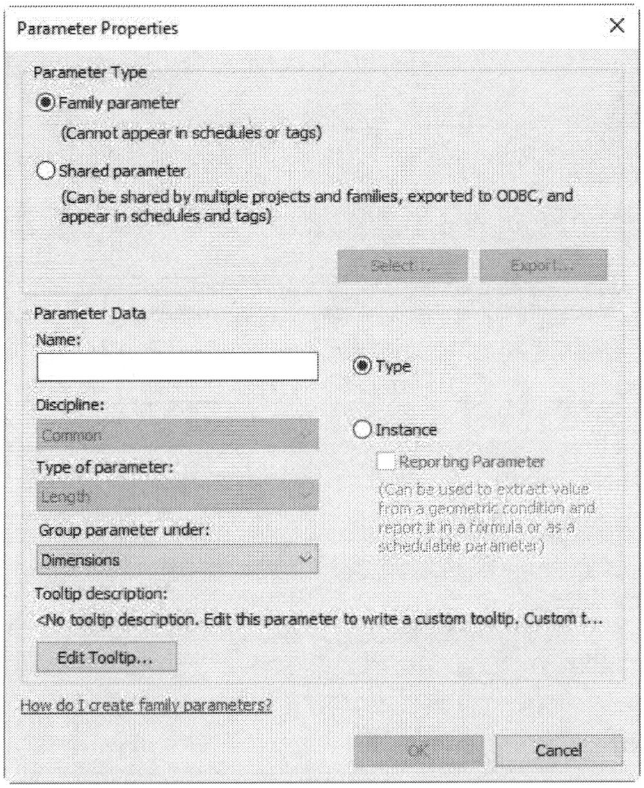

***Figure 10-30** The **Parameter Properties** dialog box*

10. In this dialog box, select the **Shared parameter** radio button in the **Parameter Type** area.

11. Next, choose the **Select** button in the **Parameter Type** area; the **Shared Parameter File Not Specified** message box is displayed, prompting you to choose one shared parameter. Choose the **Yes** button; the **Edit Shared Parameters** dialog box is displayed.

12. In this dialog box, choose the **Create** button in the **Shared parameter file** area; the **Create Shared Parameter File** dialog box is displayed.

13. Now, browse to the location **C: / rmp_2025** and create a new folder with name **c10_rmp_2025_tut**. Now, in the **File name** edit box, enter the name **Dual_Duct_VAV_box** for the Imperial system or **M_Dual_Duct_VAV_box** for the Metric system. Choose the **Save** button; the **Create Shared Parameters File** dialog box is closed.

14. In the **Edit Shared Parameters** dialog box, choose the **New** button from the **Groups** area; the **New Parameter Group** dialog box is displayed. In this dialog box, enter **Mechanical VAV box** in the **Name** edit box and choose the **OK** button; the **New Parameter Group** dialog box is closed.

15. In the **Edit Shared Parameters** dialog box, choose the **New** button from the **Parameters** area; the **Parameter Properties** dialog box is displayed.

16. In this dialog box, enter **VAV box Width** in the **Name** edit box. Select the **Common** option from the **Discipline** drop-down list and the **Length** option from the **Type of Parameter** drop-down list. Now, choose the **OK** button to close the **Parameter Properties** dialog box.

17. Repeat the procedure followed in steps 15 and 16 to create other shared parameters, as given in the table below. Refer to Figure 10-31 to view the added parameters.

Name	Discipline	Type of Parameter
VAV box Length	Common	Length
VAV box Height	Common	Length
VAV Outlet Width	Common	Length
VAV Outlet Height	Common	Length
VAV Inlet Diameter	Common	Length
VAV Max CFM	HVAC	Air Flow

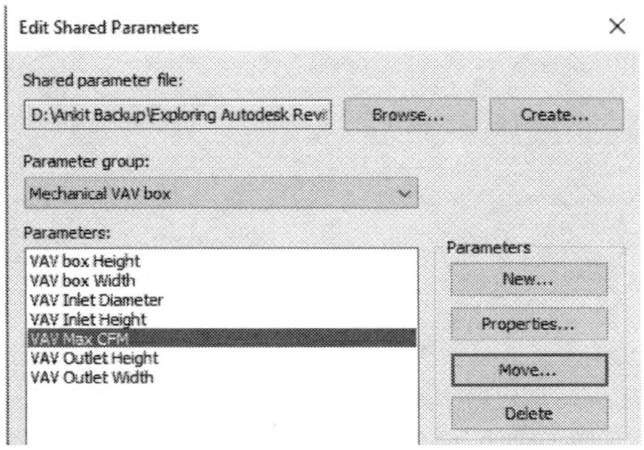

***Figure 10-31** Partial view of the **Edit Shared Parameters** dialog box displaying the created parameters*

18. Choose the **OK** button; the **Edit Shared Parameters** dialog box is closed and the **Shared Parameters** dialog box is displayed.

19. Now, select the **VAV box Width** option from the **Parameters** area of this dialog box. Choose the **OK** button in the **Shared Parameters** dialog box and then in the **Parameter Properties** dialog box to close them. Notice that the width label is assigned to the dimension line.

20. Next, select the **4' (1220 mm**) dimension line; the **Modify|Dimensions** contextual tab is displayed. Choose the **Create Parameter** button from the **Label dimesnions** panel of the **Modify|Dimensions** tab; the **Parameter Properties** dialog box is displayed.

21. In this dialog box, select the **Shared parameter** radio button and then choose the **Select** button in the **Parameter Type** area; the **Shared Parameters** dialog box is displayed. In this dialog box, select the **VAV box Length** option from the **Parameters** area.

22. Now, choose the **OK** button in the **Shared Parameters** dialog box and then in the **Parameter Properties** dialog box to close them. Notice that the length label is assigned to the selected dimension line.

23. Double-click on the **Front** subnode under the **Elevations (Elevation 1)** head; the view is changed to Front elevation. Invoke the **Reference Plane** tool from the **Datum** panel of the **Create** tab; the **Modify | Place Reference Plane** contextual tab is displayed.

24. Choose the **Pick Lines** tool from the **Draw** panel of the **Modify | Place Reference Plane** tab. In the **Options Bar**, enter **5"** (**127 mm**) in the **Offset** edit box and press ENTER.

25. Double-click on the existing reference level to create two reference planes, one above and one below it. Choose the **Modify** button to exit the selection.

26. Choose the **Annotate** tab and then invoke the **Aligned** tool from the **Dimension** panel. Next, add dimensions to the created reference planes, as shown in Figure 10-32. Choose the **Modify** button to exit the tool.

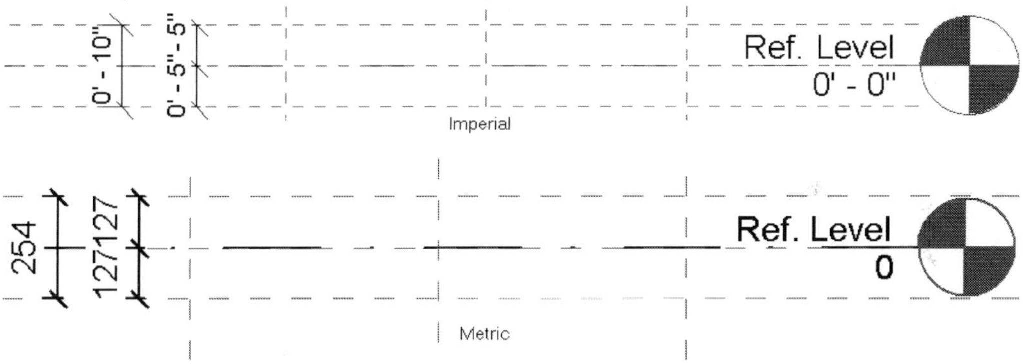

***Figure 10-32** Dimensions added to the created reference planes*

27. Select the **5"** (**127 mm**) dimension line; the **EQ** symbol appears with the selected dimension. Click on this symbol; permanent equality constraints are assigned to the created reference planes.

28. Next, select the **10" (254 mm)** dimension line; the **Modify | Dimensions** tab will be displayed. Choose the **Create Parameter** button from the **Label dimensions** panel; the **Parameter Properties** dialog box is displayed.

29. Repeat the procedure followed in steps 21 and 22 and assign the **VAV box Height** parameter to the VAV box.

30. Select the reference plane below the reference level, and then in the **Properties** palette, enter the name **VAV box bottom** in the **Name** edit box.

31. Next, invoke the **Set** tool from the **Work Plane** panel of the **Create** tab; the **Work Plane** dialog box is displayed. In this dialog box, ensure that the **Name** radio button is selected.

Then, choose the **Reference Plane: VAV box bottom** option from the drop-down list in the **Specify a new Work Plane** area, as shown in Figure 10-33. Choose the **OK** button; the dialog box is closed and the **Go To View** dialog box is displayed.

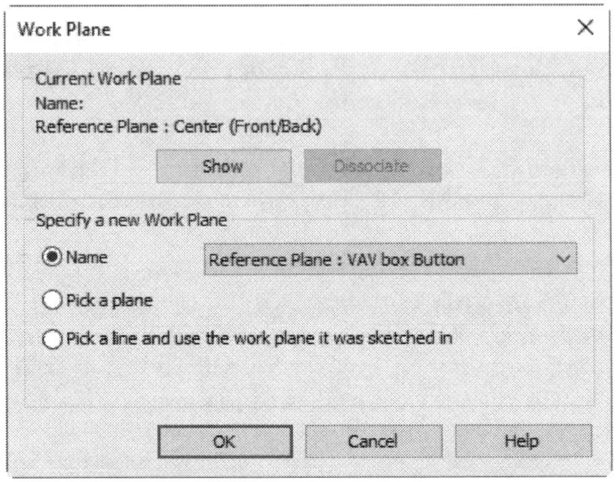

Figure 10-33 *Setting reference plane from the* ***Work Plane*** *dialog box*

32. In this dialog box, select the **Floor Plan: Ref. Level** option, if it is not selected by default, and then choose the **Open View** button; the dialog box is closed and the view is changed to plan view.

33. Next, choose the **Extrusion** tool from the **Forms** panel of the **Create** tab; the **Modify | Create Extrusion** contextual tab is displayed. In this tab, choose the **Rectangle** tool from the **Draw** panel. Now, make a rectangle with **2'** x **4'** (**610** x **1220** mm) dimensions, and with the already defined reference planes, as shown in Figure 10-34.

34. Choose the **Finish Edit Mode** button to exit the tool and to finish extrusion. Next, make sure that **Extrusion Start** value is set to **0' 0"** (**0.00** mm) in the **Properties** palette. Then, choose the **Associate Family Parameter** button next to the **Extrusion End** edit box; the **Associate Family Parameter** dialog box is displayed.

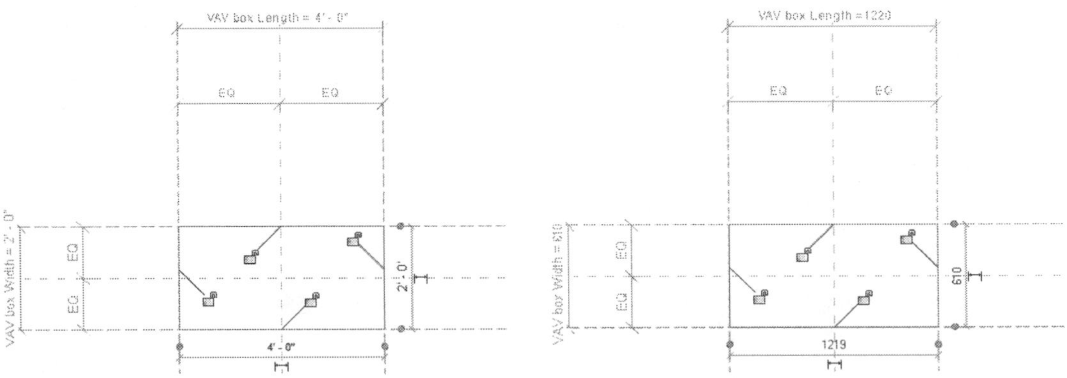

Figure 10-34 *Extrusion boundary created by using the* ***Rectangle*** *tool*

35. In this dialog box, select the **VAV box Height** option from the **Existing family parameters of compatible type** area, as shown in Figure 10-60. Next, choose the **OK** button; the **Associate Family Parameter** dialog box is closed. Notice that the end value **10"** (**254 mm**) is set in the **Extrusion End** edit box in the **Properties** palette. This value shows the height of the VAV box.

36. Next, double-click on the **View 1** subnode under the **3D Views** node in the **Project Browser**; the view is changed to 3D view and the model of the VAV box body is displayed, as shown in Figure 10-35.

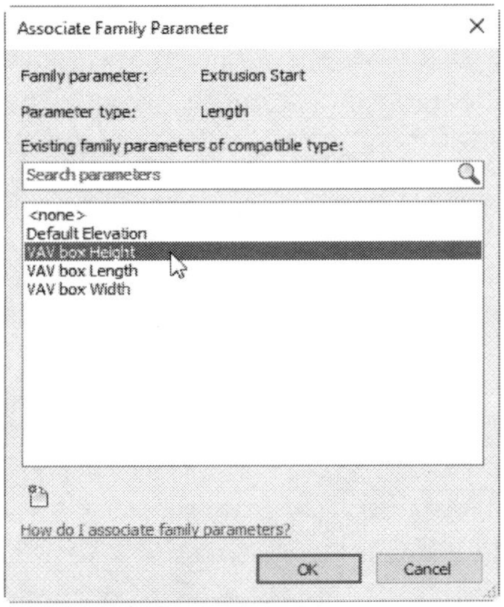

Figure 10-35 *Choosing an option from the* ***Associate Family Parameter*** *dialog box*

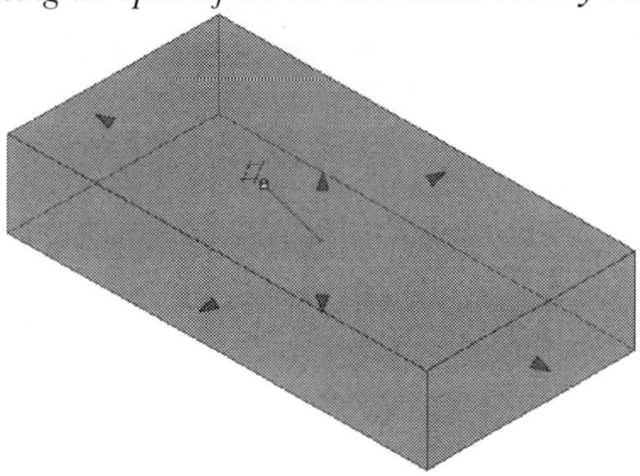

Figure 10-36 *3D model of VAV box*

Note

If the view is not similar to the one show in Figure 10-36, you can choose the ***Home*** *button near to the ViewCube.*

Creating the VAV box Outlet Sleeve

In this section, you will create the sleeve for the VAV box body.

1. Double-click on the **Right** subnode under the **Elevations (Elevation 1)** node; the view is changed to Right elevation. Next, choose the **Reference Plane** tool from the **Datum** panel of the **Create** tab; the **Modify | Place Reference Plane** contextual tab is displayed.

2. In this tab, choose the **Pick Lines** tool from the **Draw** panel. Next, enter **4" (100 mm)** in the **Offset** edit box in the **Options Bar**. Next, pick the middle vertical reference plane twice to create one reference plane on its right, and the other on its left.

3. Repeat the procedure followed in step 2 and make two reference planes, above and below the existing horizontal **Ref. Level**. Choose the **Modify** button to exit the tool. Next, annotate the created reference planes and assign the equality constraint to it, refer to Figure 10-37.

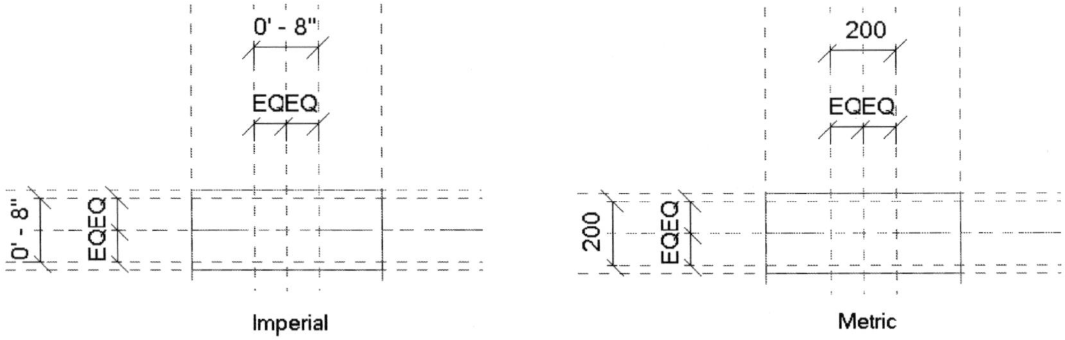

Figure 10-37 *Annotating the reference planes*

4. Now, assign the **VAV Outlet Width** and **VAV Outlet Height** parameters to the created reference planes following the procedure given in steps 20 through 22 of the previous section, refer to Figure 10-38.

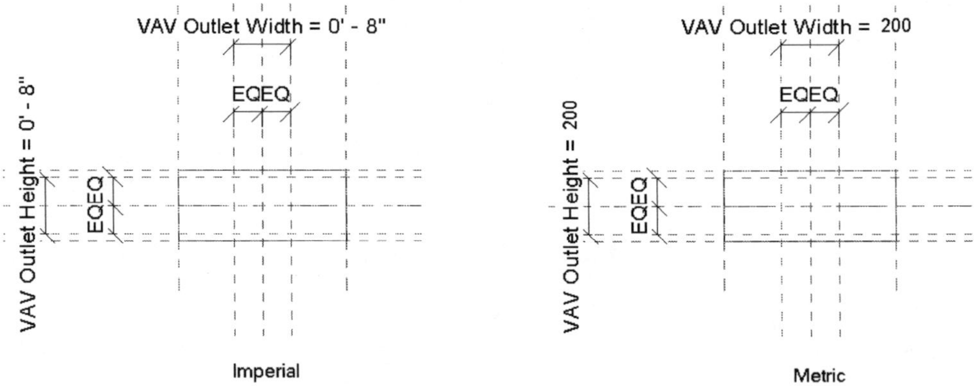

Figure 10-38 *Parameters added to the reference planes created*

5. Double-click on the **View 1** node under the **3D Views** node; the view is changed. Choose the **Set** tool from the **Work Plane** panel of the **Create** tab; the **Work Plane** dialog box is displayed. In this dialog box, choose the **Pick a plane** radio button from the **Specify a new Work Plane** area and then choose the **OK** button; the dialog box is closed.

6. Move the cursor on the right face of the VAV box body and then click on the face when it is highlighted, refer to Figure 10-39.

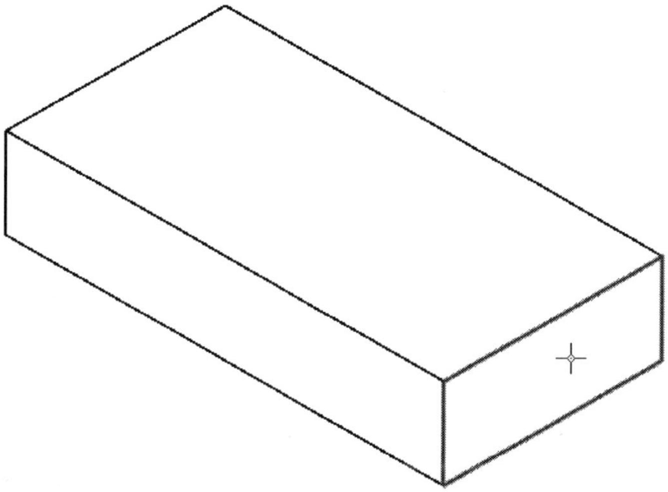

Figure 10-39 *Setting the reference plane on the face of the VAV box body*

7. Open the Right elevation from the **Project Browser**. Next, choose the **Create** tab and then choose the **Extrusion** tool from the **Forms** panel; the **Modify | Create Extrusion** contextual tab is displayed. Choose the **Rectangle** tool from the **Draw** panel of this tab.

8. Now, sketch a box with dimension **8" x 8" (200 x 200** mm) box including the boundaries of the created reference planes, as shown in Figure 10-40. Make sure in the **Properties** palette, the value for **Extrusion Start** is **2' 00" (610 mm**). Next, enter the value **2' 2" (660 mm**) in the edit box next to **Extrusion End**. Choose the **Finish Edit Mode** button to finish sketching and exit the tool.

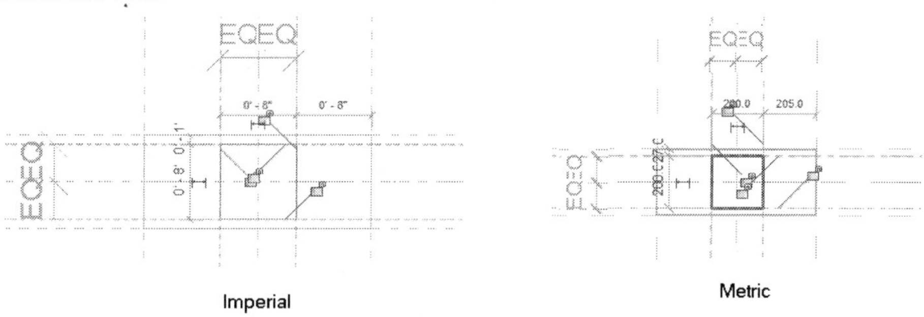

Imperial Metric

Figure 10-40 *Sketching the VAV box sleeve*

9. Choose the **Save As > Family** from the **File** menu; the **Save As** dialog box is displayed. In this dialog box, browse to *C:\rmp_2025\c10_rmp_2025_tut* and enter the name **VAV_box_Outlet_Sleeve** for Imperial system or **M_VAV_box_Outlet_Sleeve** for Metric system in the **File name** edit box. Next, choose the **Save** button; the dialog box is closed and the file is saved.

Creating the VAV box Inlet Sleeve

In this section, you will create VAV box inlet sleeve to the previously created body.

1. Choose the **Family Types** tool from the **Properties** panel of the **Create** tab; the **Family Types** dialog box is displayed, as shown in the Figure 10-41.

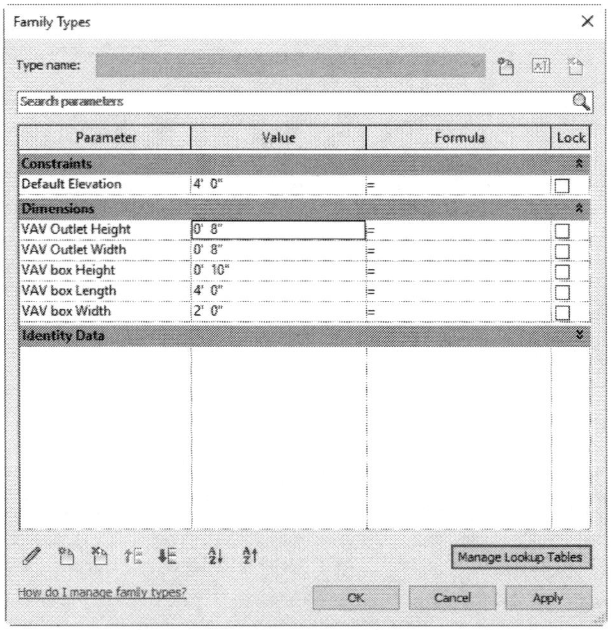

***Figure 10-41** The **Family Types** dialog box*

2. Next, choose the **New Parameter** button displayed at the bottom of the **Family Types** dialog box; the **Parameter Properties** dialog box is displayed. In this dialog box, select the **Shared parameter** radio button and then choose the **Select** button from the **Parameter Type** area; the **Shared Parameters** dialog box is displayed.

3. In this dialog box, select the **VAV Inlet Diameter** option from the **Parameters** area. Choose the **OK** button in the **Shared Parameters** dialog box and then in the **Parameter Properties** dialog box to close them. Next, in the **Family Types** dialog box, enter **4"** (**100 mm**) in the Value field for the **VAV Inlet Diameter** option.

4. Choose the **New Parameter** button from the bottom of the **Family Types** dialog box; the **Parameter Properties** dialog box is displayed. In this dialog box, enter **Inlet Radius** in the **Name** edit box in the **Parameter Data** area.

5. Now, select **Length** from the **Type of parameter** drop-down list and then select the **Other** option from the **Group parameter under** drop-down list in the **Parameter Data** area. Choose the **OK** button; the **Parameter Properties** dialog box is closed.

6. Now, in the **Family Types** dialog box, click in the **Formula** field located next to the **Inlet Radius** option and then, enter **VAV Inlet Diameter / 2**, refer to Figure 10-42. Next, choose the **Apply** button and then the **OK** button; the dialog box is closed.

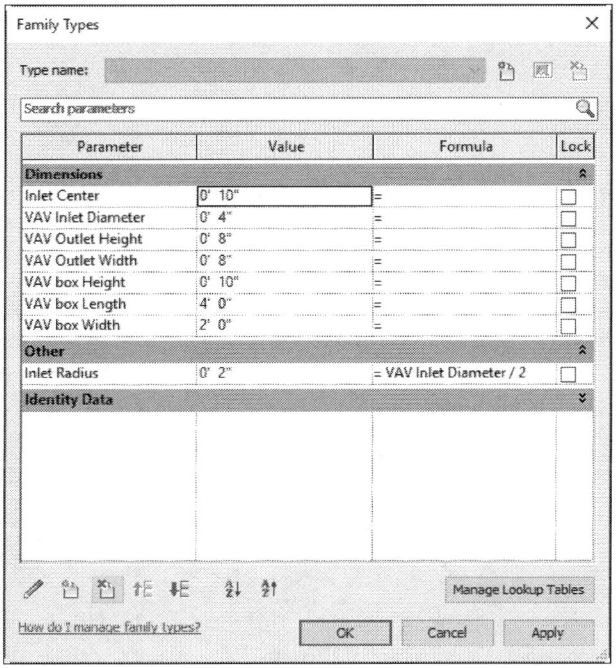

Figure 10-42 *Setting parameter values in the **Family Types** dialog box*

7. Double-click on the **Left** subnode under the **Elevations (Elevation 1)** node; the view is changed to Left elevation view. Now, choose the **Hidden Line** option from the **Visual Style** flyout in the **View Control Bar**; the VAV outlet sleeve disappears from the view.

8. Press and hold the CTRL key and select all the reference planes which define the VAV outlet sleeve. Next, choose the **Hide Element** option from the **Temporary Hide/Isolate** flyout in the **View Control Bar**.

9. Repeat the procedure followed in steps 1 and 2 of the previous section to make the two reference planes of dimension **5" (127 mm)** on either sides of the existing reference plane and assign the equality constraint to the planes created, refer to Figure 10-43.

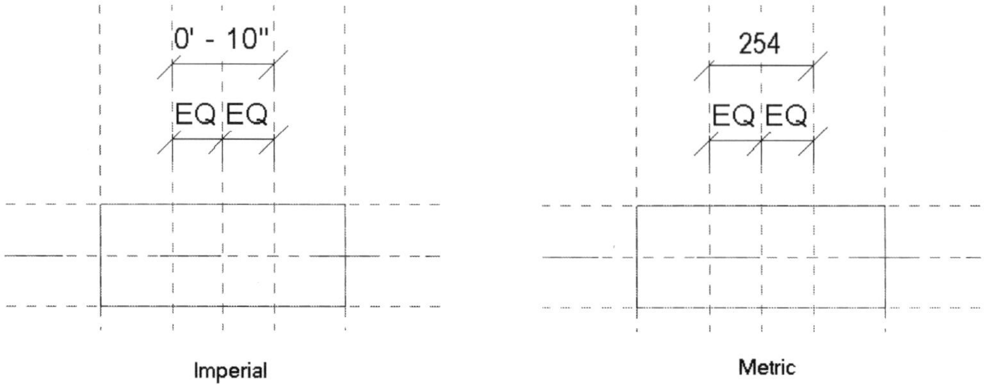

***Figure 10-43** Equality constraint assigned to the created reference planes*

10. Next, select the **10" (254 mm)** dimension line; the **Modify|Dimesnions** tab is displayed. Choose the **Create Parameter** button from the **Label dimensions** panel; the **Parameter Properties** dialog box is displayed.

11. Enter **Inlet Center** in the **Name** edit box in the **Parameter Data** area, and then choose the **OK** button; the dialog box is closed and a label is assigned to the reference planes.

12. Double-click on the **Front** subnode under **Elevations (Elevation 1)** node in the **Project Browser**; the view is changed to Front elevation view. Next, select the first vertical reference plane from the left. Now, in the **Properties** palette, click in the **Name** edit box and then enter **VAV Inlet**.

13. Now, change the view to Left elevation view and choose the **Create** tab and then invoke the **Set** tool from the **Work Plane** panel; the **Work Plane** dialog box is displayed. In this dialog box, make sure that the **Name** radio button is selected in the **Specify a new Work Plane** area. Next, select the **Reference Plane : VAV Inlet** option from the drop-down list in the **Specify a new Work Plane** area, refer to Figure 10-44. Choose the **OK** button; the dialog box is closed.

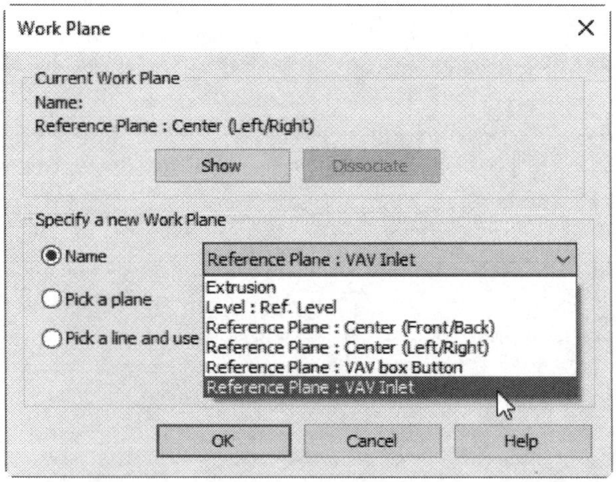

***Figure 10-44** Choosing the **Reference Plane VAV Inlet** option from the **Work Plane** dialog box*

14. Invoke the **Extrusion** tool from the **Forms** panel of the **Create** tab; the **Modify | Create Extrusion** contextual tab is displayed. Choose the **Circle** tool from the **Draw** panel of this tab. Draw two circles of **2"** (**50 mm**) radius at the intersections of the reference planes, as shown in Figure 10-45.

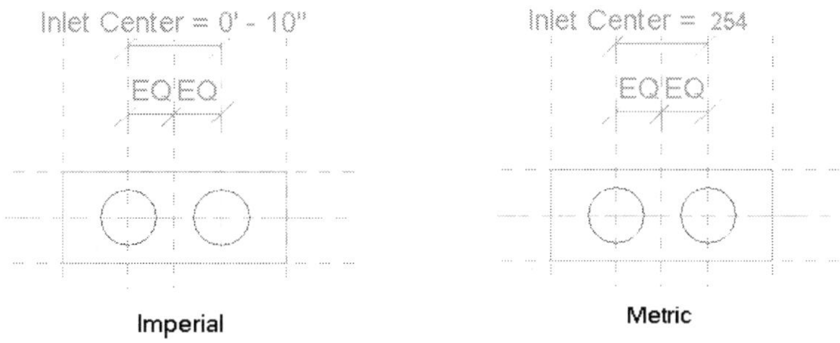

Figure 10-45 *VAV box inlet created using the* ***Circle*** *tool*

15. In the **Properties** palette, enter the value **0' 00"** (**0.00 mm**) in the **Extrusion Start** edit box and **-0' 4"** (**-100 mm**) in the **Extrusion End** edit box and then choose the **Apply** button to save changes. Next, choose the **Finish Edit Mode** button in the **Modify | Create Extrusion** contextual tab to finish the extrusion.

16. Change the view to Right elevation view and then choose the **Annotate** tab. Next, invoke the **Radial** tool from the **Dimension** panel. Next, add dimensions to the circles created and then press ESC to exit the tool.

17. Next, select both the dimension added to the circles and then choose the **Inlet Radius** option from the **Label** drop-down in the **Options Bar**. Repeat this step to assign label to both the circles, refer to Figure 10-46. Navigate to the 3D view. The geometry is created and it looks like shown in Figure 10-47.

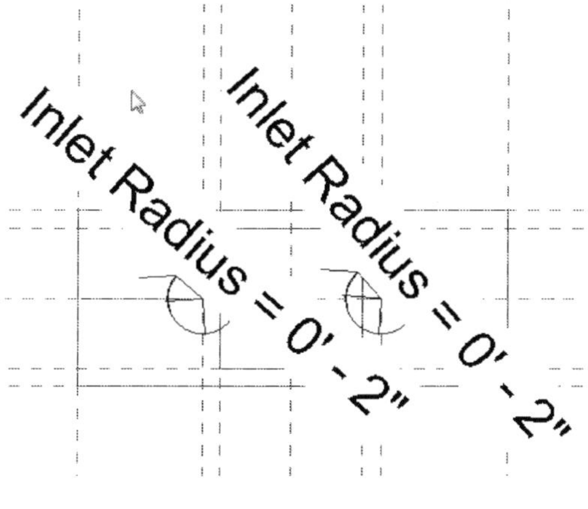

Figure 10-46 *Labels added to the created geometry*

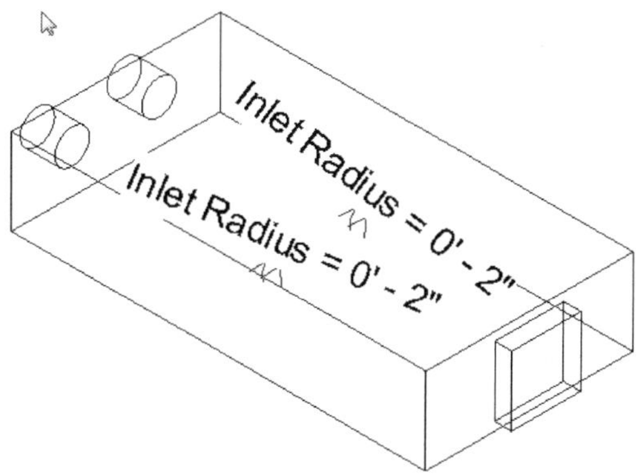

Figure 10-47 *VAV inlet sleeve*

Creating the VAV box Control Box

In this section, you will create the control boxes to the model generated in the previous sections.

1. Invoke the **Set** tool from the **Work Plane** panel of the **Create** tab; the **Work Plane** dialog box is displayed. In this dialog box, select the **Pick a plane** radio button from the **Specify a new Work plane** area and then choose the **OK** button to close the dialog box.

2. Rotate the model and move the cursor to the Front face of the VAV box and then click when it gets highlighted, as shown in Figure 10-48. Now, choose the **Show** tool from the **Work Plane** panel of the **Create** tab; the created workplane is highlighted, refer to Figure 10-49. Choose the **Show** button again to make the highlighted work plane disappear.

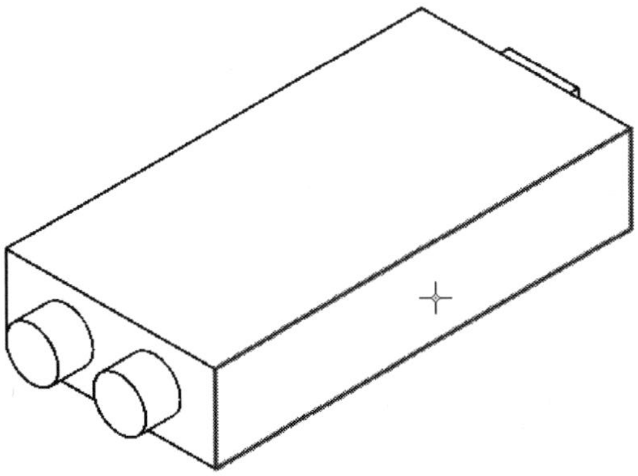

Figure 10-48 *Setting work plane to the front face*

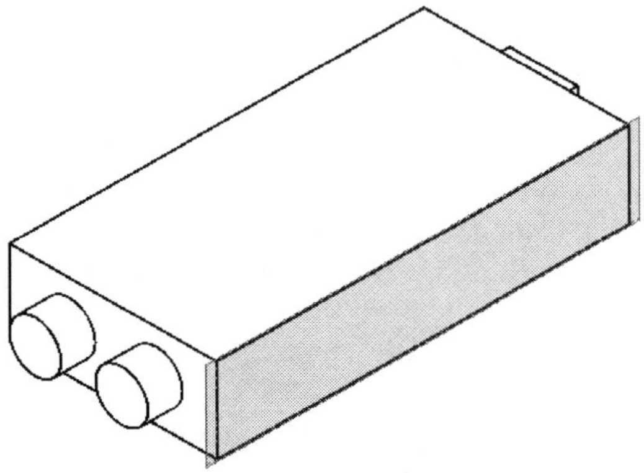

Figure 10-49 *Showing the created work plane*

3. Double-click on the **Front** view and then invoke the **Reference Plane** tool from the **Datum** panel of the **Create** tab. Next, choose the **Pick Lines** tool from the **Draw** panel. Next, enter **2" (51 mm)** in the **Offset** edit box in the **Options Bar** and then press ENTER. Pick the left reference plane to create a reference plane **2" (51 mm**) to its right, as shown in Figure 10-50.

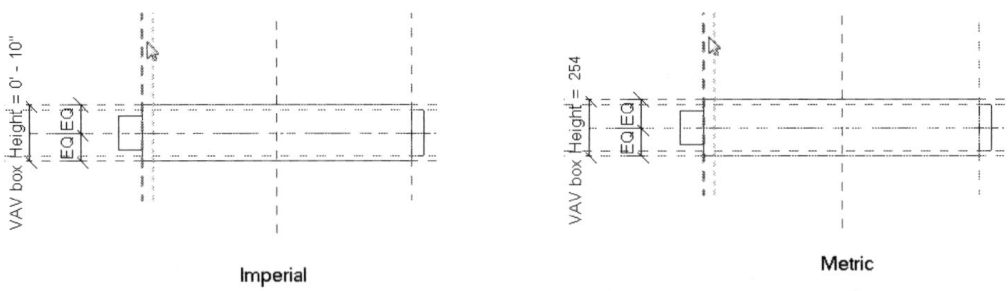

Figure 10-50 *Creating a reference plane using the* ***Pick Line*** *tool*

4. Enter **1' 00" (300 mm)** in the **Offset** edit box. Next, pick the recently created reference plane to create a reference plane on its right. Now, choose the **Modify** button to exit the tool.

5. Choose the **Create** tab and then choose the **Extrusion** tool from the **Forms** panel; the **Modify | Create Extrusion** contextual tab is displayed. Choose the **Rectangle** tool from the **Draw**

panel of this tab. Now, sketch a rectangle with **1'00"x0'10"** (**300** x **254 mm**) dimension including the boundaries of the created reference planes, as shown in Figure 10-51.

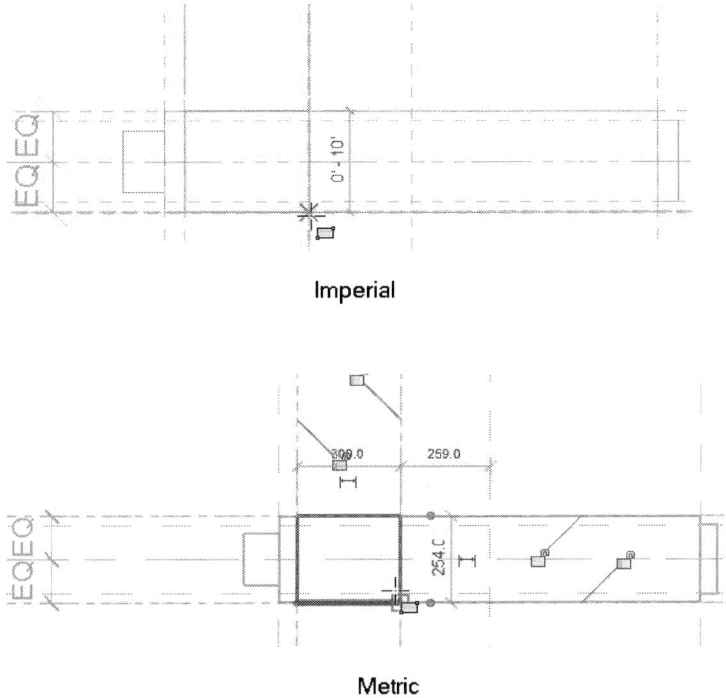

Figure 10-51 *Sketching the extrusion boundary*

6. In the **Properties** palette, click in the **Extrusion End** edit box and enter **1' 4"** (**406 mm**). Next, in the **Extrusion Start** edit box, enter **1' 00"** (**305 mm**), if it is not set by default. Next, choose the **Finish Edit Mode** button to finish sketching.

7. Repeat steps 5 and 6 to create a control box on the Back elevation with **-1' 4"** (**-406 mm**) **Extrusion End** and **-1' 00" (-305 mm) Extrusion Start** values. The final 3D view of the created geometry is shown in Figure 10-52.

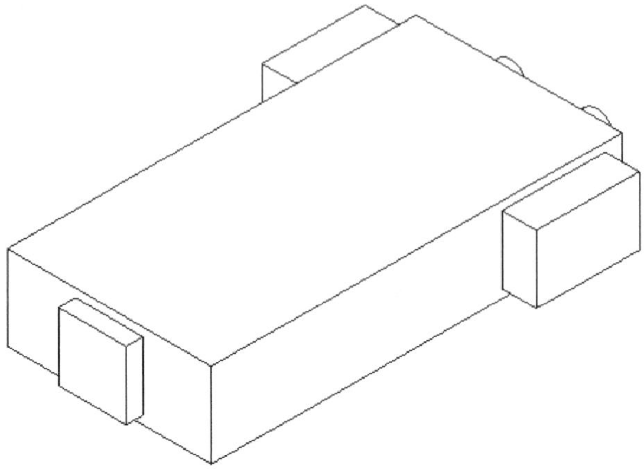

Figure 10-52 *3D view of the created control box*

Attaching the Outlet Duct Connectors

In this section, you will attach the outlet duct connector to the created geometry.

1. In the 3D view, navigate to the view in which the face of the outlet duct sleeve is visible. Next, invoke the **Duct Connector** tool from the **Connectors** panel of the **Create** tab; the **Modify | Place Duct Connector** contextual tab is displayed. Choose the **Face** tool from the **Placement** panel, if it is not chosen by default. Next, select the face of the VAV outlet sleeve to add connector to it, as shown in Figure 10-53.

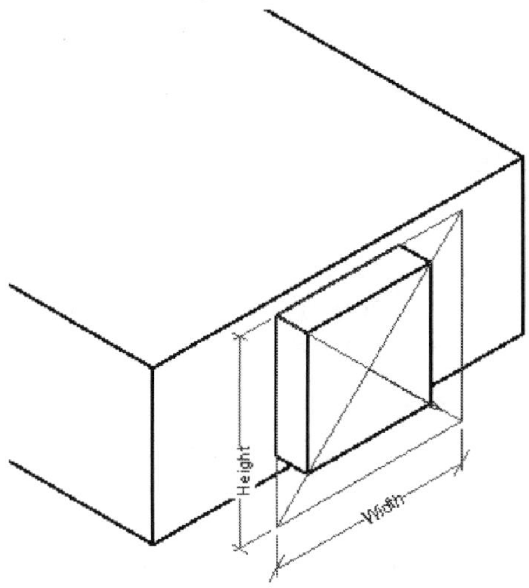

Figure 10-53 *Duct connector added to the outlet duct sleeve*

2. Choose the **Modify** button and then select the recently added duct connector. Next, click on the + sign next to the width dimension; the **Associate Family Parameter** dialog box is displayed. In this dialog box, select the **VAV Outlet Width** option and then choose the **OK** button; the **Associate Family Parameter** is closed and the parameter is added to the connector.

3. Repeat the procedure followed in step 2 and assign the **VAV Outlet Height** parameter to the height dimension.

4. In the **Properties** palette, click in the value field of the **Flow Direction** parameter and select the **Out** option from the drop-down list displayed. Next, click in the value field of the **System Classification** parameter and then select the **Supply Air** option, if it is not selected by default.

5. Choose the **Associate Family Parameter** button next to the **Flow** value field; the **Associate Family Parameter** dialog box is displayed. In this dialog box, choose the **New Parameter** button; the **Parameter Properties** dialog box is displayed. Click in the **Name** edit box in the **Parameter Data** area and enter **VAV Outflow**.

6. Choose the **Instance** radio button and then select the **Mechanical - Flow** option from the **Group parameter under** area, if it is not selected by default. Choose the **OK** button in the **Parameter Properties** dialog box and then in the **Associate Family Parameter** dialog box to close them. Choose the **Modify** button to exit the selection.

Attaching the Inlet Duct Connectors

In this section, you will attach the inlet duct connector to the created geometry.

1. Choose the **Family Types** tool from the **Properties** panel of the **Create** tab; the **Family Types** dialog box is displayed. In this dialog box, choose the **New Parameter** button from the bottom of the dialog box; the **Parameter Properties** dialog box is displayed. In this dialog box, enter **Duct Inlet Radius** in the **Name** edit box in the **Parameter Data** area.

2. Next, select the **Length** option from the **Type of parameter** drop-down list and the **Other** option from the **Group parameter under** drop-down list in the **Parameter Data** area. Choose the **OK** button; the **Parameter Properties** dialog box is closed.

3. In the **Family Type** dialog box, click in the value field of the **Formula** column of the **Duct Inlet Radius** parameter and enter the formula **Inlet Radius + 0' 1/16" (2 mm)**, refer to Figure 10-54. Next, choose the **Apply** button and then the **OK** button to save the changes and close the dialog box.

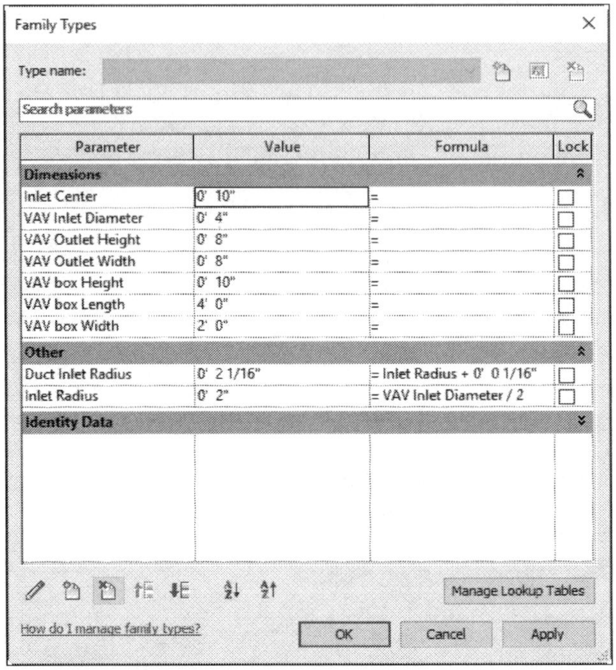

Figure 10-54 *Formulating the **Duct Inlet Radius** parameter*

4. Using the ViewCube in 3D view, navigate to the view in which the face of the inlet duct sleeve is visible, refer to Figure 10-55.

5. Next, invoke the **Duct Connector** tool from the **Connectors** panel of the **Create** tab and then choose the **Face** tool from the **Placement** panel, if it is not chosen by default. Next, select the faces of both the VAV inlet sleeves to add connectors to them, as shown in Figure 10-56.

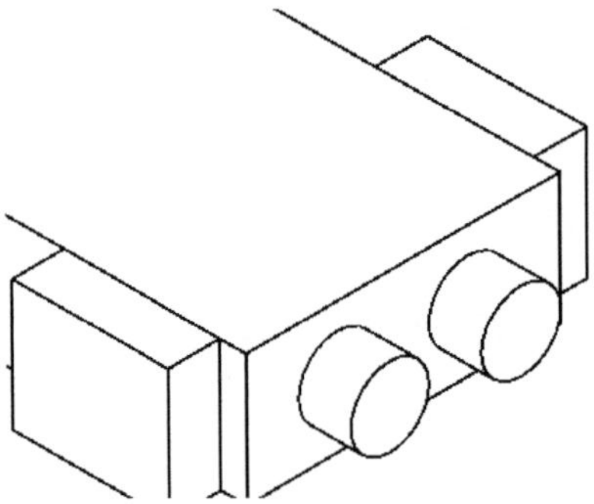

Figure 10-55 *3D view of the inlet duct sleeve*

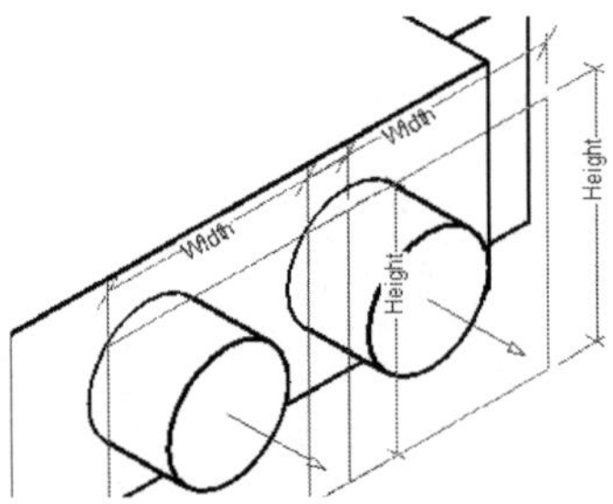

Figure 10-56 *Connectors added to the inlet duct sleeves*

6. Choose the **Modify** button to exit the tool and then select the recently added connectors. Now, in the **Properties** palette, click in the value field of the **Shape** parameter and then select the **Round** option from the drop-down list. In the **Flow Direction** parameter, select the **In** option from the drop-down list. Make sure that the **System Classification** parameter is set to **Supply Air**.

7. In the **Properties** palette, choose the **Associate Family Parameter** button next to the **Diameter** value field; the **Associate Family Parameter** dialog box is displayed. In this dialog box, select the **Duct Inlet Radius** option and then choose the **OK** button.

8. Choose the **Associate Family Parameter** button next to the **Flow** value field; the **Associate Family Parameter** dialog box is displayed again. In this dialog box, choose the **New parameter** button; the **Parameter Properties** dialog box is displayed. Click in the **Name** edit box in the **Parameter Data** area and enter **Inlet Flow**.

9. Select the **Instance** radio button and then select the **Mechanical-Flow** option from the **Group parameter under** area, if it is not selected by default. Choose the **OK** button to close the **Parameter Properties** dialog box and then the **Associate Family Parameter** dialog box. Choose the **Modify** button to exit the selection.

Saving the Project

1. Choose **Save As > Family** from the **File** menu; the **Save As** dialog box is displayed.

2. In this dialog box, browse to *C: \rmp_2025/\c10_rmp_2025_tut* and then enter **Dual_Duct_VAV** in the **File name** edit box.

3. Now, choose the **Save** button; the **Save As** dialog box closes and the project file is saved.

Closing the Project

1. To close the project, choose the **Close** option from the **File** menu.

   The file is closed. This completes the tutorial.

EXERCISE

Exercise 1 Supply Diffusers

Create an air supply diffuser in the family editor, using the file *Mechanical Equipment.rft* for Imperial system or *Metric Mechanical Equipment.rft* for Metric system. Parameters to be used are given next. Refer to Figure 10-57 to 59 to create the model geometry.

(Expected time: 45 min)

1. Project File Parameters:
   Template File-
   For Imperial *Mechanical Equipment.rft*
   For Metric *Metric Mechanical Equipment.rft*
   File Name to be assigned-
   For Imperial *c10_Supply-Diffuser_exer1.rvt*
   For Metric *M_c10_Supply-Diffuser_exer1.rvt*

2. Dimension of the air diffuser:

Base dimensions:	2'x2' (600x600 mm)
Base geometry height:	6" (150 mm)
Top dimension:	1'6"x1'6" (450x450 mm)
Top geometry height:	3" (75 mm)
Connector radius:	6" (150 mm)
Connector height:	3" (75 mm)

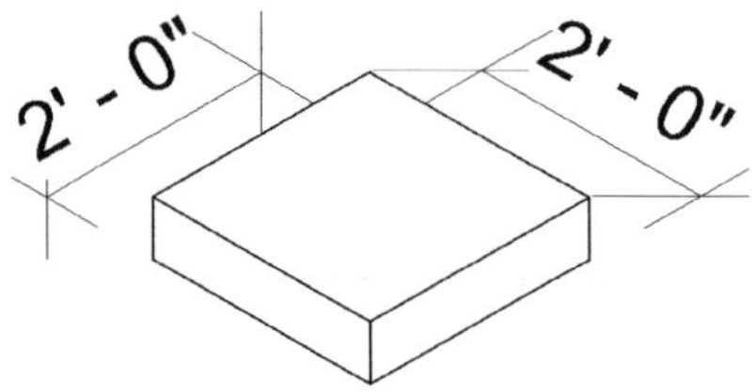

Figure 10-57 *The base geometry created*

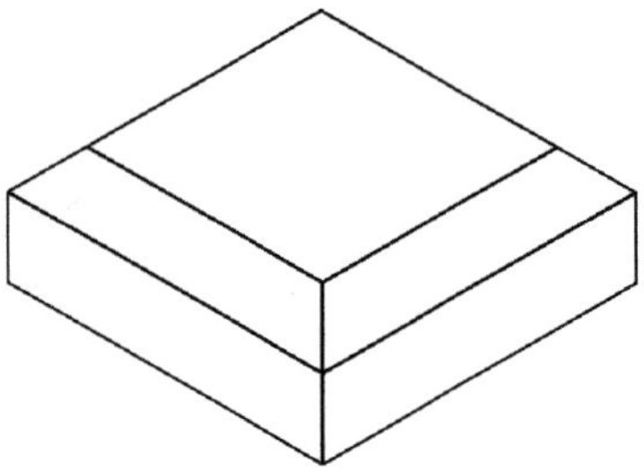

Figure 10-58 *The top geometry created on the base profile*

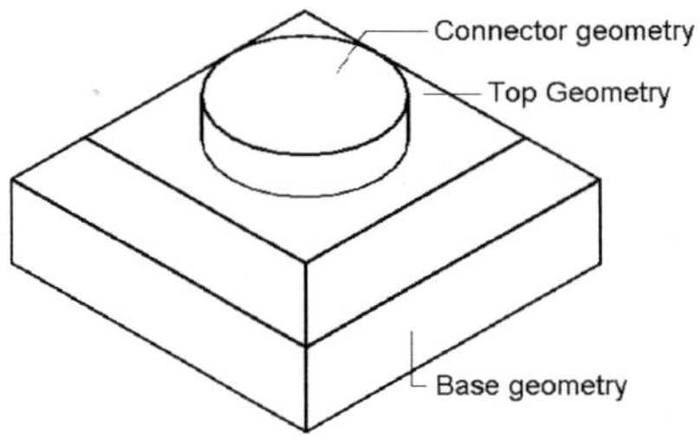

Figure 10-59 *Connector geometry created over the top profile*

Index

P

R

S

T

U

V

W

X

Z

Other Publications by CADSoft Technologies

The following is the list of some of the publications by CADSOFT Technologies. Please visit *www.cadsofttech.com* for the complete listing.

Revit Architecture for Novices Textbooks
- Revit Architecture 2025 for Novices: Learn By Doing
- Revit Architecture 2020 for Novices: Learn By Doing
- Revit Architecture 2019 for Novices: Learn By Doing

Revit Structure for Novices Textbooks
- Revit Structure 2023 for Novices: Learn By Doing
- Revit Structure 2020 for Novices: Learn By Doing
- Revit Structure 2019 for Novices: Learn By Doing

Revit MEP for Novices Textbooks
- Revit MEP 2019 for Novices: Learn By Doing

AutoCAD Civil 3D for Novices Textbooks
- AutoCAD Civil 3D 2025 for Novices: Learn By Doing

Autodesk 3dx Max for Novices Textbooks
- Autodesk 3ds Max 2020 for Novices: Learn By Doing
- Autodesk 3ds Max 2019 for Novices: Learn By Doing

Autodesk Maya for Novices Textbook
- Autodesk Maya 2019 for Novices: Learn By Doing

Pixologic ZBrush for Novices Textbook
- Pixologic ZBrush 2018 for Novices: Learn By Doing

MAXON CINEMA 4D for Novices Textbook
- MAXON CINEMA 4D R20 Studio for Novices: Learn By Doing

AutoCAD for Novices Textbooks
- AutoCAD 2025 for Novices: Learn By Doing
- AutoCAD 2019 for Novices: Learn By Doing
- AutoCAD 2018 for Novices: Learn By Doing

Inventor Professional for Novices Textbook
- Inventor 2020 for Novices: Learn By Doing

Siemens NX for Novices Textbook
- Siemens NX 2019 for Novices: (Continous Release) Learn By Doing

Creo Parametric for Novices Textbooks

- Creo Parametric 6.0 for Novices: Learn By Doing
- Creo Parametric 5.0 for Novices: Learn By Doing

SOLIDWORKS for Novices Textbook

- SOLIDWORKS 2018 for Novices: Learn By Doing

Made in the USA
Monee, IL
13 September 2026

62975269R00162